*The Red Man
in the New
World Drama*

THE RED MAN IN THE NEW WORLD DRAMA

A Politico-Legal Study with a Pageantry of American Indian History

JENNINGS C. WISE

Edited, revised, and with an Introduction by

VINE DELORIA, Jr.

THE MACMILLAN COMPANY, NEW YORK, NEW YORK

COLLIER-MACMILLAN LTD., LONDON

The Macmillan Company
866 Third Avenue, New York, N.Y. 10022
Collier-Macmillan Canada Ltd., Toronto, Ontario

Library of Congress Catalog Card Number: 72-158068

FIRST PRINTING

Printed in the United States of America

Dedicated To The Memory of

COLONEL JOHN WISE OF ENGLAND AND VIRGINIA
(1619-1691)
COMMISSIONER AND JUSTICE
OF
YE KINGDOME OF ACCAWMACKE
GUARDIAN *Ad Litem* FOR THE INDIANS
EXEMPLIFIER OF THE MOTTO ON HIS SHIELD
"Aude Sapere"

Contents

Introduction

The usual history of Indians embarks from traditional moorings and records the story from the view of the white man who invaded this continent. The perspective runs rapidly over territory that has been plowed so often by so many scholars that one study merges into another with no discernible difference. Occasionally references to aspects of Indian life reveal glimpses of the other side of the story but they rarely do justice to what has been a valiant struggle by a number of tribes against the combined forces of western civlization.

Jennings Wise, who was very active in Indian Affairs in the opening decades of this century, saw a different kind of story in the drama of the conflict of the red and white men. In large part he viewed the history of the American Indian as a part of a world drama of conflicting religions. Much of the emphasis of the early chapters of the book is on the rivalry between the Catholic and Protestant viewpoints. In the process, by emphasizing religious motivation behind territorial conquest, Wise is able to reveal a side of the Indian story that has not yet been told, or, I suspect, even contemplated by white America.

Until the fundamental differences in cultural outlook, in part inspired by different religious world views, are understood by the people with power who make decisions affecting the lives of a million American Indians, little of lasting value will be accomplished. Thus the perennial studies to find the "key" to understanding American Indians, while motivated by a genuine concern, do not begin to fathom the deep gulf that exists between Indian tribes and the rest of America.

Some of the crucial issues of the modern world, particularly the struggle over Indian lands, the crisis in civil rights of members of Indian tribes, hunting and fishing rights, the drive for self-determination, the landings at Alcatraz, the revival of Indian culture and religion, the successful fight of Taos Pueblo for the return of Blue Lake, the struggle of the Iroquois for the return of their sacred wampum belts—all these currents in today's world cannot be fully understood without some insight into the past and the nature of the relationship which has existed between Indians and white America—a relationship that ultimately derives from the completely different religious base each maintains.

History for the white man is a series of upward quantum jumps in which he is enabled to overcome his enemies, physical and human, in his unquestioned march toward immortality. This drive has created a technology which now threatens to destroy the world. As the Christian stands before God as an individual, so western man stands before the universe demanding his rights as an individual. No thought is ever given to the ultimate nature of either the universe or society. Thus, while America has produced great businessmen and scientists, it has been unable to produce one great philosopher or theologian.

The Red Man in the New World Drama is a pivotal book for today. It should be considered the opening wedge by which questions of ancient import are introduced into the contemporary world. It chronicles the history of the American Indians in a religious-political sense, thus adding a dimension to the understanding of Indian people that has not previously existed. I do not support all of Wise's contentions in this book. The mere suggestion that Dekanawida might have been a Jesuit is enough to get me banned forever in Iroquois country. Let me say at the outset that I believe that he was an Iroquois and one of the outstanding figures of world history—although learned anthropologists will tell you he is somewhat shadowy since they have been unable to find his birth certificate.

I am not certain that the Jesuits were as active and influential as Wise feels that they were. Organized religions and their agents have not always been as efficient as Wise would appear to credit the Society of Jesus. Thus I would appear to be neutral on this point since it is not a crucial issue in my view.

In certain instances I have rearranged the material to conform with new chapter headings. In the original manuscript Wise linked the California Indians with the rising militancy of the Sioux in Minnesota. This did not seem to follow and so I have moved the Sioux material backward to a chapter that contains the primary material on the Sioux. I have also expanded the material on Geronimo since he certainly deserves more space than two pages and should appear, in my opinion, with Sitting Bull as the last of the real Indians fighting for a homeland.

What I hope to present to the readers of this book is Wise's viewpoint modified by the present situation. There should be a book that views the Indian story from an entirely different viewpoint and encourages young Indians to take a different vantage point when they learn about themselves and their tribal histories. For the white reader there are innumerable incidents that Wise gallantly brought to view from their years of hiding and neglect. What is the true story of the Norsemen? Were the Hurons really descendants of the Scandinavians who had assimilated into eastern Indian tribes within a century and a half after the destruction of their colonies?

In conclusion, I hope that bringing this book once again to the attention of the reading public, particularly younger Indians, will inspire people to reject conventional histories of the American experience and to search for a theory of history that will encompass the struggles of all groups in American society. The ultimate history of America must be written by an Indian encompassing the thousands of years that man has spent on this continent and putting the years of the white man in a different perspective. If *The Red Man in the New World Drama* can shake people out of their traditional way of looking at the world, then my task as revisor and editor and the task of Jennings Wise, who originally wrote the book, will have been useful. It is this hope, more than any other, that has driven me to try and update this book in view of the present turmoil in American society and the desperate need for us all to take a new look at the drama in which we are embroiled.

VINE DELORIA, JR.

Bellingham, Washington
February, 1971

Foreword

Historians of the United States almost invariably commence their story with a picture of the American wilderness inhabited by fiercely cruel, implacable savages, against whom the European emigrants to the New World were compelled to contend at every step forward from the Atlantic seaboard. Once having started white civilization on its way, they look to Europe for the influences which dictated almost every act of the American people. They mention the Indians about as casually as the periodic plagues of the Middle Ages are referred to in European history. Yet, to a thinking person it must be manifest that a true history of the United States can no more ignore the influence exerted upon the conquerors of the wilderness of the New World by its inhabitants, than one of England can disregard that of the Celtic peoples of the British Isles upon English civilization. This work, however, does not purport to be a history such as the works of Adair, Schoolcraft, Catlin, McKenney, Drake, Williams, Morgan and others. Nor is it a mere indictment of the American people. Primarily it is designed to disclose the real inspiration of much in the making of the nation by establishing the political nexus of the events in which the red race has been involved. Yet, with the utmost candor, this new story of the Indians has been told in the hope that it will cause the civilization of today to pause as did the migrant trains of the West, even if but for a single Sabbath, and, pausing, ponder what lies behind before choosing a trail into the future.

In the picture that has been presented, deep shadows as well as highlights appear, nor can it be viewed without those shadows, falling as they do like indelible blots upon the canvas of history, giving to the so-called romance of the New World the aspect of pure tragedy. With thoughts transported back from the "Golden West" to the gloom of Central Asia, again we shall see wild tribes pouring down from the withering grasslands of the steppes, and ask ourselves wherein the skin-deep humanism of the American people, in the remorseless surge of mankind forward into the realms of time, has differed from that of those whom in our vanity we call barbarians.

While much may be said in extenuation of the sins of those who, confident in the valiance of their youth, learning nothing from the experience of Spain and France and Holland, bringing religious bigotry to the support of the greed with which they were consumed, claiming to be the founders of human freedom, yet ignoring utterly the rights of a dependent people, it is useless for the historian to go on pouring balm upon the national conscience with the smug conclusion that these things were inevitable. The fact remains that even within the lifetime of those living, the Indian fatherland has been saturated with the blood of a helpless race; that under our very noses even now practices are going on as certainly destructive of that race as the violence of our forefathers.

Surely the time has come for the American people to ponder seriously the noble words of Grant—

Our superiority of strength and advantages of civilization should make us lenient toward the Indian. The wrong inflicted upon him should be taken into account, and the balance placed to his credit. The moral view of the question should be considered and the question asked, cannot the Indian be made a useful and productive member of society by proper teaching and treatment? If the effort is made in good faith, we will stand better before the civilized nations of the earth and in our own consciences for having made it.*

JENNINGS C. WISE

Washington, D. C.
September 19, 1931

* Second Inaugural Address, March 4, 1873.

1

The Norse Discovery and Colonization of Ancient America

THE YEAR 1583 is memorable in the history of mankind. It saw the appointment by His Holiness, the Pope, of Torquemada as the Grand Inquisitor of Rome with power to extirpate heresy from the earth. He, of course, knew nothing of charity. But it also saw the birth in Holland of a great intellectual missionary among the nations—Hugo Grotius, or De-Groot—whose internationalistic philosophy was characterized by that charity to which in the urgent upbuilding of political society little thought had been given.

He was a learned man. After visiting Greece, Italy, Portugal, France and Denmark, he found himself in the employ of Queen Christina of Sweden. Having known well Francis Bacon, the great English scholar and philosopher who died in 1626, and also Shakespeare, who had epitomized in English verse the history and the philosophies of the world, he himself ranked as one of the greatest living masters of history. Known today as the father of international law, a monument to his memory exists in the Peace Palace at the Hague which, erected with a wealth drawn by Andrew Carnegie from the soil of America, rears its glorious cathedrallike tower to the high heaven of all races and creeds. And it was this seer who, with access to the existing records of Scandinavia, agreed with Bacon that the Norse civilization had contributed much to ancient America.

After examining their conclusions, another philosopher and one of the most searching students of his day, writing in 1812, said:

When the Spaniards encountered its inhabitants they were still in the infancy of the social state; none of their faculties were wholly developed; they were weak physically as well as morally; it could be distinctly seen they belonged to a race different from the White and the Black. They belonged to the Red Race but were not pure. They were the result of a primal mixture at a very remote epoch when the White race did not yet exist and of a second mixture much less ancient when this race had existed for some time. These indigenous peoples had lost the trace of their origin; only a vague tradition survived amongst them which declared their ancestors descended from the highest mountains of that hemisphere. The Mexicans claimed that their first legislators came from a country situated at the northeast of their empire. If attention is given here, the two principal epochs of mankind will be found in these two traditions; the first dates back to the disaster of Atlantis, whose memory is perpetuated among all nations; the second belongs to an emigration of the Borean race which was effected from Iceland to Greenland and from Greenland to Labrador, as far as Mexico, traversing the countries which today bear the names of Canada and Louisiana. The second epoch is separated from the other by several thousand years.*

These in themselves are amazing words. They can no more be dismissed with a shrug, even in the light of modern learning, than the statement of Montezuma, made before his execution by Cortez. In a verbal last will and testament he bequeathed the Aztec fatherland to his race after reminding them that he himself was descended from a prince who came with the rising sun in a great vessel with white wings—a graphic description of a Viking ancestor.

In the face of all this, assuredly it behooves the historian of the red race to examine, with a care not common, the facts of Scandinavian history relating to Ancient America.

The extraneous influences which came to bear upon the inhabitants of ancient America appear to have been many and strange. In view of the physical formation of the North American continent, the proximity of Kamchatka and Alaska, and the stepping stones of the Pacific islands, it would have been extraordinary if now and then Japanese, Chinese and Polynesian junks had not touched upon its western shores to sow some germ of an Asiatic culture more recent than that of the period of Indian migration. But while the evidence is slight as to such influences, when that relating to ancient America is collected and analyzed, no reasonable mind can doubt that during the five centuries preceding the arrival of the Spaniards in the West Indies there had been an extensive admixture of European and red blood with a correlative influence upon native thought and institutions. Nor is it possible for one to understand much that occurred in the succeeding drama without a knowledge of the prelude.

* *The Hermeneutic Interpretation of the Origin of the Social State of Mankind,* Fabre d'Olivet, Putnam Translation.

Iceland was discovered in the eighth century by pagan Celts from Scotia and Hibernia. Yet, there is no record that the island was populated until after it had been rediscovered accidentally in 870. Then, under petty rulers called Vikings, Scandinavians from Norway began to migrate to the island in great numbers, by reason of the tyranny of their king. It was an immense movement. At the end of four years Iceland had a population of 50,000 despite the forbidding name which its first settlers gave the island.

That this large migration across the North Atlantic ended absolutely at Iceland is hard to believe. From the number of accidental arrivals of Europeans on the continent beyond, which were recorded after written records began, it seems certain that many Gaels and Norsemen must have been there of whom there is no knowledge whatever. Nevertheless, the fact is that although the great island beyond Iceland is not nearly so far away as Norway, the first recorded visit to it was made by Gunjorn in 874. Having learned the psychological effect of the name Iceland upon a people who were spreading over the world in search of a new home, Eric the Red named the island which he was eventually sent to govern Greenland. Thus is explained the apparent anomaly of Greenland's icy mountains. Eric was but the originator of the first land fraud in the Western World! Often he chuckled over his shrewdness as he watched settlers arrive in ever-increasing numbers in a land which loomed out of the cold blue chilliness of the Arctic like a great iceberg.

A century passed before any record was made of a visitation to the Continent. In 986, Bjorn Herwolfsn reported to Eric, who had now returned to Norway, that he had coasted along the shores of the present Labrador. But let it be repeated, it is by no means certain that he was the first to behold the vast continental forest there.

Here we must visualize what was going on among the Scandinavians. The Danes had conquered the British Isles and Denmark was soon to furnish England with a king. The Norwegians had visited North Africa, Sicily and Egypt, and the Swedes had made large settlements, including Novgorod in Russia, and passing down the Volga had reached the Caspian Sea, Persia and Palestine. Meantime, Iceland had adopted an independent constitution, and like Greenland was part of the Archbishopric of Bremen. With their rulers squarely behind the Vatican, the inhabitants of both islands were being Christianized by fiat of law. Moreover, after the baptism of Eric the Red, allegiance to Rome or death were the alternatives.

So far the natives whom the Norsemen had encountered belonged to a race of yellow type, common to the Arctic regions of both hemispheres, known today by the name which the Algonquin Abenakis of Newfoundland gave them—Ashkimeq, or Eskimo, meaning "raw flesh eaters," but who called themselves Innuit and Aleut in America, and Yuit or YuKouk in Asia. The early Norse, who had known them in Northern Asia long before reaching Greenland, had called them Skrellings and despised them as cannibals.

Many people, because of their ignorance of Norse history and the records extant in Iceland, Scandinavia and the monasteries of Normandy and Wales, still assume that the Norse visitations to the American conti: nents verge upon the realm of pure myth. The fact is, however, an im mense amount of contemporaneous documentary evidence exists which, though largely ignored by Fiske, leaves no doubt of just what occurred in North America over five centuries before the coming of Columbus.*

Inspired by Bjorn's accounts of the realm he had beheld, fourteen years after the old pirate's discovery, Leif Erikson, the son of Eric the Red (who had sent him to rule Greenland), decided to take possession of it in the name of God and found there a colony. Therefore, in 1000 A.D., with Bjorn as his guide, he set out with a single ship bearing on its prow a great cross. Soon arriving at Labrador he named the region Helluland because of its forbidding aspect. Then he came to the present Bay of Gaspé and landed in Newfoundland which he named Markland. After exploring the present Acadia and naming Cape Careen and Cape Alderton he proceeded around Cape Cod to the Pocassett River in Rhode Island Bay. Here log cabins were built in which the explorers spent the winter, naming their settlement Leif's Boudir.

Soon after their arrival, an event occurred which suggested to Leif a deception which has puzzled the historians for over ten centuries. A German sailor named Tryker had discovered the unpalatable grapes indigenous to the region and made a quantity of strong wine. Thereupon, for exactly the same reason that his father named the great island to the north Greenland, Leif named the country he had found Vinland, and in June, 1001, with an immense accumulation of furs, returned to Greenland. Ever since, overlooking the nature of his treasure, the historians have envisaged a land appropriate to the name which Leif had given to the region of Leif's Boudir.

Greatly enriched, he himself knew no need to brave the sea and the continental wilds of Vinland again. It was his brother Thorwald who, to enrich himself, set out in the spring of 1002, for Vinland. After spending the winter at Leif's Boudir, one party of his company explored what was probably Long Island while another continued southward. In the absence of the latter in 1004, Leif returned to Cape Careen and Cape Alderton obviously to explore more thoroughly the peculiarly mild region of Acadia. This itself is conclusive that Leif's Boudir was not in a southern

---

* *The Vinland Voyages,* Thordarson, N. Y. Am. Ger. Soc., 1930; *Narratives of the Discovery of America, Adventures in the Wilderness,* Yale University Press, 1925.

clime and that the settlers were searching in both directions for a more hospitable region.

There now occurred with dramatic fitness the opening event in the great human tragedy of the ten succeeding centuries—the first meeting of the red and white races.

It is not mere sentimentality but an appreciation of psychologic influences upon history that gives pause to the narrative of events. Let us envisage the picture. In the vast solitude of the semi-arctic wilderness craggy peaks rear themselves out of the primeval forest. Blue in the cool sunshine, bejeweled with drifting icebergs gleaming with the irridescence of huge pearls, the reaching fingers of the sea clutch at the sylvan shadows which enshroud the land. With its prow resting upon the rocky strand, its sails flapping in the breeze, a strange craft is moored to the huge firs which bathe their feet in the tide. Upon its beak a massive cross is fixed. Now and then the scream of wild fowl breaks the widespread silence as invading voices come from the covert where Leif and his companions are drinking deep of Tryker's wine. Unconsciously, with ever-roving eyes, these peerless mariners glimpse down the forest lanes the rippling bosom of the sea. And lo! In the distance they behold three tiny birchen shallops skimming along the shore toward them.

In an instant they cast aside their flagons, unbuckle the huge blades slung upon their backs. Grasping immense ashen bows charged with the steel-pointed shafts which have all but conquered the world, breathlessly they peer upon the possible enemy.

"'Tis naught but beastly Skrellings," muttered one at last.

"Yea," agreed Thorwald. "Let us fall upon these heathen cannibals for God hath willed that we shall possess their land."

So as they crouch there, not the Merciful Father of Christ Jesus but the bloody Yahva of Joshua speaks to them—these recently converted pagans who, even more savage, are no less heathen than those whom they mean to slay. The cross gleaming there in God's sunshine is but a meaningless thing, no more than a new armorial bearing which the fiat of Christ's vicar has compelled them to substitute for the winged dragon. Odin still rules their souls.

Apparently without fear the paddlers draw their tiny craft upon the strand. Boldly, unsuspecting in their curiosity the fate which awaits them, chattering excitedly they approach the "great canoe." Then they enter the silent wood as innocently as children in search of its crew. In the damp air twang a score of bows. Death, with unerring wings, has taken its first murderous flight! Now flash the great two-handed blades of the Norsemen. It is all over in a twinkling.

So Christian white men have slain their red brethren at the first sight of them! The sin of Cain is on their heads, for in truth their native victims are not cannibal Eskimo, but the Algonquin Micmacs of Newfoundland. Unfortunately for Thorwald and his companions, one of what was obviously but a reconnoitering party, has escaped. So, while the Norsemen are gloating over their prowess with many "skalds" of Tryker's wine, even before they have cleansed the gore from their murderous blades, upon the bay, hugging the shore in all directions, appears the main body of the Micmac fleet. Skilled by centuries of ceaseless warfare with the semi-cannibal Mohawks to the West, skillfully, resolutely, in native battle array they approach.

Rushing to their ship the Norsemen fear not the attacking fleet. But what is this? The forest emits a veritable cloud of skirmishers also armed with bows and arrows. Then is the Norse metal tested to the limit.

In the end the assailants are driven off by the superior weapons of their opponents whose bodies are protected by the great shields which they bear, steel casques and metal-studded doublets. Yet, it is with strangely prompt retributive justice that a penalty has been exacted of Thorwald— the one most responsible for the collective crime. A quartz arrowhead lies deeply imbedded in his lungs. Its poison has entered his blood. The hemorrhage of death is upon him. So as if with symbolic import, did the stone-age race of America seek to defend itself against the already finely tempered steel of the Europeans who had come to claim its fatherland.

Solemnly the fallen leader's remains are laid to rest in a stone tomb erected to receive them. Helmet and sword are placed upon the body. Now aware of their danger, with a final message from Thorwald to his wife in Greenland, the survivors hoist their sails and depart for Leif's Boudir. It is with strange irony that they call the site of the first conflict between white and red men, the Cape of Crosses.*

Here indeed in the incident related is to be found food for the psychologist! For many years a group of literati were wont to assemble annually at the Inn of William the Conqueror at Dives in Normandy. To each in turn fell the task of relating an informing story. Among those recorded is one which is peculiarly apropos here. A whaling vessel came upon an undiscovered arctic island. About the crew the unaffrighted penguins crowded with a show of confidence that appealed to the sympathies even of the rough seamen. The men conceived a great fondness for the dumb creatures who seemed almost human in their affections. Unfortunately the ship's dog got loose in the night before the ship sailed and killed an entire brood of helpless young penguins. The next year the sailors longing to see their bird friends again returned to the island. Not a

---

* Centuries later the tomb of Thorwald, still containing his bones and arms, was found and still those who call themselves historians deal fearfully with Vinland as if it were more myth than fact.

penguin was seen until they were discovered in their distant hiding place. The dog had destroyed their trust. Fear had been born among them. Now they only desired to be let alone in their solitude. Man had nothing to offer them which they prized as much as security.

In the mind of the Micmacs, though themselves warlike under the stress of the hard conditions of aboriginal life, there was a clear distinction between warfare and murder. Trustingly they had approached a white-faced race of strangers whose canoe bore a great shining cross. Murder had followed. Fear had replaced trust among them as among the penguins. As the Norsemen sped southward from the Cape of Crosses where they had placed their strange emblem over Thorwald's tomb, was not their great ship observed from countless native lookouts? Nor can one familiar with native nature doubt that ahead of it passed the warning "Ware! Big canoe. Pale faces with the mark of Cain."

Meantime, the other party of Thorwald's expedition had taken part in another tragedy. Coming after a long voyage upon a realm of soft air and fair skies whose inhabitants were gayly befeathered, it appears from their descriptions of the natives, the fruit, flowers and foliage that they had probably reached Florida.

Although they declared that the inhabitants of this pleasant, fruitful country were very friendly there, too, a conflict had followed their arrival in which the explorers captured two native children. From this alone one can read between the lines. These they brought back with them to Leif's Boudir and named them Valthof and Vimar. But the death of Thorwald had cast a gloom over the whole company and in 1005 they returned to Greenland. The dragon's teeth had already been sown from Newfoundland to Florida.

Gudrid, the widow of Thorwald, soon married Thornfinn Karlsefn, a famous sea rover who had visited both Asia and Africa. About her there is much suggestive of a Valkyr. Together she and Thornfinn organized an expedition to avenge Thorwald and conquer Vinland. For a while they feasted amid oath takings which boded the natives no good. Finally, they set forth in the summer of 1007, accompanied by Thorwald's sister Freydize, a veritable "she devil," in two ships which mustered a company of 132 people including men and women. Passing Helluland, Markland, Bear Island and Careen Cape, instead of looking for Thorwald's tomb as originally planned, the happily married couple in command continued southward. This angered Freydize, who deemed it a betrayal of her brother's memory. After exploring Nauset, Chatham and Monomy Bays, upon reaching Buzzard's Bay the party landed with their cattle at Bird Island where a temporary settlement was made. Here Freydize attempted to poison Gudrid. Upon the latter's recovery, on account of the harsh climate the whole company removed to the more sheltered locality of Leif's Boudir. One group occupied the old settlement while Thornfinn and Gudrid with another built a new town beyond. About both settlements a large area of land was soon brought under cultivation and the prosperity of the colony seemed assured.

Unfortunately for Gudrid, and Thornfinn's colonists, the native lookouts had observed their movements along the coasts. Fearful of visitors who appeared to be the same as those with whom the battle had occurred at the Cape of Crosses a few years before, for a while the aboriginal inhabitants had held aloof. Meantime, however, the diminutive Algonquin Etchemins, or Canoemen, had been assembled in full force and now their fleet appeared upon the river. Freydize was for falling upon them at once and avenging Thorwald, but to her disgust Gudrid, who had just given birth to a son—the first European child born on the continent—in the spirit of a true empire builder, counselled peace. Finally, though not without difficulty she persuaded her husband that this was the best policy.

After reconnoitering the situation the natives disappeared, soon to return. The record is quite plain that for over a year a thriving trade in furs was driven with them in which bright colored frippery was the currency of the Norsemen. As usual, according to the latter, they treated the natives well, giving them food and much drink. It was not long, however, before they began to charge their visitors with being unreasonable. The fact that they began to fortify the towns with great timber palisades after the Norse fashion, is conclusive that Freydize and her party were antagonizing them despite Gudrid's counsel. Nor is there any doubt that soon the lusty Norsemen were trying to debauch the native women. Now, as always thereafter, this was fatal to good relations with the warrior race who, venerating the female sex, placed it above illicit cohabitation. In other words, the exact conditions prevailed in Vinland which ever characterized the American frontier—rough, unmarried men, liquor, and an irresponsible craving for women which was indulged without regard to consequences.

The result was inevitable. After Freydize had turned a maddened bull loose upon the native visitors to drive them off, they returned from their nearby village much offended. Next the venerable "Queen" of the tribe fled for protection to Gudrid's cabin just after the birth of the latter's second son, while the Etchemin who sought to protect the old woman was slain. Then followed the conflict which had long been brewing.

The native tactics were typical. Freydize herself led the defenders. Although the assailants were finally driven off, a majority of the colonists, including Freydize, were slain. Thereupon, Thornfinn and Gudrid placed a stone on the banks of the Cohasset River near the present town of Taunton to mark the site of their colony, with a complete account of the battle recorded thereupon, showing that Thornfinn and 131 others had occupied the land.* Then, upon the counsel of Gudrid, the survivors moved to Buzzard's Bay for safety. After spending two months exploring the country to the south, including the Chesapeake and the Potomac, Thornfinn returned to find the women in a great feud, and a virtual revolution on his hands. Therefore, in 1011, taking their share of the accumulated wealth, he and Gudrid returned to Norway.

During the next several decades great numbers of people from Iceland and Greenland visited the continent, scattering in every direction. Meantime, too, Valthof and Vimar, who had been educated by their captors, had excited much interest by their reports of the land to the south of Vinland, where they said the natives were ruled by a great white chief. There, they declared, was a great health-giving fountain.† Consequently in 1501 Horveld and a party with the two Indians as their guides set out to find this region, and near the falls of the Potomac actually came upon a tribe of natives headed by a venerable Norseman who still possessed a horse which years before had brought from Iceland. This was one Bjorn Kappi, a banished knight who, penitent in spirit, had come to the continent to live out his exile. He told his visitors that Ary Marson, a Viking, had ruled the local natives before him, and that believing he was Marson resurrected, they revered him as a god. After giving them a ring to return to his friends at home, he pleaded with them to depart lest harm befall them. Such were the strange things going on in this land of mingled romance and tragedy.

Upon returning to Iceland, Horveld told his story to a monk who having travelled far afield on the continent, had known well both Marson and Bjorn Kappi. With her curiosity aroused by a land of "white people" concerning which Valthof and Vimar continued to talk, Syasi, the wife of Horveld, now despatched the two Indians to negotiate an agreement with the natives that would permit her to found a colony among them. This having been arranged, with a large company she founded Syasi's Boudir in what is probably the present North Carolina. There among fruit and flowers a thriving colony had been established with large plantations when a fierce tribe of natives from the north—probably the Iroquoian Tuscaroras—who were traditional enemies of the local Algonquins, appeared. Now was anticipated the tragedy of Roanoke Island in 1585. Syasi's Boudir was completely destroyed, but although the Norsemen's ship was burned, the survivors eventually reached Leif's abandoned settlement on the Taunton River, where they now established themselves.

---

* Today this marker is known as the "Dighton Rock" because it is near the town of Dighton.

† The same fountain, no doubt, of which Ponce de Leon also heard reports, probably the present popular hot springs in Georgia.

Meantime Gudrid had visited Rome and aroused great interest in the New World, so that in 1057 the Catholic priest Jonus was despatched to Leif's Boudir. The colonists, however, were semi-pagans at best and, resenting his assertions of authority, promptly executed him. Nevertheless, by this time Vinland had grown into a prosperous country with many trading settlements based upon the main one at Rhode Island, so that immediately Erik Upsi, Bishop of Gardar, abandoning his diocese, took up the work of Jonus. Making his headquarters at what is now Newport, he built an Episcopal residence whence to carry on the work of baptism. The remains of this are today found in the "Old Mill" at Newport. Thence, too, missionaries were despatched to the region of the James River and Potomac where other settlements had been established.

Such is the story of early Vinland plainly recorded in the Icelandic and Norse literature. But it is evident that the expeditions made to the continent after the Norsemen had once visited southern Vinland, were not prompted solely by longings for adventure, or by the valuable furs and timber to be had in the New World. The mortuary mines of the Middle and South Atlantic states show that gold in the free or nugget form was in possession of the natives at the time of these expeditions. Unquestionably it was soon discovered by the Norsemen who probably resorted to every conceivable artifice to take it from its owners. Furthermore, as in the case of the Spaniards when they first came to Florida, undoubtedly when questioned concerning the source of this gold, the natives pointed northwestward and southwestward, so before the end of the eleventh century the Norsemen were scouring the continent for the putative Eldorado just as the Spaniards, French, Dutch and English each in their turn were to do five centuries later. Certain it is, once they had caught sight of the native gold, these dauntless sea rovers would have gone far afield in their quest.

Until 1103, Iceland, Greenland and the Diocese of Gardar in Vinland were part of the Archbishopric of Bremen, but in that year were transferred to the Archbishopric of Lund. At this time there were intimate connections between the great monasteries of Scandinavia and Hibernia with Rome. What was known in one country was, of course, soon known in the other through the medium of the itinerant proselytizing priesthood. Moreover, the Norsemen themselves were scattered all over the world moving from one port to another, so that there was no secret whatever about Vinland and the gold to be had there. In the Flateyarbok many statements of visitors to the Continent disclose the great wealth they were acquiring. Nor were Gudrid's reports in Rome unknown in the monasteries of Wales. The evidence is conclusive that in 1170 a westward migration of Welsh occurred. Nevertheless, it was not until 1262, or almost coincident with the formation of the Swiss and Henseatic leagues, that the European rulers began to pay much attention to Iceland, Greenland and Vinland. Then, having already formed a union, the islands were induced to attach themselves to the Crown of Norway as the best means of protecting the large trade they were conducting between the continent and Europe. And it is particularly interesting to note that three years later was born the great humanist, Pierre du Bois of France, who was soon preaching throughout Europe of a league of nations to enforce peace.

By 1267 Norsemen had almost reached Alaska from Greenland. As shown by remains recently discovered, they were then in Melville Sound, obviously searching always westward through the inland waterways of Canada for the continental Eldorado. What they found does not appear. At any rate Vinland was now prized by Rome. Thus, in 1276 the Vatican appointed Archbishop Jou to Christianize the natives of the continent, and at once he sent representatives to the Diocese of Gardar in Vinland and "the neighboring lands and islands."

Although Iceland was far more advanced at this time than Greenland, early in the fourteenth century the latter boasted two large settlements—the East Byd with 190 villages, a cathedral and four monasteries, and the West Byd with ninety villages and four churches. Before the middle of the century there were twelve churches in the West Byd alone. These facts are indicative of a large number of visitors to Vinland.

It was in 1347, or just ten years after the French fleet, under a Norman captain, defeated the English despite the desertion of Philip the Bold by the Genoese admiral and his ships, that the last recorded expedition from Iceland to the continent was made. Nor is it difficult to understand this. At this time the "Black Death," which was ravaging Europe, spread to Iceland and Greenland, and no doubt to the continent. While it was still raging two years later, the report was widely current in Iceland and Scandinavia that the people of the West Byd of Greenland had suffered terrible losses in a great conflict with the Skrellings. Naturally, therefore, the Icelanders ceased to put out for the continent as in the past. And in 1379 a similar report reached Iceland about the people of the East Byd. Thus, it is seen that even before the middle of the fourteenth century the continental colonists were engaged in a serious warfare with the natives.

In 1380 Iceland and Greenland both passed to the Danish Crown, and now commenced a period of terrible adversity. A series of tremendous volcanic eruptions was attended by serious climatic changes and repeated scourges of the "Black Death."

At this time the Danes generally were looked upon in Europe as no more than pirates, so that in 1379, Henry Sinclair, the Norman Count of the Orkneys, undertook to wipe them from the seas. The following year Nicolo Zeno was appointed Grand Admiral of Venice and assigned the same task. With his brother, Antonio, Nicolo Zeno soon joined Sinclair, and together they descended upon Greenland, which the Danes were now making a base to prey on the trade with the continent from which, apparently, much gold was coming. After ravaging the Byds, they came upon the Monastery of St. Thomas in Greenland which was named after the resident monk, Thomas Moel. From him they learned much about the continent where St. Thomas himself had been a prisoner seventeen years. Passed from one native tribe to another, it seems certain from his descriptions that he had traveled from the St. Lawrence River to Florida and back. Moreover, after telling the visitors of the fabulous wealth of the southern country, he was induced to guide them upon a search for it. Although Moel died immediately after they reached the continent, and Nicolo Zeno was compelled to return to Venice on account of illness, at the end of four days the others came to the region which the dead guide had described as Estoitland, or Eastland. This was Acadia or the Markland of Leif. Here they found neither naked nor fur-clad savages but a people whose clothing was made largely of feathers and with the Latin cross tatooed on their shoulders. Moreover, a Norseman in the company who was familiar with Latin translated their prayers.

"May God give us life and health, for ourselves, and our families, food for the day, victory over our enemies." These were the words they spoke —purely the semi-pagan religion of the early Christians. Some of the older members of the tribe actually described the white men who many years before had been among them—Scandinavian priests from Rome. The visitors also found the native burial grounds full of crosses. Evidently this was the settlement of whose destruction reports had reached Iceland in 1379—the continental East Byd.

But it was not the land of fabulous wealth which St. Thomas had promised, so that Sinclair and Zeno soon continued southward without exploring the St. Lawrence River, where it will be shown, there was a large Norse colony. Caught in a tempest, eventually they were driven to a southern volcanic island where they found the natives "half savage but very peaceable." Here the crews revolted and upon Sinclair's refusing to leave the land, Zeno returned to Europe to record his explorations.

Finally, in 1409, or the year the last vessel from Greenland reached Norway, the report reached Iceland that the Skrellings had completely destroyed the Greenlanders, and that both the East and West Byds had been demolished. Thereupon it was concluded that Greenland itself had been invaded by the natives which, of course, was impossible. Manifestly, they could not have travelled a week on the open sea in open canoes even had they dared assail the most fearless warriors of medieval Europe in their own land. The Norse pirates who were still ravaging Greenland were simply confused with the Skrellings on the mainland. In fact the cathedrals were destroyed by a great earthquake that year. Naturally with Greenland completely depopulated by migration, disease, piratical massacres and earthquakes, and her colonies being assailed by the natives, the Icelanders during the period of their own great distress did not visit Greenland and the continent, so that the curtain of a century and a quarter now lowered over Greenland and Vinland.*

* Nothing more was heard of Greenland until 1535 when John Davis visited the island where he found the ruins of the cathedrals and the sites of the once prosperous Byds tenanted only by a handful of Eskimo. From this it was concluded that the Byds had been destroyed by the Eskimo and strange as it may seem, even so great a scholar as John Fiske took Davis' word for this. Nor was it until long after the beginning of the Reformation in Europe that the Danish king thought to look after his island possessions. Then he found that the colony of Greenland was extinct. As to what happened in ancient America, Fiske is unreliable. Compare his *Discovery of America* with *Danish Greenland*, Rink, *History of Greenland*, Crantz, and authorities already cited, including the most recent ones which are thoroughly harmonized in the foregoing brief account. It seems certain that in the archives of the Vatican there exists a wealth of historical matter not yet published, including the contemporaneous documents of the Archbishops, Bishops and the numerous monks who ruled the monasteries in Greenland and established the Missions in Vinland.

2

The Tragedy of Westland

Iɴ ᴛʜᴇ preceding chapter it has been shown that Eastland, or the eastern colony of Greenland on the Continent, had been destroyed when in 1380 Sinclair and Zeno found there the evidences of a preexisting Christian settlement. No doubt by this time all the outlying settlements scattered over the continent from Newfoundland to Florida had also been destroyed in the great conflicts reported in 1349 and 1379 which were probably the result of general uprisings among the natives such as those which occurred periodically during the succeeding centuries. Eastland would logically have included the settlements of New England. That there was a terrific struggle there between the Norsemen and the natives is indicated by the fact that when Gosnold and Pring visited Massachusetts in 1602 and 1603, and a colony of English was planted on the Kennebec River in 1605, the original tribes from Newfoundland to the Hudson River were broken up into countless small groups bearing little resemblance to the large tribes in other sections. This would naturally have occurred where the Norse settlements were the oldest and most dense as in the original Vinland which had come to be called Eastland.

Inasmuch as there was an Eastland which was destroyed in 1379 according to report current in Iceland, presumably there was a Westland. At any rate we know that the western Norsemen were assailed in 1349. Moreover, since Eastland was not assailed until 1379 but had been destroyed by 1380, and there was a great conflict in 1409, the settlement of Westland must have been in existence until the latter date.

When the first report reached Iceland of a great conflict between the

Norsemen of Greenland and the Skrellings, according to Morgan, the wave of aborigines known to the ethnologists as Huron-Iroquois, was moving eastward in successive swarms. Proving their superiority in bravery and intelligence over the Algonquins who had preceded them, when they reached Niagara the Iroquois divided. From there one wave passing north of Lake Ontario rolled eastward along the St. Lawrence and the other down the valley of the Susquehanna River which, rising in central New York, meanders through Pennsylvania to Chesapeake Bay or "the great river" of the Algonquins. Thus, the Mohawks, Oneidas, Onondagas, Cayugas, and Senecas came to rest in the order named up against the Algonquin Crees who occupied the plain to the north of Lake Ontario. This left the Mohawks contending with the Algonquin Micmacs in the region to the east where Thorwald found them in A.D. 1004. Of the southern wave, the Cherokees came to rest in Georgia, the Shawnees in Kentucky, the Tuscaroras in North Carolina and the Conestogas and Susquehannas in western Pennsylvania. The two last tribes who were notorious for their large stature lay up against the Algonquin Lenapi of New Jersey, Maryland and Delaware, along the Delaware River, while the Conestoga claimed both sides of Chesapeake Bay. On the peninsula between Lake Huron and Lake Ontario remained only the Hurons, while the Eries occupied the region of Ohio, or "the land of the beautiful river," south of the lake which took their name.

Behind the Iroquois swarmed the Athabaskan-Ojibway family. The Athabaskans were halted by the Crees to the north of Lake Superior while the Ojibways threw the Ottawas and Potawatomis into Michigan, and their most numerous tribe, or the Chippewas, into Wisconsin. Pressing hard upon the Chippewas came the Siouans, or the most numerous linguistic family next to the Algonquins. Throwing the Winnebagoes into Wisconsin they formed contact in the north with the Athabaskans in the region of Lake Winnipeg and the Lake of the Woods. In the great triangle between the Ohio and Mississippi remained numerous Algonquin tribes.

Our first knowledge of the Iroquis comes from Jacques Cartier who visited Canada in 1534. According to him they were divided into two hostile groups from the Bay of Gaspé all the way to Lake Huron. Those to the north of the St. Lawrence River and Lake Ontario, who were called Hurons, were at war with the five great tribes to the south. It was soon discovered by him that the Huron peninsula which Morgan calls Huronia, was the bone of contention. The reason for this is plain. Although Mooney declared that not a word of foreign origin is found in the native vocabularies it should be manifest from the name of the occupants.

Huron is not a native word nor was it ever used by the aborigines. Derived from the French, or old Norman, or Norse, huré, it means "bristly," "bristles," "rough hair" (of the head) of man or beast, or a wild boar's head. In old French it meant, "muzzle of the wolf or lion," "the scalp," "a wig." Thus, the peasants who rebelled against King John in England in 1358, or while the Norse were in Vinland, were called Hurons. Father Lalemont in the Jesuit Relation for 1639, in attempting to explain the application of this European term to the natives of Canada said that when these people came to the French trading posts on the St. Lawrence, a French soldier or sailor about 1600, noting their peculiar hair, gave them the name because their heads suggested wild boars. "Others," he said, "attribute it to some other though similar origin."

No better description of the Norsemen could be found, for they, like all the ancient Celts, according to their clans, wore helmets fashioned to resemble wolf and boar heads, and were also generally bearded, or bristled. It would appear, therefore, that they used the term to distinguish themselves from the natives.

On the other hand, the hostile Iroquois designated the Hurons by the name Wendat, eventually perverted by the French into Guyandotte, or Wyandot, which meant "people from an island." Because the Hurons did not live on an island in Canada this name has puzzled the ethnologists. But surely it should not be surprising that the natives applied it to the Icelanders and Greenlanders.

Then too the meaning of Canada itself—"the land of cabins"—also indicates that the Hurons who occupied it were different from the natives. Moreover, it is well established that the Hurons not only dwelt in wooden structures, or cabins, but built palisades around their villages often thirty-five feet high with platforms and ladders exactly like the early Celtic defenses, and sometimes dug moats around them with high escarpments. So, too, the hostile Iroquois were notably house dwellers from an early day, which is untrue of natives who were not in intimate contact with the Hurons.

The most cursory examination of a map will indicate why Huronia was the center of this ceaseless conflict. Commanding every approach by the inland waterways of Canada from the north, it also stood at the head of the great basin of the St. Lawrence River. Thus, it was readily approachable from the East. Plainly it was the strategic point at which the Norsemen would naturally have established themselves in an effort not only to reach the Eldorado of the Black Hills in Minnesota, or "the land of sky tinted water," whither the native guides were no doubt always pointing, but to work the great fields of free copper which lay in the region of Lake Superior.

Undoubtedly from the time the Norsemen entered the St. Lawrence River they met with strong opposition. This was due not merely to the hostility which Thorwald's murderous assault on the Micmacs and Thornfinn's battle in Connecticut had engendered, but to their quest of the native gold.

It was with trained military eyes that the Norsemen selected the site of Quebec for their fortified base. This they named Stadcona, no doubt after the Celtic Strathcona of Scotia. Other fortified points along the route to Huronia were Hagonchenda, or the town of Hagon, Hochalay, or some high place, Tutonagay, or the Norse Tutonia, and finally the grand site of Montreal, which they named Hochelaga, or the high fort.

Here then, from Quebec to Huronia, was Westland, a region in which the native tribes are known to have been involved for centuries in what is ordinarily deemed a great family fight among the Iroquois.

The nomenclature applied to the tribes also indicates the opposition they encountered. Thus, the Hurons called the hostile Iroquois Trudamans, or Trudimani, or Toudamans, probably from old Norse "trude," meaning woman, and "man" meaning man. Trudamans was obviously an appropriate name for those who claimed descent from their mother and had the same implication as squaw-man, or one who subordinates himself to a woman. The Hurons also called the Iroquois Agonionda which in Onondaga means "those who attack us." This shows that Trudamans could have implied no lack of valor. Certainly the meaning of Iroquois— adder—as well as the Norse experience with the tribes of this family, confirms this view.

Finally, the seven tribes of the Siouan family which lay between Huronia and the Black Hills, or the guardians of the gold fields, were called by the Chippewas, or the guardians of the copper region, Nadowe-is-in, meaning snakes, adders, enemies. Corrupted by the French into Nadowessioux or Sioux, it was taken to mean "real adders" or the worst enemies of all. This indicates that upon reaching Huronia the Norsemen encountered a new resistance probably from the Sioux and the Chippewas combined. And it was undoubtedly to overcome this resistance and enable them to reach their objective that the Huron Confederacy, known to have been the first Indian confederacy in America, was formed.

It consisted of four equal and several dependent groups. The former were the White-Eared People, the Cord People, the Rock People and the Bear People. In these descriptions it is not difficult to see the original Norse warriors with their metal dragon-winged casques, or white ears; the sailors or riggers who dragged the boats over the portages; the miners; and the voyageurs, or explorers.

The dependent tribes were called Bowl People, and because they took no part in the fighting against the Southern Iroquois, or their own kin, were called the Neutral Nation. In all probability they were the natives whom the Norsemen reduced to a virtual state of vassalage in order to provide laborers for the mines, or to rock the bowls or gravel sifters along the streams.

These dependents did not call the Norsemen either Hurons or Wendats, but spoke of them as Attiwendoronk, or "stammerers" or "people who spoke a strange language." Here it is to be noted that the Sioux have always designated Norwegians, Swedes, Finns, Poles and Hungarians as "tangled-tongues." Cartier, like the keen-eyed Jesuits who followed him, noted not only the differences between the Hurons and other natives, but their undoubted superiority even over the hostile Iroquois.

Exactly when the Norsemen reached Huronia cannot be said. They could certainly have reached it as soon as they reached Melville Sound, or the year 1267, nor is there much doubt that they were there and already federated when the last large expedition left Iceland in 1347. No doubt this new reenforcement of Norsemen had joined the confederacy before the conflict reported in 1349. These conclusions are well supported. The hostile Iroquois told Cartier that their own confederacy, which unquestionably was induced by the Norse invasion, was from 150 to 200 years old when they first saw the French. This would fix its formation at a date not later than 1385, which accords with the list of Tuscarora Sachems recorded by David Cusick, the Tuscarora historian.* Since it is known that the Huron confederacy was formed before the Confederacy of the Five Nations, and thus the hostile federation was formed not later than 1385, it appears that the Norsemen must have been established in Westland when the reports of the great attack of 1379 occurred, if not before that of 1349.

The governmental system of the Huron Confederacy has been thoroughly described. It embodied representative political forms and institutions which ethnologists have recognized as without precedent in the culture of the American aborigines. The executive power was vested in chiefs who, after the manner of the Norsemen, were chosen by the suffrage of the child-bearing women and constituted a central council for legislative and judicial purposes. There were five distinct political units of ascending magnitude—the family, clan, phratry, tribe and confederacy. The functions of the civil officials were entirely separate from those of the military, though sometimes the same person held both a civil and a military office. Ownership of property in severalty and the right of inheritance were both recognized, contrary to ordinary aboriginal conceptions. Adultery and murder were crimes. The Danish blood feud and the giving of male names and family relationships were also recognized by a system of law and equity which was not of Indian origin. Burial by interment was practiced, while the deity worshipped with public and periodic thanksgivings was not the Great Spirit of the aborigines, but an anthropomorphic deity like Odin or the demiurge of the semi-pagan Norsemen.

Nor is it certain exactly when the Huron Confederacy was finally destroyed by the Iroquois, Chippewas and Sioux. According to the Jesuit

* Dating back erroneously from the time of Champlain, Morgan fixed the date at from 1400 to 1450.

Relation for 1639, the Bear, Rock, Cord and White-Eared People claimed they knew with certainty that their tribes had been in Huronia at least two hundred years before, which indicates that the Norsemen were still there when the great crash of 1409 occurred, and it was probably then that the Norsemen in Huronia were finally overwhelmed and the survivors dispersed. Where they had come from is clearly indicated by the fact that the White-Eared, Cord, Rock and Bear people fled back to Hochelaga and the other fortified towns along the St. Lawrence River, while the Neutral Nation, or the Bowl People—the native laborers—remained in Huronia.

That the Norsemen had actually located the copper fields and taught the natives how to work them is indicated by structures not of native origin which the early French found there. Moreover, upon reaching Canada Cartier obtained from the Hurons a supply of what he deemed gold ore only to learn upon reaching France that it was "fool's gold."

With no suspicion that the Norsemen had been in Huronia centuries before 1603, Lescarbot was much puzzled by not finding the tongues recorded by Cartier. Nevertheless, he concluded that the changes in the languages of Canada were due to some great "destruction of people." In this, of course, he was right. Nor is it extraordinary that even in 1534 when Cartier came upon the Hurons, there was nothing to indicate they were descended from the Norsemen although it was obvious that they were different from the Iroquois. In the course of three generations the survivors of Huronia, cut off from the escape to Greenland, would naturally have been almost completely absorbed by the natives. Moreover, about red blood, there is something hostile to the beard. Thus, even half breeds often have no hair on their faces, so that at the end of 124 years the Hurons or Wyandots would naturally have been beardless. On the other hand, these people without written records would have preserved but a dim tradition of what had happened over a century before. All they knew was that a ceaseless feud between the Hurons and the Five Nations had been going on for centuries. Here it is to be recalled that when Davis visited Greenland in 1535, he could get no facts whatever from the Eskimo as to what had happened there in 1409. They did not even know that an earthquake had destroyed the cathedrals. So too, the Croatans knew little of their ancestors a century after the disappearance of Raleigh's colony. On the other hand, Father Christian Le Clerq, a Franciscan missionary, found Acadia just as described by St. Thomas Moel to Sinclair and Zeno —a land where there were cemeteries with crosses and many other evidences of an ancient Catholic mission.*

* It is easy to account for the ultimate complete absorption of the Norse blood. In 1615 Champlain reported 20,000 Wyandots, or the mixed descendants of the Norsemen, living in eighteen fortified villages in the general neighborhood of Huronia, while Sagard and Brebeuf put their number at 30,000 and 35,000 respectively. Be-

It is not unlikely that other Europeans beside the Norsemen were involved in the tragedy which has been described. Knowledge of a Welsh migration across the Atlantic first came to Europe from Meredith, the Welsh Bard, in 1486. Columbus himself had heard of this. Moreover, in the Welsh monasteries there is a record that an expedition was fitted out in 1170 by Prince Madoc who left his homeland with 120 men and soon returned with accounts of a region beyond Iceland and Greenland which he had visited. Recruiting a larger expedition of ten ships, he soon departed with about 300 men and women never to be heard of again. According to Gabriel Gravier, the great scholar of Rouen and President of the Norman Historical Society, who devoted his life to the study of the Norse sagas, Prince Madoc and his colonists went to continental America to join the Norsemen of Greenland and Iceland, and in a work on the Pre-Columbian visitations to America, he set forth a mass of evidence which cannot be ignored.

When, after returning to England, Captain John Smith set about recording his adventures in the New World, he wrote many things which, as in the case of Ptolemy, Heroditus, Marco Polo, and Raleigh were long cast aside as mere fiction, but which time has had a way of verifying year by year, until today his work is highly prized by the historian as a mine of reliable information. At the very beginning he refers to the tradition extant in Europe that a large expedition of Welshmen had set sail for the west in A.D. 1170. The fact that he added "no man knew whither they went," indicates he had heard nothing from the Indians of their being in America. So we commence our search for the Welsh with the historical fact that they migrated to some place in the west.

Early in the eighteenth century the Reverend Morgan Jones published an account of his captivity by the Tuscaroras of North Carolina in which he claimed that he conversed with them in Welsh. A few years later another Welshman, the Reverend Charles Beatty, declared that these Indians had a Welsh parchment Bible in their possession which they could not read, but which he was able to explain to them partly in Welsh. About the same time, Griffith, a Shawnee captive in 1764, claimed to have met a Welsh-speaking tribe. Soon, in his journal of 1774, David Jones undertook to make Welsh identities for the language of the Ohio tribes. Next, the great explorer, George Catlin, upon visiting the white Mandan and Minataree Indians a few years after Lewis and Clark had found them, and who are generally declared to be the highest types of the then aboriginal

---

cause they were allied with the French against whom the Five Nations were bitterly hostile, in 1656 they were crushed by the latter who had claimed Huronia since the overthrow of the Norsemen. Now the Bear People were incorporated by the Mohawks, the Rock People by the Onondagas, while most of the others, widely dispersed, remained with the French. This absorption would alone account for much in the peculiar cosmogony and character of the Iroquois.

race, was so much impressed by their superior culture that he deemed them the descendants of the Welsh immigrants. To their forebears, therefore, he attributed the circular moated earthworks in West Virginia, Kentucky, Ohio and Missouri, which in plan and profile, he showed to be the counterparts of the Pict works in Europe. Catlin's theory was soon expanded in an interesting work in which the Aztec state was traced to the Welsh whom the author declared were driven westward after landing in Virginia. Finally, General Custer, who knew the Sioux well, believed that they were of Gaelic origin. Certainly there were Sioux in South Carolina, because they were found there in 1557 by the Spanish.

"Various persons," says Mooney, the ethnologist, in referring to the Welsh, "have attempted to identify this mythic tribe with the Nottoway, Croatan, Modoc, Noki (Hopi), Padonca (Comanche), Pawnee, Kansa, Oto, and most of all with the Mandan." And Fiske, though admitting there was a Welsh migration in 1170, like Mooney, relegates the theories of Catlin, Bowen, Custer and others to the realm of limbo. Science, however, recognizes no such place, and it is obvious that the Pict-like mounds in America do not belong in such a "mythical" region. Furthermore, Prince Madoc and the Welsh must have gone somewhere, and the question of where they went is not solved by sneering at those who try to answer it. Certainly if Celts were in Iceland in the eighth century, and in Alaska in the thirteenth, there is no physical reason why they could not have visited America in the twelfth. Today the Indians do not speak Scandinavian but that does not prove the Vikings were not in America.

For a long time it was the habit to attribute the works described by Catlin to a prehistoric race of mound builders just as everything strange in Europe was attributed to the Pelasgians. But now we are told by the ethnologists that there was no such race in America. So that we still have the mounds, and no prehistoric race nor realm of limbo to which they can be assigned.

In truth these so-called mounds were no more than villages surrounded by earth banks, moats and wooden palisades with apertures for the discharge of arrows. Ordinarily they were located in some very strong defensive position which only people accustomed to European warfare would have selected. The dwellings were within the earth banks so that they were proof against fire arrows. There was a central area for games and ceremonies with a Communal Council House. From one of them a silver button with the image of a mermaid, which was a Welsh crest, and a European sword have been exhumed. It is highly significant too that the first English in Virginia found the Powhatans dwelling athwart a river named Appomattox, and the ruler's daughter bearing the name of Matoaka. The first at once suggests Ap Madoc, or the Welsh for "son of Madoc." Certainly there was no Welsh word the natives would have preserved longer as a name for the reigning family than that of the Welsh chieftain, were he ever in fact among them. Then too, the Manahoacs of Virginia, a small unidentified group, bitterly hostile to their Algonquin neighbors, had a name strangely suggestive of Madoc. It has been suggested by the ethnologists who cannot account for them otherwise, that they "may have been of the Siouan family."

All this gives us a clue. The Pict-like works led directly from Virginia to Huronia. It is not to be doubted that the Welsh would have learned of the presence of gold and copper in the latter region just as surely as did the Norsemen, and that just as surely, being natural miners, would have set out to find it, and it is right along the line of these works that the Conestogas and Susquehannas, in every way peculiar physically, were found.

There is nothing to show whether the Norsemen from Greenland or the builders of the earthworks arrived at Huronia first. Did the latter arrive there from Virginia and amalgamate with the Norsemen? This is not likely because they seem to have retained their identity after leaving Ohio, since the works from Huronia to Oregon are identical with those from Virginia to Huronia.

Did the Norsemen find them in Huronia and drive them out, or did they find the Norsemen there and aid the Sioux, Chippewas and Iroquois in their assaults upon the Norsemen?

While these things are purely conjectural, the works which continued from Ohio westward, are not. Plainly the builders retained their Caucasian identity better than the Wendats or Norsemen who survived the crash of 1409. This was due, no doubt, to the fact that they clung to their communal village system which prevented their absorption.

The movement westward is not difficult to understand. In the end, the natives of all tribes would probably have tired of these visitors just as they did of the Norsemen, and sooner or later would have driven them off. At any rate, after a wide gap, the fortified villages reappear in Missouri, always in strong positions at some bend of the Missouri River in the immemorial domain of the Sioux.

By their new hosts they were called Moatakni, or Metootahak, which was the nearest they could come to Modock or Madoc. These names had no other significance than "southerners" or "villagers of the south," or "the builders of villages in the south." Certainly it was an appropriate description for migrants from Virginia and Ohio. On the other hand, Mandan is but a corruption of Mawatami, meaning a "people who dwell on the banks of a river," while the name Minataree has not been identified. This latter white tribe merely called themselves "the people," just as those deeming themselves superior to their neighbors would naturally have done.*

---

* The White Indians, or Modocs, Mandans and Minatarees all claim they came from the seacoast in the east.

Here it is to be noted that after leaving Ohio and reappearing on the Missouri River, the village builders as shown by their works for a while moved straight toward the Black Hills and the sacred Red Pipe Stone Quarry in Minnesota. Of the quarry, which was deemed the Garden of Eden of the red race, the Sisseton Sioux are known to have been the traditional guardians. Thus, when Father Hennepin visited them in 1683, and Le Seur in 1708, they were still residing along the east bank of the Missouri River, about the present site of Sioux City. So also, in 1804 Lewis and Clark found the Sioux with a cordon around the region embracing portions of the present states of Iowa, Minnesota, North and South Dakota, Nebraska, Wyoming and Montana, within which lay both the sacred shrine and the Black Hills. Nor was it until after treaty relations with the Sioux had been established by the United States, and the Dakota, or League of Friends, had been more or less dissolved, that white men were allowed to enter this territory. Thus, it was peculiarly appropriate that the Sioux who were guardians of the gold region and the quarry were called by the Hurons "the worst enemies of all." In a sense they were the Levites of the race.

Whatever further evidence the ethnologists may find with respect to the connection between Europe and ancient America—and the present scratchings in Yucatan will no doubt yield much—enough already exists to dictate the rewriting of American history. It must not only include the tremendous influence which the red race has exerted upon the nation, but it must begin with the story of Iceland, Greenland and Vinland, instead of with San Salvador, Roanoke Island, Jamestown, Plymouth and Manhattan. Certainly Leif Erikson and Gudrid, and possibly Prince Madoc must be given their rightful place in the annals of the American people instead of being relegated to oblivion.

3

The Rediscovery of America

THE NORSE colonization of Greenland covered a period of just four and a quarter centuries. During all of this time, as shown, Europeans were passing back and forth from Scandinavia to the continent by way of the Orkneys, Iceland and Greenland. It is difficult to grasp the fact that from the discovery of Greenland to the great earthquake of 1409 which completed the destruction of the Byds, was about the same length of time as from the discovery of San Salvador by Columbus to the present day. Only a few of the events which had happened in America during this period are known. Nevertheless, the darkness of the four centuries following Leif Erikson's first settlement in Vinland is nothing as compared to that of the fifteenth century.

The reason for this is plain. While Greenland and Iceland were struggling with the "Black Death" and being convulsed by great volcanic disturbances, while the natives of the Continent were destroying the white invaders, Europe too was shaken to its foundations by titanic forces. Following the Crusades, the explorations of Asia by Marco Polo and the Italian Renaissance, Europe was determined to share in the wealth of the east. First Genoa and Venice must wipe Norse piracy from the seas. Then Portugal led the way around Africa to an unbounded Eldorado. Meantime, however, the red race with a faith and hopes of its own, seeing nothing among the Norsemen of the charity about which their priests had preached, had risen in its might to destroy the Norse civilization which Rome had encouraged among them, while the earth continued to belch flames upon the monasteries of Greenland and Iceland. It was as if the tragedy of Sodom and Gomorrah were being reenacted. The Vatican was so much engaged in a struggle to maintain itself against the new thought of the Old World, that it could no longer undertake to spread old thought to the New World.

Plainly too, Rome must keep abreast of the great economic movement in Europe. For many reasons which need not be discussed here, the politicians of the Vatican saw in the potential of the Spanish nation forming through the union of Aragon and Castile the best prospect of developing a political power that could be made to serve their purposes. It was, therefore, not Portugal, but this new Spain which the Vatican decided to temper into a sword for Rome. It was to be a veritable Toledo blade in the struggle against a spreading heresy and for the economic supremacy of the world. Accordingly, with its eyes dazzled by the glare of the new-found treasure, with a moral blindness which obscured utterly the meaning of Jesus, the Vatican set out to extirpate heresy by crucifying mankind upon a cross of gold. So it was that Torquemada came to Spain to assist the joint sovereigns in the task of expelling the Moors and the Jews along with a culture richer than all the gold of the East. Little thought it had now of the frozen wilds of the arctic region where lay the ruins of the old Norse civilization.

Nor was it until Cristoforo Colombo, a Genoese map maker, began to talk of a short route to the east by way of the Atlantic, that the interest of the Vatican in the old Vinland was rearoused. Having taken part in some of the Portuguese enterprises, this obscure man had studied with keen interest the record left by the Zenos of their expedition to Greenland and Vinland, and of Sinclair's voluntary exile on an island twenty days west of Europe. Thus, in 1477 he was in Iceland looking into all that was to be found about the land to the west which Rome had abandoned in 1409. Nor is there any doubt that he actually conversed with people there and with Normans, Bretons, Basques and Portuguese who claimed to have visited this region. Even now the Portuguese in the Azores were seeking a royal patent under which they might claim a western archipelago of which they knew.

Why was the Portuguese King immovable? Was it not because Rome had no idea of letting him lift the curtain of oblivion from the potential Eldorado of the Norsemen? At any rate, Colombo proceeded at once to Spain. This was not merely to solicit the aid of the poor and ignorant court of Ferdinand and Isabella, as commonly supposed. Obviously it was through Torquemada, who had arrived there two years before, that he must gain the assent and cooperation of Rome. Historians who take themselves seriously still write about eggs and the accidental interest of a king and queen in a poor Italian adventurer, and declare that Colombo was referred to learned mathematicians in order that he might convince Spain of the soundness of his theories. The truth is, the Vatican knew much more than Columbus about the earth. So instead of collaborating with mathematicians and cartographers, in fact Colombo was trying to convince the Vatican that unless it proceeded at once it would be forestalled in the race for the gold to be had in the abandoned Vinland as well as in the East.

There is no doubt that in England the likelihood of Colombo securing a commission through Rome from the Spanish king was foreseen. Was it merely accidental that in 1486 the Welsh bard, Meredith, or the compatriot of Henry VII began to rouse the English from their economic torpor?

Was it accidental that in 1488, without authority either from the King of France or Rome, Jean Cousin of Normandy set forth upon a voyage which carried him to Brazil?

Was it accidental that the report of Cousin's exploit, like that of the Zenos, was virtually suppressed and that even today it is hardly known?

Finally, was it accidental that suddenly the reluctance of Ferdinand and Isabella was overcome, and that one now called Cristobal Colon for obvious political reasons, was authorized to proceed on an expedition equipped by Spain?

It seems plain what had happened. While English, French and Portuguese visitations to Vinland had been frowned upon, the Vatican was at last compelled to permit the known wealth of the western continent to be exploited on behalf of Rome by its own chosen agent lest it be claimed by some power not under its control. Moreover, the vicar of Christ was sorely in need of gold to carry on the struggle with the heretics. With the Eldorado of the old Norse Vinland and the region described by Sinclair and Cousin to the south of it secured to Rome, Torquemada would be free to expel the Jews from Spain regardless of their wealth. Nevertheless, Colon was to say nothing about Vinland and was not to touch there since already it had been preempted. Unquestionably Rome knew that even now the remnants of a Christian civilization were to be found there upon which it could base its claim of exclusive jurisdiction. Therefore, steering straight west, on October 12, 1492, coincident with the royal decree expelling the Jews from Spain, Colon came upon an island which he named San Salvador. Of its exact identity the evidence is less clear than as to the first landing of the Norsemen in Vinland. Nevertheless, confident that he had found the region visited by Sinclair in 1380 and by Cousin in 1488, he explored the Caribbean archipelago which he named the West Indies.*

In the face of what has been said, it is unlikely that Columbus really believed he had come upon the fringe of India. The Vatican knew, and Columbus knew that the Norse Skrellings of Vinland were no part of an

---

* Since this was written the press has published a notice of the finding of Colombo's commission in the Vatican.

Asiatic civilization, that the Indians of Asia were not of a red race, and that those whom he found at San Salvador were of the same stock as the Skrellings. In all likelihood, therefore, the name West Indies was but a blind to hide from Europe the fact that a new hemisphere had been discovered until Rome could organize the grand scheme necessary to preempt it.*

"They are not idolators," declared Columbus, speaking of the people upon whom he had come at San Salvador, "nor have they any sort of religion, except believing that power and goodness are in heaven, from which place they entertained a firm persuasion that I had come with my ships and men. Wherever we went, they would run from house to house, and from village to village, crying out, 'Come and see the men from heaven!' so that all the inhabitants hastened toward us, bringing victuals and drink.

"In none of the islands I have visited have I found any people of monstrous appearance; they are all of pleasing aspect, with straight hair and complexion not very dark. After they have shaken off their fear of us, they display a frankness and liberality in their behavior which no one would believe without witnessing it. No request of anything from them was ever refused, but they rather invite acceptance of whatever they possess, and manifest such generosity that they would give away their own hearts."†

Such was the character of the American aborigines according to the contemporaneous testimony of the European who discovered them. How different is this description from that of those who were soon to cite the ferocity of these kindly people as justification for their destruction!

By reason of the name which had been given for the reasons shown to the western archipelago, the Europeans naturally called these people Indios, which in Italian and Portuguese became Indio, in French, Indian, Dutch, Indiane, and German, Indianer.‡

---

* This, some day, the archives of the Vatican will probably disclose.
† Letter of Columbus to Ferdinand and Isabella, February, 1493.
‡ "The American Race" was the title given by Dr. D. G. Brinton to his notable monograph on the natives of America (1891).
Creole did not necessarily denote a person of mixed blood as often supposed, though a mulatto, or a white-black cross, and a mestizo, or a white-red cross, might be a creole, that is, if the white parent were born in America. The red-black cross was called a zambo, English sambo. In popular parlance the red man is called "Red-man" and "Redskin," the latter corresponding to the German Rathante and French Peaux-rouge; in the French vernacular of Canada "Siwash" from bottle sauvage (moccasin) and traine sauvage (toboggan); in the United States "Mr. Lo" by reason of Pope's famous words: "Lo! the poor Indian, whose untutored mind, sees God in clouds, or hears Him in the wind."
The awkwardness of the name American Indian which is also scientifically incorrect, caused Major J. W. Powell, director of the Bureau of American Ethnology, to

A graphic picture of ancient America, based on the ethnological studies of Morgan and Powell, has been painted by Fiske which, although historically unsound because it almost completely ignores Norse influence, presents the social and economic facts of native life. The aborigines generally were found organized in communities called by Voltaire "tribal republics." As in the case of all primitive peoples they were bound together in the typical social unit of the tribe by ties of consanguinity and affinity, and by certain esoteric ideas derived from their philosophy concerning the genesis of the environing cosmos. With different languages, institutions, and rituals, their culture varied widely though a system of ownership that was universal existed among them. Thus, while personal effects were owned individually, these people, who still in part belonged to the stone age, had not yet attained a conception of property in land which, like the air they breathed, though necessary, was not understood as something to be appropriated to individual use.

The communalism which existed among them, in fundamental respects differed in no wise from that which the students of the Middle Ages will find everywhere in Europe. There, too, it seems to have been an all-pervading principle. Because communities rather than individuals appear as the chief units in the governmental systems of early Europe, just as among the aborigines of America, innumerable writers have endeavored to trace the origin of legal ownership there to systems of property holding by the tribe, the clan, the village or the family. So, too, the social structures of India and Russia have been set side by side with those of Western Europe, and the controversy which has raged around the village community has not yet subsided. Its progress may be followed through the pages of Maine and Seebohm, Von Maurer and Fustel de Coulanges, and in the more recent works of Round and Professors Vingradoff and Maitland.

The evidence is overwhelming that in their native state the Indians were a gentle, kindly people; that they cherished peace and were not as prone to strife among themselves as the peoples of Asia, Africa, and Europe of whom we first have knowledge—possibly not as much so as the Italians of the fifteenth century.*

---

suggest "Amerino," as a substitute which would lend itself easily to the formation of convenient ethnological words such as pre-Amerino, post-Amerino, psuedo-Amerino, Amerindish, Amerindize, etc. Although objected to by purists, this arbitrarily coined term has a certain vogue, like Nordic, which is widely employed to designate the superior Celtic stocks.

* The description of Columbus was confirmed by that of Cabaza de Vaca, the first European to enter Texas—"the land of peace"; by the Huguenots who visited Brazil in 1555 and South Carolina in 1562; by La Salle who descended the Mississippi River to Texas in 1585; by John Smith and the pilgrims of New England; by Father Hennepin and Le Seur who were among the first to visit the Northwest; by William Penn in 1682; and by Lewis and Clark in 1804. Penn's description of them was almost identical with that of Columbus.

That they were not more cruel by nature than the Europeans of a similar stage of development, is also abundantly clear. The system of slavery which existed among them was of the mild type of Greece and Rome and far more lenient than that extant among Europeans.

Despite the vitiating influence of the Norse, in their personal relations they were still a noble race among whom many of the vices common to the early civilizations of the Old World were unknown. The sacredness of the family ties such as they were, sexual probity, a staunch honesty and good faith in all their dealings, a broad generosity toward their fellowmen, a personal courage born of these qualities, were the outstanding characteristics of their untainted nature.

Morgan states that next to the Aztecs the Iroquois presented the highest stage of political development in ancient America.* And it is admitted that the Iroquois were the first to pass to the conception of an altruistic deity. These facts are readily accounted for by their initial contact with the medieval Christians which has been described. No one can study American ethnology without noting the striking similarity between the solar and lunar and the male and female cults among the aborigines and the early Norse Celts. Moreover, in a study of Nanabozha, or the demiurge of the Algonquins, Iroquois, Ojibways, Siouans and Mandans— the great ethnic groups with which the Norsemen first came in contact, one finds much suggestive of both the pagan mythology of the *Nibelungenlied* and the medieval Christian conceptions admixture of the Norse gods and those of Olympus, of Valhalla and the Indian Happy Hunting Grounds, of God and Satan, Heaven and Hell, the garden of Eden, Adam and Eve, the flood, Noah's ark, the Son of God with the orenda to work miracles, His death, ascent to Heaven, and resurrection, is readily apparent. Moreover, the cosmogony of these groups was little more primitive in conception than that of the Celts of Iceland and Greenland who, though nominally converted Christians, necessarily retained a pagan mentality for many generations. Certainly one cannot read Penn, Colden, Adair, Jefferson, Gallatin, Heckewelder, Schoolcraft, McKenney, Drake and Williams, and study the great American ethnologists—Shaw, Morgan, Powell, Curtain, Cushing, Matthews, Mooney and Hewitt—without being struck by much in common between the Aryans and the red race.

While it is not to be doubted that the Norse occupation of America had exerted a profound influence upon native thought and institutions before the Western Hemisphere was rediscovered, from what has been said, it is apparent that the efforts of the Vatican to uplift the red race from paganism by the agency of the Church, which it had established in North America, had been utterly unavailing. The reasons for this are plain. Freed from

* *Ancient Society*, Morgan, Harvard University Press.

the restraints of civilization, the continental Norsemen were themselves too close to the borderline of savagery and paganism to civilize a less advanced people. Indeed, if we disregard the moral code which a few priests had sought to enforce upon them, the natives were more moral even according to Christian conceptions than the Scandinavian Celts. Thus, there is not a recorded instance of an Indian chief or priest poisoning a brother, or an Indian guest raping the wife or daughter of his host, or robbing or burning a sheltering home in the night. These things happened so often among the white settlers of Vinland that the absence of murder, poisoning, robbery, rape, arson and the sexual vices of the so-called Christians among the hated Trudemans, can only be taken as a tribute to the latter. It seems surprising indeed that along with the epidemic diseases of civilization they did not contract its moral habits. But the truth is that native nature was repelled rather than seduced by the early civilization of Europe.

Unquestionably, there is romance enough for any poet about Gudrid, Freydize, Syazi, Valthof and Vimar, Ary Marson, Bjorn Kappi, St. Thomas Moel, Jonus, Erik Upsi, and the self-exile of Sinclair in the Wilderness of the New World. But romance is one thing and civilizing influence is another. Verily the first Europeans had raised tremendous obstacles in the path of the next. Nevertheless, the Spaniards found nothing hostile to Christianity so that, as might be expected of a people who were highly moral even according to European standards, the fundamental principles of Christian theology were peculiarly congenial to them. This is a fact to which the almost unanimous and amazing success of the Catholic missions testifies.

Their religion was not only not idolatrous, but in their minds the creative force was not anthropic. As declared by Pope in his exquisite lines, they saw God in nature and heard the Divine Voice in every wind. They were fatalists only to the extent of accounting for the specific form of the universe by the assumption that events which had happened in the past had settled once and for all the forms in which the same kind of events would continue to occur. They cherished a code of morals, strict adherence to which, they believed, would win them favor with the wonderful forces of nature which constituted the Great Spirit of Life; they believed, too, that by prayer and certain ritualistic practices this Spirit could be propitiated. The fact is, despite their taboos, charms, magic, symbolic actions, offerings, sacrifices, incantations and punishments, which when regarded alone appear as evidence of the rudest paganism, they had developed a very high intellectual conception of Divinity which was in every way consonant with the theory of evolution and the natural selection of species. Their rites no more affected their fundamental religious philosophy than the various ritualistic ceremonials of Christian sects—often relics of paganism—are to be taken to argue the basic concepts of the Christians, and it is the universal testimony that they were naturally more religious than the whites.

In the Indian there was an instinctive spirit of religion. Alone on the open plains or in the primeval forests, under the blue sky of God's world, he was nearer the deep springs of life than the European who came to convert him from his so-called heathenism. It was natural that, ever aware of nature, he should always be conscious of the eternal purpose of the elements. His God—the Great Spirit—was always in his daily life. When he went out of his tepee in the morning, he said his prayer to the Spirit who sent the sun; when he smoked his pipe he raised it to the four quarters of the globe, murmuring a prayer to the Spirit who watched over and provided for him. In the deepest shade the mingled green and blue of the mallard's down upon his pipestem reminded him of the earth and sky. God was always near him—the spiritual in all things taking precedence over the material.

To repeat, to the mind of the Indian, nature was not actuated by an anthropomorphic deity—a mental idol to which he should make obeisance. This being so, all the more credit is due him that out of the qualities of his mind and heart he could evolve the conviction that happiness, present and future, was only to be attained by obedience to a self-imposed code of ethics. It is in his strict adherence to the latter that is to be found the real evidence of the intellectual character of his religion and the true effect of it upon his morals. It is this aspect of his philosophy that is to be compared with that of the Christians who were now to seek to enforce their civilization upon him.

How many of these people were there?

As to this there has been much speculation. Extremists have imagined a native population of millions. On the other hand, it has been claimed, and persistently repeated, despite the removals, wars, epidemics, lowered breeding propensity and the patent fact that the aboriginal population of large regions has completely vanished during the past four centuries, that the Indians are more numerous today than at any time in the past.

The first error is due in part to the common tendency to magnify the glory of a vanished past, and in part to the mistaken idea that the numerous ancient remains scattered throughout the land were built or occupied at practically the same period. The contrary error—that the Indian has increased—is due to several causes, chief of which is the mistake of starting the calculation at too recent a period, usually at the establishment of treaty relations and governmental supervision.

Another factor of apparent increase is found in the mixed-blood element, since many persons are now officially rated as Indians who possess as little as one sixty-fourth of Indian blood. A careful study by the late

James Mooney, in 1910, probably the foremost authority on many Indian subjects, including population, led to the conclusion that at the time of the coming of the white man, there were approximately 10,000 Indians in Greenland, 72,000 in Alaska, 220,000 in British America, and 846,000 within the present territorial limits of the United States. The most thorough scientific analysis of these estimates fixes them as accurate within ten percent, so that it is conservative to take 1,150,000 as the figure representing the aboriginal population of the Western Hemisphere north of Mexico in 1492.

Since the thirteenth century many territorial readjustments had occurred. Nevertheless, in the successive human waves which had rolled eastward across the Rockies, there were still to be recognized five distinct families—Algonquins, Huron-Iroquois, Ojibwas, Siouans, and Caddoans —besides the Mandans.

Commencing with the Micmacs of Newfoundland and New Brunswick, and extending down the coast as far as Cape Fear, the Algonquins constituted about half of the native population east of the Mississippi and south of the St. Lawrence. The Etchemins, or Canoemen, dwelt along the St. John's and St. Croix rivers, extending as far south as Mount Desert, while the Abnaki furnished the Penobscot, Androscoggin and Norridgewock tribes. The forest west of the Saco in New Hampshire as far as Salem was claimed by the Pawtuckets, while the Massachusetts were already broken up in small scattered bands before the English arrived.

The Pokanets dwelt around Mount Hope and claimed Nantucket, Martha's Vineyard and a part of Cape Cod; the Narragansetts, Rhode Island, sharing Long Island and Connecticut with the Pequods, a branch of the Mohegans of the Hudson River country to whom were kin the Manhattans. Behind the Mohegans, always pressing up into Canada until Champlain finally drove them south of the St. Lawrence and Lake Ontario, were the Five Nations of the Iroquois, or the Mohawks, Oneidas, Onondagas, Cayugas and Senecas in the order named from the upper St. Lawrence to Ohio and extending down into Pennsylvania. The Guyandottes of the French, or Wyandots of the English, now occupying Huronia, were nothing more than an Iroquoian tribe known as the "Tobacco People" among whom remained only a few descendants of the Wendats of former days from whom they took their name. To the south of them, along the Wabash, were the Algonquin Miamis, or Twitchwees, beyond whom were the Algonquin Menominees, Illinois, Kicapoos, Mascoutens, Sauks and Foxes, with the Ojibway Chippewas, Ottawas and Potawatomis to the north. New Jersey, Delaware and Pennsylvania contained the Iroquoian Lenni-Lenape, or Delawares, the Conestogas, and the Susquehannocks, while the Algonquin Nanticokes occupied the eastern shore of Maryland and the Accomacks the portion of the peninsula to the south, being dependents of the Virginia-Maryland Confederacy extending from the James to the Patuxent. Along the coast between Virginia and Cape Hatteras dwelt the Algonquin Pamlicos, and around Cape Fear the Corees, or Coramines, while in the Midlands of Carolina were the Woccons, Catawbas and Meherrins, possibly kin to the Dakotas. Farther south were the Chowans, Nottoways and Tuscaroras, all kin to the Iroquois, the last being the dominant tribe in this region just as were the Five Nations in the region of the Great Lakes.

The Iroquoian Cherokees occupied the upper valley of the Tennessee River as far west as Muscle Shoals, and the highlands of Carolina, Georgia and Alabama. Southeast of them between the Chattahoochee and the Augusta were the Uchees and Natchez; the last, like the Yamasees and Savannahs, along the coast of South Carolina and Georgia, being attached to the great Mobilian or Muskogee-Choctaw family which included the Seminoles of Florida, and the Creeks of Georgia, the Choctaws of Alabama and Mississippi, and the Chickasaws north of them. Along the Cumberland River the Iroquoian Shawnees, who had been expelled by the Cherokees, connected the Algonquins of Virginia with those in the Ohio country, or the Miamis of the Wabash.

"It is not easy," says Bancroft, who followed Albert Gallatin, "to estimate their probable numbers at the period of their discovery. Many of them—the Narrangansetts, the Illinois—boasted of the superior strength of their former condition; and, from wonder, from fear, from the ambition of exciting surprise, early travellers often repeated the exaggerations of savage vanity. The Hurons of Upper Canada were thought to number many more than thirty thousand, perhaps even fifty thousand, souls; yet, according to the more exact enumerations of 1639, they could not have exceeded ten thousand. In the heart of a wilderness a few cabins seemed like a city; and to the pilgrims, who had walked for weeks without meeting a human being, a territory would appear well peopled where, every few days, a wigwam could be encountered. Vermont and northwestern Massachusetts and much of New Hampshire were solitudes; Ohio, a part of Indiana, the largest part of Michigan, remained open to Indian emigration long after America began to be colonized by Europeans. From the portage between the Fox and the Wisconsin to the Des Moines, Marquette saw neither the countenance nor the footsep of man. In Illinois, so friendly to the habits of savage life, the Franciscan, Zenobe Mambre, whose journal is preserved by Le Clercq, describes the 'only large village' as containing seven or eight thousand souls; Father Rasles imagined he had seen in one place twelve hundred fires, kindled for more than two thousand families; other missionaries who made their abode there describe their journeys as through appalling solitudes; they represent their vocation as a chase after a savage, that was scarce ever to be found; and they could establish hardly five, or even three, villages in the whole region. Kentucky, after the expulsion of the Shawnees, remained the park of the Cherokees. The banished tribe easily fled up the valley of the Cumberland River to find a vacant region in the highlands of Carolina; and a part of them for years roved to and fro in wildernesses west of the Cherokees. On early maps the low country from the Mobile to Florida is marked as vacant. The oldest reports from Georgia dwell on the absence of Indians from the vicinity of Savannah, and will not admit that there were more than a few within four hundred miles. There are hearsay and vague accounts of Indian war-parties composed of many hundreds; those who wrote from knowledge furnish means of comparison and correction. The population of the Five Nations may have varied from ten to thirteen thousand; and their warriors roamed as conquerors from Hudson's Bay to Carolina, from the Kennebec to the Tennessee. Very great uncertainty must indeed attend any estimate of the original number of Indians east of the Mississippi and south of the St. Lawrence and the chain of lakes. We shall approach, and perhaps rather exceed, a just estimate of their numbers at the spring-time of English colonization, if to the various tribes of the Algonquin race we allow about ninety thousand; to the eastern Dakotas less than three thousand; to the Iroquois, including their southern kindred, about seventeen thousand; to the Catawbas, three thousand; to the Cherokees, twelve thousand; to the Muskogean tribes—that is, to the Chickasaws, Choctaws, and Creeks, including the Seminoles—fifty thousand; to the Uchees, one thousand; to the Natchez, four thousand; in all, it may be, not far from one hundred and eighty thousand souls."*

* History of the United States, Bancroft.

4

The European Politico-Religious Conspiracies

AFTER EXPLORING THE West Indian archipelago, Columbus returned to Barcelona in 1493 bringing with him a number of specimens of the red race so that Rome found itself confronted not only with a great politico-economic but also a religious and racial problem.

The first step taken by the Vatican was the papal bull of this year by which it undertook to divide the so-called New World between Spain and Portugal. Plainly it was determined to balk England's designs upon the continent to the north. The next step was an attempt to consolidate the opinion of the Christian states behind His Holiness Alexander VI. This task was entrusted to the Dominican monk, Fray Francisco de Vitoria, protegé of Ferdinand and Torquemada, and professor of theology at the University of Salamanca. At once he began invoking the old Catholic doctrine of Pierre du Bois of France, now called Saint Pierre. Plainly, a league of the Christian nations with a supergovernment empowered to enforce peace upon recalcitrant potentates such as Henry VII of England was threatening to become was not only the best way to extirpate heresy but to insure a monopoly of the Western World for Rome. And in this far-flung scheme the aborigines, now called Indians instead of Skrellings, were to play a momentous part.

These "savages" he argued, were not subject to any king by human right. Their dealings with potentates were not to be determined by human law. Only a divine law, moving a political state as a theodicy, had brought about a political relation between the Indians and the king of Spain. Only divine law was competent to rule these creatures of God. Expanding his thesis, he insisted that the world was one society—a *societas naturalis*—a natural society of nations. The link between political states is God. Between the peoples of this one society peaceful intercourse may not be forbidden. What is true as to European people is equally true as between them and the natives of America. Therefore, the Indians cannot exclude the Spaniards. Since there exists the right of immigration—*jus communicationis*—Spain is justified in exercising the right of American penetration. The Spaniard also has the right to preach the Gospel to the Indian. The Indian has the right to preach "heathenism" to the Spaniard. Either may resist conversion just as, despite the *jus commercii*, he may decline to purchase proffered goods.

This argument was subtle, leading to unescapable conclusions. Since no state might prevent a stranger from settling on its lands, nor even from becoming a lawful national, the Spaniards were legally at home in the Americas. But strong powers owed it to God to defend by arms the menaced liberties of smaller states. That is the prerogative of a true society of nations. How much more readily, therefore, should strong powers defend the menaced liberties of individuals in every state! Just as Spain should protect the innocent of Europe from "religious sacrifice," Spain should protect the natives of America from "cannibalism." If need be, to protect the innocent, a state might subjugate wholly an unjust nation. Plainly Rome was threatening England.

Thus, the theologian undertook to lead Spain to her "cosmic place" in the Americas as Christ's vicar in the society of nations, and of his philosophy was logically evolved the papal bull of 1497 by which Alexander VI, with "divine justice" assigned to Spain and Portugal, as God's best tools, the mission of Christianizing the American aborigines within the separate spheres of influence already set apart to them. They were now to take up the task assigned in A.D. 1000 to the Archbishop of Bremen, in 1103 to the Archbishop of Lund and in 1276 to the Archbishop of Jou.

The commercial minded English, however, with a Welsh king and a much weakened attachment to Rome, had no idea of allowing the Pope to deny them a share in the Eldorado of Vinland. Spain might have the West Indies which Columbus had discovered, but not the continental region which both Rome and Denmark had long since virtually abandoned. Moreover, Henry VII himself was a Welshman. Had Meredith not shown that Madoc was his own countryman? Therefore, boldly disregarding the known purpose of the Vatican but keeping within the letter of the Roman law, he commissioned John Cabot to set up his standard on the northern continent. Thereupon Cabot proceeded in the utmost haste upon an ill-equipped expedition to Labrador. In order to justify the claim that Denmark had abandoned Vinland it was alleged that he found no trace whatever of the occupation of the region by Norsemen.

Over this the Vatican was naturally enraged. How did Henry VII dare lay claim to a region which belonged to Denmark by right of prior settlement and over which centuries before Rome had asserted its pontifical jurisdiction? Thus, we see, that long before Henry VII ever dreamed of a break with Rome over his marital problems, a widening break between England and Rome existed which the historians of both sides to the controversy have found it expedient to cover up with all manner of diversions.

In 1498 on a second voyage Columbus reached the southern continent while Cabot and his son proceeded to Newfoundland, obviously still searching not only for a northwest passage to Europe as ordinarily stated, but for the continental Eldorado which Henry VII knew Rome was determined to monopolize.

In the scheme of Rome the Italian navigator Amerigo Vespucci was to play an important part. Deputed by the Vatican to set up the Spanish standard in the northern continent, plainly it was for no accidental reason as supposed by many historians that his name instead of that of Columbus was given to the continent which England was claiming. Moreover, in Rome the Vinland of old had no connection whatever with the regions which Columbus had lately discovered. On the other hand, Columbus went to his grave knowing that he had not discovered a new world. It was Amerigo who, after the death of Columbus, showed that the old Vinland and the realm discovered by Columbus belonged to the same hemisphere. While Amerigo was off exploring, Caspar and Miguel Corte-Real were despatched in 1500 to the Gulf of St. Lawrence where they soon perished.

The semi-official pronunciamentos of Vitoria had no immediate effect upon Spain, much less other countries. At this time the Spaniards were among the most intolerant people in the world, and soon showed themselves to be as savage as the early Norsemen. Long they had been taught to hate and despise both the dusky Moors, who during five centuries had ruled a part of Spain, and the Jews. Naturally there was no room in the Spanish heart for the naked befeathered peoples inhabiting the Caribbean Islands. These they deemed no more than a species of animal having nothing whatever in common with the human race. Consequently, even during the time of Columbus, dogs, valued according to their ferocity, were employed with which to hunt down the natives of Haiti, all of whom had gold in their possession. There a large population was destroyed along with 40,000 unfortunates imported from the Lucayos Islands to replace it. Now, too, the Crown began to make large grants of land.

According to Vitoria not even the Pope had the human right to partition the property of the Indians. With all the Dominicans behind him, he stood firmly against both Pope and King, insisting in 1500 that "the Indian has as much right to possess property as the Catholic peasant." The potential equals in the natural society of man, they also had the right to plebiscite. A majority of their votes alone could justify the annexation of their territory to the Empire of Spain. Beyond the divine and human privileges that are general in the society of nations, "Spain must commit no act in the New World, except by treaty."*

Preceding as it did the ameliorating influence of Grotius by a century, though following that of Erasmus, this philosophy was so radical in its relation to the times that it may be said to have been the cornerstone in the foundation of European morality with respect to the people of the New World.

Numerous Basque and Breton fishermen were still making regular visitations to the Gulf of St. Lawrence, and in 1506, the year that Ximines succeeded Torquemada, Denis of Hornfleur arrived there to be followed two years later by Aubert of Dieppe.

What was Henry VII now to do without himself openly repudiating the Pope (for which England was not ready)? The answer came quickly. Perhaps the foremost intellectual in Europe at this time was Erasmus of Holland who after studying at Oxford had become an intimate of Grocyn, the champion of English liberal thought. The first European to print a translation of the Bible in a modern language, he rendered the Holy Scriptures available to the whole world and thereby did much to dispel the mysteries with which the priestcraft had enshrouded them. Having been bitterly assailed by the Catholic Church because of his liberal tendencies and obvious association with Henry VII through Grocyn, as the intellectual successor of Wycliffe, he now undertook in collaboration with them to defend himself by a tremendous assault upon Rome. Very cleverly, he did not direct this against the Pope himself but upon the priesthood and the priest cult which had grown up within the Holy Catholic Church. Against the control of the Vatican which was the complete expression of these alleged evils, the vicar of Christ was powerless to contend. By this time the whole Christian world was in an intellectual convulsion.

Such was the situation when in 1509 Henry VIII, with all the acquisitive instinct and shrewdness of his Welsh father, succeeded to the throne of England to find himself instantly beset by the politico-religious conspirators of both sides of the controversy which was raging between the Vatican and the reformers. Nor did he hesitate to take an active though secret part in the conspiracy of the latter. Thus, at once he despatched Sir Thomas More, the disciple of Grocyn and Erasmus, and the Sheriff of London, ostensibly to work out a trade agreement with the Dutch, but in fact a secret understanding with their bankers who were bitterly hostile to Rome. Plainly it was with Dutch money that he must lay the keel of the sea power by which alone he could defy Rome in a struggle with Spain for

* Here was the beginning of Indian treaties.

the possession of North America. The Vatican, of course, was fully apprised of his purpose. Therefore, what was no more than a second political "gold rush" to the so-called new world now commenced. The year 1513 saw Balboa exploring the Pacific and Ponce de Leon searching Florida ostensibly for the fountain of perpetual youth of which Valthof and Vimar had informed the Norse priesthood in 1051, but in fact for the pool of gold which was to give renewed vigor to Rome!

Having witnessed the horrors going on in the West Indies, and freed his own Indian slaves at the instance of the great Dominican monk Montesino, Bartolomé de Las Casas, who had just been elevated to the priesthood, proceeded to the Court of Spain in 1515 to plead with the king for the Indians, presenting the arguments contained in his subsequently published works. These were perhaps the most terrible indictments of European civilization ever written. But although the publication by him of the facts was forbidden for the time being, he succeeded in securing the ear of Ximines and the Pope who, having promised Indian reforms, flattered him with the title of "Protector of the Indians." The truth is they were no more able in 1515 to abolish the enslavement and murder of Indians than Lincoln could emancipate Negro slaves without a war, for the simple reason that there was no humanity left in a people like those of Spain. Europe had been completely demoralized by the gold nuggets found in the hands of the Indians, all of whom were pointing to the west as the source of it. Now that the aboriginal civilization of the West Indies had been destroyed, Las Casas himself was compelled to accede to the introduction of Negro slavery in America in order to prevent the attempt to further extend slavery among the red race by providing the Spaniards with labor to work the land they had seized. Instantly large importations from Guinea of the less advanced black race of Africa commenced. In this the murderous civilization of Europe was soon to find the penalty which the continuing sin of Thorwald was to bring down upon it.

In 1516 De Léry made a settlement at Sable Island in Canada, and the next year another Frenchman planted a colony on the St. Lawrence River at the mouth of the Saguenay. Plainly Henry VIII must act if anything of the life-giving wealth of North America were to be had for England. Therefore, in 1516, the very year that the two French colonizers arrived in Canada, Sir Thomas More, the successor of Grocyn at Oxford and the associate of Erasmus, and of Martin Luther, who was now openly assailing Rome, published his *Utopia*. In fact no more than a modern version of Plato's *Timaeus*, it was plainly designed to discredit Rome by adroitly comparing the reign of terror being sanctioned by the Pope in America with the ideal society which Plato had depicted. In other words, it was a shrewd appeal to the conscience of the Christian world in favor of Henry VIII against the dominion of Rome under which the Indians had been barbarously slaughtered and Negro slavery instituted in the New World. The implication was that the empire which the English had in mind would be one in which complete tolerance and amity would prevail so that men might think and worship and live and love as they pleased. And over and above all this was an even more shrewd appeal to the economic instincts of a world becoming more and more commercialized. In Utopia the communal system prevailed which as portrayed by Plato and More, avoided all the unpleasantness incident to poverty, tyrannical governments, taxes, and the like. Since there was no governmental or religious oppression, no war, and a wealth of happiness for everyone, the world might draw its own conclusions as to the desirability of such a society.

In a sense then, More did no less than arouse the natural socialistic aspirations of the downtrodden masses who from the appearance of *Utopia* were encouraged to associate the idea of liberty and freedom from economic want with a home in the New World. On the other hand, Rome was to be ceaselessly represented as the sole obstacle in the way of the fulfillment of the Utopian dream. Certainly it was all very cleverly contrived as a means of making Protestantism stand for something more in the undiscerning mind of the common people than a mere academic protest against religious dogma. Into the great religious debate had been introduced a hidden economic factor skillfully designed to break down old traditions and old loyalties and thus make it possible for Henry VIII to create the British Empire which he had in mind, backed by a navy that would enable him to claim for England the North American Eldorado! The latter was in fact the impelling motive behind all the wrangling of the Protestants with Rome until the dream of an exclusive Roman empire in the New World had been dissipated by the English conquest of Canada in 1760.

Aside from this there was another factor which now entered into European religious politics. The communal principle upon which More based his Utopian society was utterly inconsistent with the economic order of which Rome was necessarily the champion. The politico-religious feudalism of Rome and communism simply could not be reconciled. Once the latter was established the governance of the Vatican would be doomed. Therefore it was natural that the priesthood should secretly look upon Henry VIII in much the same light as the capitalists of later days came to regard Rousseau, Saint-Simon, Comte, Owen, Louis Blanc, Marx, Kropotkin, Bakunin, Lenin, and the socialists, communists, anarchists and social revolutionaries generally. There was all the more reason, therefore, why the Americas should be secured to Rome by every possible means.

Despite the promises made to Las Casas, the year 1518 saw the arrival of Fernando Cortez in Yucatan, already noted for the ruthless manner in which he had suppressed Indian uprisings in Santo Domingo and Cuba.

Crushing the natives at once in a bloody battle, the following year he took unto himself the famous concubine Marina whom he soon made the arch traitoress of her race and a valuable aid in the rape of Mexico. It was utterly vain now for Las Casas to establish a colony in Venezuela and attempt to Christianize the natives, who saw only too plainly the unChrist-like aspect of Negro slavery. They quickly destroyed his settlement.

To counteract the English-Dutch conspirators and in the hope of weaning Henry VIII away from them, in 1520 he and Francis I, of whom the former was known to be intensely jealous because of his American designs, were brought together upon the famous "Field of the Cloth of Gold," just as a meeting between Napoleon and the Russian Czar Alexander I was arranged three centuries later when, restive under the control of Rome, they were both trying to conquer the world.

But the league of Christian states under the hegemony of Rome of which Vitoria had been preaching meant only one thing to the counsellors of Henry VIII and the reformers. They deemed it but an obvious trick to place England under the complete domination of the Vatican. Therefore, having been excommunicated and sentenced to death by the Pope, the very day in 1520 that Rome pierced the golden heart of Mexico with the sword of Cortez, Martin Luther nailed his public protest against Rome upon the door of the Wittenberg cathedral. These things, of course, were not accidental.

In retrospect it seems plain in view of what was at stake which side of the contest Henry VIII would officially espouse. For the moment sheer expediency caused him to hold aloof from open participation in the great revolt which Luther, the Protestant, had instituted. In fact, the Vatican and the kings of Spain and France, now in complete cooperation, left him no choice as the champion of English nationalistic and commercial aspirations.

After Cortez had forced Montezuma's own subjects to massacre him, his successor, Guatimozin, was placed upon live coals to force him to disclose the place where the Aztec treasure was hidden. The conquest of Mexico, like that of Peru soon to follow, was represented as a prodigy of valor, in order to cover the real purpose of it. In fact it was no more than wholesale murder for gold with the tacit approval of those whose support Cortez had.

It was now that the real nature of the Aztec state came to light. According to the Aztecs themselves, there appeared among their ancestors a man favored by Heaven who engaged several wandering tribes to settle in the country of Anabac, the most fertile and most pleasant in the land, and there establish themselves under a regular government. This state, at first somewhat limited, is known to have extended gradually by the agglomeration of several tribes who became united and formed the flourishing empire of which Mexico City became the center about the time the Norse were in Melville Sound. Montezuma was the ninth emperor. The capital was quite large and thickly populated, but the structures, even the greatest, such as temples and palaces were so badly built they indicated an architecture still in its infancy. While the religion of the Aztecs, gloomy and ferocious as that of the ancient Celts, permitted human sacrifices, forms of the feudal system were found in the government. The emperor had under his dominion thirty nobles of highest rank, each of whom had in his own territory about 100,000 citizens, among whom were 300 nobles of an inferior class. The caste of the Mayeques or Mayas was similar to that of the ancient serfs. In the cities as in the country the ranks were distinguished, and to each man was set apart a rank according to his profession.

The Mexicans had only a crude knowledge of nearly all the arts without perfecting any. Their writing consisted only of hieroglyphic pictures. They had, nevertheless, a sort of post by means of which the orders of the emperor or important news was sent forward from the center to the extremities of the empire. Their year was divided into eighteen months of twenty days each, to which they added five complementary days, which indicates some astronomical knowledge. Their agriculture was imperfect. As they did not understand money, the taxes were paid in kind. Each thing, according to its kind, was arranged in storehouses, from which they were drawn for the service of the state. The right of territorial property was known in Mexico; every free man possessed there a certain extent of land; but the social ties, still uncertain, showed a social state at its dawn.

The Empire of Peru was also in its infancy, but possessed more agreeable forms than that of Mexico. The more gentle religion, the more brilliant cult, gave more charm and *éclat* to the government. The Peruvians worshipped the sun and the moon, paying certain homage to their ancestors which indicated that their legislator was of Asiatic origin. According to the Peruvian traditions, this legislator, named Manco-Capac, appeared with his wife Mama-Ocollo upon the shore of Lake Titicaca and announced himself as the son of the Sun. Assembling the wandering tribes, he persuaded them to practice the agriculture he taught them. After this first step, the most difficult of all, he initiated them into useful arts, gave them laws, and had himself recognized as their theocratic sovereign. It was on religion that he founded every social edifice. The Peruvian Inca was not only legislator and monarch but revered as son of the Sun. His person and his family were sacred. The princes of the theocratic family espoused their own sisters to avoid mixture with any other blood, as the Egyptian monarchs had done in former times.

All this does not necessarily imply an Asiatic origin for the institutions of Mexico and Peru. The son of the Sun would probably have come with the rising sun. The descendants of those who accompanied him, had they been but a few Norsemen, naturally would have intermarried closely to preserve their racial character and superior powers as Europeans. Moreover, the Norsemen possessed the mentality of pagans even though they had been legally Christianized.

By this time every explorer and cartographer in Europe had looked into what there was to find about the Norse occupation of Vinland and like Cabot, Denis, Aubert and de Léry, they all discovered that latterly the Norsemen had concerned themselves with the northern hinterland. Therefore, in 1524, under the commission of Francis I, coasting northward from Florida, Verrazano tried to penetrate into the northern interior by way of Chesapeake Bay. Foiled in this, he finally arrived at the mouth of the St. Lawrence where he found those who had preceded him, but no gold.* Two years later De Allyon, a Spaniard, planted the standard of Spain in North Carolina. After ravishing the natives there and carrying off a whole community into captivity, he founded the Spanish settlement of San Miguel on the James River at the present site of Jamestown. Thus, it is seen that in the search for the northern Eldorado the same policy which was employed by Cortez was employed by De Allyon with the result that henceforth the natives of the middle Atlantic region not unnaturally had the same dread of Europeans which those in the North had inherited. In 1528, after Cortez had disgraced civilization by the ruthless gutting of Mexico, moving from Florida along the Gulf coast, de Vaca arrived in Texias, or Tejas, the Caddo "land of peace," which was scourged throughout the next eight years. The injustices done the natives by his companions are recorded in great detail.† Almost coincidentally with him came de Guzman who also found numerous Tejas or Texas Indians.

Meantime Montesino and Las Casas had convinced Rome that Cortez had gone too far. In estranging the natives by resorting to torture to wring from them their gold, with wholesale massacre as the penalty for not yielding it, the Spaniards were defeating their own ends by arousing the whole native population of North and South America. There was a better way than that to secure the western Eldorado. Therefore, upon the counsel of Ximines, the account written by Las Casas of the destruction of the Indians was published in order to gain popular support in Spain for a new Indian policy and in order to prepare the world at large for what was soon to come. The church now took up the issue under instructions from its leaders and promoted a great reaction against the murder of Indians. The

---

* Adams, in his *American Epic,* makes the mistake of assuming the sight of Verrazano's vessel was a novelty to the natives.

† It was a sad commentary on what happened that the name Apache, or that of a tribe which inhabited Texas, the "land of peace" later should have been applied to the human devils of Paris, just as Mohawk was the name given to the thugs and ruffians of London.

truth is, Ximines and others had seen that if the Americas were to be secured to Rome, the aid of the natives must be had. The next step in his program was, therefore, the formation by Ignatius Loyola in 1528 of the Society of Jesus for the direct purpose of providing papal agents with which to Christianize the Indians and organize them politically behind Rome. Therefore, after Mexico had been looted and had nothing more to yield, explicit instructions were issued to Cortez as Captain General of New Spain for the institution of the entirely new system of governance which was to be applied under the direction of Ximines and Loyola to the Indians wherever the jurisdiction of Rome could be asserted. Accordingly, in his commission of 1529, he was directed to "give his principal care to the conversion of the Indians; to see that no Indians were given to the Spaniards to serve them; that they paid only such tribute to His Majesty as they might easily afford; that there should be good correspondence maintained between the Spaniards and the Indians and no wrong offered to the latter either in their goods, families, or persons."

According to Antonio Herrera, Bishop Don Sebastian Ramirez, acting governor under Cortez, earnestly endeavored to put into practice these humane orders. He not only abrogated the enslavement of the Indians, but took care that none of them should be made to carry burdens about the country, "looking upon it as a labor fit only for beasts." Nor was he less exact in the execution of all the ordinances sent by the Council of Spain for the ease, improvement and conversion of the natives. "By all means," adds the old historian, "the Country was much improv'd and all Things carried on with Equity, to the general Satisfaction of all good men." And the fact is that the laws of Spain soon came to recognize certain rights of the Indians to the land actually occupied by them though not in unoccupied areas which were designated "waste lands" or Crown domains, while it became the practice to compensate them for occupied lands that were taken, usually by grants of waste land. Soon, too, the gentle Franciscans were earnestly engaged in Christianizing the Indians who everywhere responded promptly to their efforts despite all that had gone before.

The Indian policy proclaimed for New Spain was the answer of Rome to *Utopia*, but it was too much for Henry VIII, who had witnessed with jealous longing the great gold-bearing galleons returning from Mexico and Brazil. Accordingly, in 1629, or the very year that the account of Verrazano's voyages was published, the English King with his navy near completion, broke with Francis I and Rome. An open fight between them was now on. Instantly Hawkins, the greatest seaman in Henry's new imperial navy, appeared off the coast of Africa to break the Portuguese and Spanish monopoly of the slave trade. In this highly profitable business, too, the Utopian idealists of England wished to share!

5

Roberval and the Prophet Dekanawida

I T IS SIGNIFICANT that after the debacle of 1409 in Greenland nothing more was said in the Icelandic literature about Vinland. On the map
published in 1529 by Verrazano, however, appeared the name Aranbega
to designate the region of the New England coast. Thus, it is apparent that
a new name had been substituted for Vinland.

But what was, and where was Aranbega?

Plainly, the Cabots, Denis, Aubert, de Léry, Verrazano and others had
looked for it in Canada. Yet, the accounts left by them negate the idea
that there was a town with such a name either in New England, Newfoundland or Canada. This indicates that the term was applied to a region
instead of a town. Certainly it had the searchers puzzled. Thus, when in
1531 Aubert of Dieppe published an anonymous account of his visit to
America in 1508, he used the name Norumbega to designate the whole
coast from Cape Breton to Florida, parts of which had already been
searched by Amerigo, the Corte-Reals, the Cabots, Ponce de Leon, Denis
of Hornfleur, de Léry, Verrazano and De Allyon, himself, each in turn,
and by all with the same result—the finding of a little gold in the hands of
the natives but with apparently no local supply of it.

It was now time, however, for Spain to seize the wealth of Peru which
had long since been prospected. When Pizzaro, a worthy pupil of Cortez, arrived there in 1531, a successor of Manco Capac named Huano-Capac
was on the throne. He left a son, Ata-hualpa, to whom he wished to give
only half of his empire, the kingdom of Quito, declaring his brother,
Huascar, whom he loved dearly, heir to the kingdom of Cuzco. This

unprecedented division caused a general discontent, kindling a civil war. Of this the perfidious Pizzaro took advantage to offer aid to Ata-hualpa, to approach him and carry him off from the midst of his subjects in an odious manner. A priest named Valverde loaned his services for this execrable act and dared even to confirm the sentence of death which was pronounced by the ferocious Spaniard against this unfortunate monarch. Ata-hualpa was strangled by special grace instead of being burned alive as the sentence ordered. Thereupon, a new flow of gold to Spain commenced.

The years 1534-35 saw Cartier searching for Norumbega in Canada while John Davis, representing Henry VIII, proceded on a similar quest to Greenland. The speed with which Cartier hastened home after his discovery of what he deemed gold shows that as early as 1634 the French believed they had discovered the Eldorado of Norumbega at last. But when the mistake was discovered a great search was instituted by Spain which was to leave no part of the continent untouched. Thus, in 1639, de Soto arrived in Florida and carrying fire and sword among the Muskogees, slaughtered over 2,000 of them before arriving in 1542 at the "Great River" near the present site of Memphis in the land of the Natchez. Thence he passed through Texas to Mexico. Coincidently Cabrillo searched the Pacific coast where he planted the first Jesuits among a hitherto untouched native population. At the same time Coronado swept through western Texas, New Mexico, Colorado, Utah, possibly Montana, Wyoming, Nebraska, Iowa, and Kansas, and back to Texas. And while all this was going on the greatest cartographers of the age placed on Mercator's map for 1541 the name Anorumbega in the region of the Hudson River! Thus it is seen that, already aware of the real purpose of the Spaniards whose priests inquired over-much about their gold, the Indians, with no outside aid as yet, were deliberately shielding the Red Pipe Stone Quarry and the gold fields around which the cordon of the Sioux, or the "worst enemies of all," was still thrown just as in the days of the Norse invasion. From this fact alone it is manifest that they already possessed something of a racial organization. This was only possible through the unifying control of the native shamans at the shrine subsequently found at the quarry.

Although disappointed by Cartier's "fool's gold," the French had not failed to recognize the value of the great copper deposits in the region of Lake Superior. These they proposed to monopolize. But that is not all. Undoubtedly Cartier had learned of the great mineral wealth of North America, and the certain presence of gold in the Northwest. Therefore the French knew that Norumbega, or the region where the Norsemen had operated, had been found at last.

From this moment it was not Cartier but the Jesuits who were to dictate

French colonial policy. Henceforth Spain was to abandon all her claims beyond Florida, Mexico and California to France as the agent of Rome for the purpose of monopolizing North America.

It has been said that if you scratch a Russian you find a Tartar. It is certainly true as shown by their whole history that those who wore the cassock of Jesuit missionaries among the Indians were political agents of Rome in the great scheme of French Empire. Certain it is, no one can understand American Indian history unless it be read in the light of the primary objects of the Jesuits.*

The endearment of the Indians to Rome by Christianization and respect for their rights had been the guiding principle of the Society of Jesus from the first; therefore, when the purpose of Henry VIII to overthrow the sea power of Spain and to contest the American empire of Francis I became fixed, the Jesuits had seen only too plainly the immense possibilities of employing the Indians to make British colonization impossible. But they also learned from Cartier's experiences in Canada of the great obstacles in their path. First, the ancient Iroquois Confederacy completely dominated Canada. Certainly there would be no chance of founding a French empire there without controlling it since the Iroquois had reduced all the neighboring tribes to a virtual dependency. Second, the traditional hatred of this powerful dominating force for the Wyandots or Hurons made it impossible for outsiders to befriend the latter without antagonizing the former. Third, the Hurons and the hostile Iroquois alike remained far from disposed to abandon their own peculiarly congenial form of mysticism. In the Indian cults the medicine man or shaman ordinarily embodied the noblest personal attributes of the social democracy which was common to the tribal communes. He was a philosopher, historian, teacher, moral exemplar, leader, and often a personal friend, all in one.

Nevertheless, their recognition of these facts led the Jesuits to the next steps. What if the tribes in every great ethnic group in America should be organized into a confederacy after the Iroquois pattern? Was it not obvious that their power to resist the British would be enormously enhanced, and that it would be easier to control them through their supergovernments than by trying to deal with them singly? In this the Jesuit and Franciscan missionaries in Canada, Florida, Mexico and California could be brought into cooperation. Once this had been achieved the founding of

* On one occasion these things were being discussed with a man of great learning. Perhaps it was in Washington, or Scandinavia, or in the wilds of Canada, or at the great Indian shrine of Minnesota, for one of the two had searched for an answer in all of these places. Whether the other was a Catholic, a Protestant, a white man, an Indian, or a "cross-breed" need not be said. "Perhaps," he remarked, "if those who call themselves ethnologists, anthropologists, archeologists, historians and what not, had access to the documents in the Vatican, they would understand much which now seems obscure."

a Protestant empire in the New World such as that which More and Coligny had envisaged would be impossible.

For the carrying out of this far-reaching plan a large force of carefully selected intrepid men was to be sent to Canada at once with no women. In every possible way they were to be encouraged to intermarry with the natives in order, with the aid of priests, to form a medium of union between France and the tribes who had overwhelmed the Norsemen. Meantime the rights of the Indians were to be scrupulously respected while the quest for the Eldorado was going on. In this way the shrewd French proposed to search out the innermost secrets of the red race while endearing it to themselves by a policy in marked contrast with that of Spain, or the terrorizing of the natives which had failed utterly. Accordingly, Cartier returned to Canada as governor in 1540 with one party and in 1542 was joined by Jean François de la Roche, Sieur de Roberval, of Picardy, as lieutenant governor, with another. Instantly those who had been carefully selected for the task spread out all over Canada and began to take unto themselves Indian wives. Thereupon Cartier returned to France leaving Roberval and the priests in charge.

No priest could work long among the northern tribes without soon learning of the great mystic shrine "in the mountains of the prairie" of Minnesota. According to common belief it was there that the Great Spirit —Gitche Manito—spoke to his red children as Jehovah spoke to Moses from Mount Sinai. And although Frazer had not yet contributed to the world *The Golden Bough*, no one knew better than the Jesuits the history of religious cults and understood better the principles of taboos and magic universally found in primitive religions.

Who conceived of the scheme which was now devised to make use of the native shamans for the political purpose of federating the tribes as well as leading them gradually toward Christianity through the use of the Indian oracle, cannot be said with certainty. Yet, it is easy to narrow the possibilities down to a very few persons if knowledge of the Indians and chronology both be considered. Whoever it was had to know about the workings of the shrine at the Quarry, and he had to evolve his plan between 1534 when Cartier first arrived in Canada and 1545 when the first evidences of it occurred. Therefore, we need only search among the French who were present in Canada during this single decade.

Cartier's own account of his experiences in Canada on his first visit was widely published. In it there is no indication that he could then have entertained this scheme but much to indicate the reverse. The same is true of his second visit. This is all the more strange in that meantime Roberval had vanished in thin smoke. According to his friend Thevet he was later killed in Paris, but according to others, after leaving the Saugenay settlement in 1643 he never returned to France and certainly no more was heard of him in Canada. The conflicting evidence as to his fate suggests much.

Roberval was a learned man. He ruled the country well and before his disappearance learned a vast amount about its people. Nor is it beyond the range of possibility that in 1543 he deliberately associated himself with the Indians, submerging his identity among them for a great political purpose, just as Sam Houston did two centuries later. If in fact he did this in concert with the priests in Canada, the secrecy of Cartier's third visit is readily explicable; also the strange fact that no searching inquiry by Cartier or anyone else as to Roberval's whereabouts is recorded. To say the least it was most extraordinary that the Sieur de Roberval, Deputy Governor, could vanish with no contemporaneous discussion of the matter by Cartier, who discussed almost everything else.

When in 1832 George Catlin was attracted to the realm of the Dakotas, or League of Friends, by rumors of the secrecy surrounding the Red Pipe Stone Quarry, as the first white man known to have visited the Quarry, he learned many strange things.*

The region of the Quarry was deemed sacred because it was supposed to be analogous to the Christian garden of Eden. According to the prevailing Indian tradition, the red pipestone which this quarry alone yields had been stained by the blood of the ancestral race which had perished there in a great battle among the red nations who had survived only through three maidens impregnated by the Great Spirit. Mystic rites were still being conducted amid awesome phenomena easily traceable to medicine men hidden away in the caverns of the wooded cliff near the quarry, while it was out of the murmuring falls of Winnewassa that Gitche Manito was supposed to speak to the Indian pilgrims to the shrine.†

All this has a strangely familiar Christian note. Countless similar shrines exist in Christian countries even today. One cannot fail to wonder if some druidical-minded Gael or early Norseman were not the founder of these mystic rites long before the Jesuits in Canada learned of them. However that may be, soon after Roberval's disappearance the Indian oracle began to speak, and before long the belief was prevalent among all the northern tribes that in the past Gitche Manito had assembled the nations there, and after molding a calumet, and smoking it as a signal, after commanding them to cease their warring, and unite to enforce universal peace, had promised them a prophet to teach them a new system of life.

* We know that Father Hennepin and Le Seur both visited the region of the shrine many years before.
† It was upon the facts recorded by Catlin and the legends reported to Longfellow by Mrs. Eastman, that the great poet based much of his *Song of Hiawatha*. The red pipestone was named Catlinite after its white discoverer.

Above all else the Indian was a poet-artist. It took natural surroundings like this to touch his soul—a veritable mountain oasis rising with sudden beauty sheer and green out of a prairie desert, a torrential stream apparently coming from nowhere, hurling itself over a wooded ridge, gurgling a while with foaming violence, calming into the murmur of a pellucid pool audible in its crystal clearness above the silence of a vast solitude, returning as suddenly as it came to mother earth. Oh, yes! It was possible there for the red man and even for Roberval to commune with the Great Life Giver, to hear His voice, with thoughts uplifted, with spiritual arms outspread to Him, to penetrate with a clarified vision into the mysteries of life. Peace on earth and good will among men! That was the message—the sermon on the mount—that now came to these so-called red barbarians while the French and the Spanish and the English were murdering them and each other in the name of Christ.

At any rate, with such a unifying legend abroad among the Indians, which soon attained to the dignity of a religious conviction, obviously a splendid moral background had been prepared for the federation of the red race under the spiritual guidance of Rome through the native philosophers, or shamans, and the tribal leaders. Surely it was a grand conception, one eminently worthy of Roberval, one which would have done justice to the most astute Jesuitical mind.

The prophet whom in the way shown the northern tribes had been led to expect, soon appeared among the Hurons in the person of Dekanawida. Inasmuch as he was living after 1570, at which time he was deemed a venerable seer, presumably his birth did not occur before 1500. This is the year Roberval was born. Not only that but the prophet commenced his labors about 1545 which concurs almost exactly with Roberval's mysterious disappearance.

The name taken by the prophet is particularly significant—"one in whom two rivers meet." This would be an eminently appropriate designation for one who was seeking to bring two great blood streams together, or unite two races. Finally, although Dekanawida appeared among the Hurons just where Roberval would most likely have commenced his task, his origin is uncertain. About it was thrown a great mystery. His alleged mother claimed to have drowned the son who had come back to her because of a premonition of evil to the Hurons on his account. This, of course, would have fitted in exactly with Roberval's plans since he would have had to claim that he was Indian born and assume a false identity beyond the risk of detection. On the other hand there was no difficulty about a dark Frenchman claiming to be a Huron. Furthermore the widespread belief that Dekanawida was a resurrected spirit is exactly what Roberval would have desired most in order to give him the standing of a prophet of supernatural birth possessed of the orenda to work miracles.

Whether Dekanawida was or was not Roberval, certain it is he had resorted to an old well-known Christian trick with which also the Jesuits were thoroughly familiar. Hundreds of them have been elevated by their flocks to the spiritual rank of an inspired sainthood. Moreover, according to the ethnologists the prophet was supposed to have been driven out of Huronia by the Hurons, which at once clothed him with the aspect of a martyr. Plainly Roberval would have been shrewd enough to work among a tribe different from that of which he claimed to be a member since the chance of detection would have been less. On the other hand, it was the Iroquois and not the Hurons he would have wished to dominate.*

At any rate, soon the prophet was preaching among the Eries, and then came to the Senecas whom he found bitterly hostile, not so much to the radical social reforms he proposed as to peace with the hated Hurons or Wendats. Among the latter he continued to labor many years inasmuch as the conversion of the Senecas was vital to the success of his scheme.

---

* In view of all these facts the author after years of study and discussion with the most learned Indians of the Sioux tribe, in the absence of proof to the contrary, is convinced that Dekanawida and Jean François De La Roche, Sieur de Roberval, were one and the same.

6

Dekanawida, Hiawatha and the Hodenosaunnee

THE DEATH OF HENRY VIII, whose navy alone had saved England, occurred in 1547. Edward VI was succeeded by Mary in 1553. Meantime, Coligny, the Grand Admiral of France, as the great leader of the Huguenot, or French Protestant, party and the ally of Luther, Melancthon, and Calvin, had been completely fascinated by More's Utopian theories. Therefore, with no part in the Catholic scheme of Canadian colonization, he undertook to establish a colony in Brazil as the germ of a great republic of free worship in America based on the communal system. The immediate marriage of Mary to Philip I of Spain, her deliberate destruction of the expanding English sea power, the massacre by Menendez of Coligny's experimental community, the reunion of England with Rome by act of Parliament in 1555, the burning of English Protestants, including Archbishops Cranmer, Ridley and Latimer, were but successive steps in the process of the two conspiracies between which the helpless Indians were now being ground to death with only their native instinct to protect them.

Succeeding to the throne of England in 1558, for a while Elizabeth sought to compromise with the opposing forces centering in Rome and Geneva. But this was impossible. The Vatican was fully aware of the purpose of the theocratic state set up by Calvin in Switzerland, and of the growing power of the Huguenots under Coligny in France and of the Dutch Protestants under Coligny's son-in-law, William of Orange. Of their undying purpose to found a Protestant state in America, it had no doubt. One who did not obey the Pope was an enemy of the Vatican. Eventually therefore, led by Bishop Parker and strongly reenforced by the large Dutch element of England, Elizabeth was compelled to choose between the economic destruction of England and peace with Rome. But before the final choice, which was inevitable on her part as the ordained foundeı of a British Empire free of Rome's control, her father's navy must be restored in order to protect England against Rome, to say nothing of seizing North America. Her great minister William Cecil points out to her that this cannot be done through the mere fostering of fisheries and foreign trade. Lest they be mistaken for Papists, the English Protestants and the immense Dutch population in England, which the persecutions of Philip I in the low countries is constantly recruiting, will not eat fish! Moreover the commercial capital of the country is almost entirely monopolized by the Dutch and Italian bankers and trade interests which cannot be estranged. Therefore, but one thing remains. Piracy must be legitimatized through the legal fiction of privateering in order to encourage the impoverished sheep-growing gentry to turn to maritime enterprises. Thus are the bold seamen of Devon like Hawkins, Drake and Gilbert induced to take to the Spanish Main for the direct purpose of enriching England by plundering the Spanish trade with America. The gold-bearing galleons from Mexico and Peru which had irresistibly aroused the cupidity of Henry VIII and were the real cause of his break with Rome, are now the legitimate prizes of England! In other words every tie of blood and religion known to the old civilization of Christian Europe has failed under the temptation of "King Gold." The papal bull of 1497 imposing a moral duty upon Spain to save the soul of the heathen red race had proved but a scrap of paper.

In 1561 De Villefane ravaged the north of Florida, or the present South Carolina, which he claimed for Spain. Thereupon Coligny, with Protestant France, William of Orange, the Stadtholder of Holland, and the British squarely behind him, undertook to dispute this claim. The experiment in Brazil had been too brief to test the communistic theories of More. Recruited in Geneva and France, a new colony was dispatched to northern Florida under Ribout in 1652 which was reenforced two years later. Founded on a principle with which the colonists had no experience whatever, unfortunately it included an excessive number of fanatic preachers who knew no more than Luther, Calvin and Knox of the all-embracing tolerance of Coligny. Despite the gentle Laudoniére, soon life among them was intolerable, for each was convinced that he alone had discovered the secret of life. In one thing only did these pioneers in the field of human tolerance find a community of interest—the maltreatment of the Indians. Inasmuch as none were willing to work they all robbed the natives. This continued down to 1565 when, aided by previous experience as the Lord High Executioner of Rome and the Capitalistic Order, Menendez arrived on the scene and ended the strife with a wholesale massacre of all sects, including the Catholic communists. Thus, in the twentieth century the Bolsheviki of Russia were readily able to find historical precedents for the most expeditious method of maintaining their own economic theories.

Upon the murderous warfare among the Europeans the natives looked amazed. Apparently what the Indian deemed murder was permissible among the visitors when committed in the name of their God, the Prince of Peace. It was not strange that subject to economic conditions which made their own communal system possible, they feared a civilization which in its irrespressible craving for gold and individual wealth sought to maintain itself by wholesale political murders and the individual murders of an Inquisition which the vicar of Christ justified as necessary to vindicate the true faith. As between the intolerance of Rome and Geneva their stone-age intellects could make no choice. By this time an Indian would not believe a Christian of either sect upon an oath sworn on his own Bible.

Such was the Europe which, under the guidance of astute Jesuits and commercialized Puritans, with an arrant hypocrisy common to both, continued with increasing impatience to search America for the wealth which equally they craved with no real interest in humanity and particularly in the red race.

A generation had passed since the appearance in Canada of Cartier and Roberval with the progenitors of what was now, perhaps, the most wonderful race of men in the world—"the friendly ones," or the Franco-Indian habitants of the St. Lawrence River and the region of the Great Lakes. With all the knowledge of Europeans they possessed the native qualities as well. Thus, they were born conquerors of a wilderness which could conceal nothing from their eyes. They had discovered the gold fields. And so on Mercator's map for 1569 appeared a picture of Norumbega as a fortified place with high towers, indicating only too plainly the far-flung indefinite region of the north which plainly France meant to hold against the world. Obviously this picture was nothing more than a trademark for France—a sign warning others: "Keep out. Police protection here." And still the historical debate goes on over Norumbega!

In his *Discoveries of America Down to 1528*, Weise derives this name from the old French *L'Anormee Berge*, meaning the Grand Scarpe or the Palisades, or a place which had been fortified, while the Horsford's identified it as the word Norvege, or Norway. The derivation of the name from a native word meaning "still water or place of a great city," also has been suggested. But what difference pray does it make whether the name of this region had a native, Scandinavian, Norman, or purely French origin? There can be no reasonable doubt that Norumbega and the vanished Eastland and Westland of the Norsemen were one and the same.

But were Frenchmen and the Indians one and the same? Not by any means. The Canadian or French-Indian bloodcross spiritually as well as physically and intellectually was an Indian, for the average halfbreed in these days far preferred the purity, the integrity, the unwritten code of honor and the freedom of the American forest to the cant, the hypocrisy and the economic hardships of European life. The Jesuits had made a woeful miscalculation in dealing with the red race. Again captive Greece was leading captive Rome!

The concurrence of all this with the labors of Dekanawida would be surpassing strange if the great prophet were not in fact a Frenchman. Meantime, he had accomplished little with the Senecas, so that now he went among the Oneidas to urge them to take the lead in forming a great league to enforce peace based upon the existing Confederacy of the Five Nations but with an entirely new system of law. While favorable to his plan, the Oneidas at once pointed out their inability to form such a league without the cooperation of the other tribes. At the same time they told him that the Onondaga chief had destroyed his whole family with the exception of a twin brother because they too had advocated reforms similar to those being proposed to them. If, however, he could enlist the cooperation of the Mohawks, or the great warrior tribe of the Confederacy, it might be possible to accomplish his end. Meantime they would stand like a "great tree," or as the benevolent spirit which in many primitive cults a tree represents, and which now became the peculiar symbol of the Oneida tribe.*

Whether Dekanawida then went in search of the Onondaga reformer who had sought voluntary exile from his tribe in the wilderness, or whether the latter sought out the Huron prophet, is not clear. At any rate, what either a Huron or Roberval would have most needed at this juncture because of the traditional hatred of the Mohawks for the Wendats was an Iroquois aid. This need Dekanawida plainly felt, for he and the Onondaga were soon working in the closest cooperation. Moreover, their plan was for the latter also to assume the role of an inspired prophet under the name of Hiawatha, "The Sifter." Hewitt sees in this name no significance. Obviously, however, it was eminently appropriate for one whose mission it was to sift the knowledge of life in order to find the finer particles of truth.

Soon the two were in the Mohawk country. "But if you are a prophet," said the sachems to the sifter, "surely you fear not death." Therefore, they demanded that he submit to the ordeal of death in order that they might test his powers, for like the disciples of Jesus they had little faith in one not of divine origin. The prophets were equal to the occasion. It was

* *The Golden Bough*, Frazer, New York: The Macmillan Company.

planned that Hiawatha should agree to ascend a large tree which when cut down by the skeptics would hurl him into a river from which there could be no escape. Beneath the bank Dekanawida was to place a canoe and substitute himself for Hiawatha in the tree. Everything worked according to plan. When the Mohawk sachems reached their counsel lodge to discuss the death of the impostor, they found Hiawatha smoking the peace pipe, or the symbol of his doctrines.

The almost instant result of the cunning artifice to which the prophets had resorted was the deification of Hiawatha and his associate despite the fact the latter was a Huron. Once they had established a belief in their miraculous birth and power to work miracles, the Iroquois generally were ready to accept their guidance. Yet it took them many years to overcome reactionary chieftains who saw their own leadership threatened. It was a bitter struggle between old and new ideas which followed. In the end, however, as the disciple of Dekanawida, Hiawatha succeeded in founding the Hodenosaunnee, or League of the Five Nations, about 1570.

This league was called by Morgan the Long House, a name which he attributed to the great central lodge or the federal council chamber. But in fact Hodenosaunnee had no such significance. Hewitt has shown that it implied just what had happened—the extension of the powers vested by the tribes in the council of the confederation.

The structure of the league, as shown by both Morgan and Hewitt, was not only without precedent among the Indian tribes but entirely novel to the political science of Europe. Its peculiarity was that the constituents retained in themselves all the sovereign powers not expressly delegated to the federal council of the superstate which, with respect to the powers delegated to it, was sovereign. Each of the five tribes retained its local self-government and the exclusive enjoyment of its domain, but the supreme government of the Confederacy was conducted by a council of sachems of equal rank and authority. Fifty in number, they represented the clans of the tribes and were appointed for life when once selected. Vacancies were filled by the clan council which alone could remove a federal counsellor from office. The qualification was a sachemship or membership in one of the tribal councils. The federal council could not convene itself but could be called into session by any tribal council. Regular sessions without call were held annually and were opened by thanks to the Great Spirit. The super-council voted by tribes under the unit system, and unanimity was required for any action. There was no chief executive, but for military purposes leadership was vested in two permanent chieftainships, one from the Wolf and one from the Turtle clan of the Senecas to whom the guardianship of the march on the Huron border was entrusted. With no authority whatsoever as to civil affairs these chieftains were elected by their clans for life, or good conduct. There was also a council of matrons charged with the duty like the vestal virgins of keeping alive the council fire. This, of course, was merely symbolic of the natural duty of the female sex to continue life. While the fundamental purpose of the league was to enforce a Pax Romana, among other objects it was designed to accomplish was reverence and the doing away with the eating of human flesh to which the Mohawks appeared to have been given to some extent. Provision was also made for the absorption instead of the killing of captives. And above all was the provision that while the individual tribal domains were to remain subject to the occupancy of their tenants they were to be the common property of the Confederacy and not subject to alienation in whole or in part by any tribe or member thereof. In this way did these wise people undertake to secure unto themselves forever their fatherland by making it impossible for Europeans with their gold and blandishments to whittle away the territory of the Hodenosaunnee. Moreover, what would seem strangest of all was the European as well as the Christian influence behind this design not apparent, is the written constitution of this stone-age people. This was embodied in rich wampums entrusted to the guardianship of the federal council—the first documents to appear in the native culture, yet writings which embodied a wisdom superior to that of the foremost political scientist of the so-called civilized world.*

Such then was the origin and the character of the political state which as shown by Hewitt was to be the pattern for the republic in which the dreams of Coligny and Raleigh were to be realized 219 years after the labors of Dekanawida and Hiawatha were crowned with success—a republic which Immanuel Kant, with little knowledge of the Hodenosaunnee, was to describe as the hope of humanity.

While Dekanawida and Hiawatha were unquestionably wise in their

---

* In the course of time these wampums came by methods which need not here be discussed into the hands of a citizen of New York for alleged valuable consideration at a time when the Hodenosaunnee was under the most irresistible duress of the State. Thereafter the Onondaga Nation brought an unsuccessful suit in the state court to recover its constitution. Upon appeal to the Supreme Court of the United States in the case of Onondaga Nation vs. United States, 189 U. S. 306, the possessor of the wam pums was confirmed in his ownership. In the record the following description of them appears: "One belt of dark wampum beads representing the confederation organization of the Five Nations created by Hiawatha; one belt representing the first treaty stipulation between the Six Nations and General George Washington in 1784, in which is depicted in bead work the council house, General Washington, the Odotaho, or president of the tribes, and thirteen representatives of the Colonies; also two fragments of other belts, one representing the first approach of the Indians of the 'people with white faces,' and the other a narrow belt representing the unity of the Five Nations." The wampums have been translated into English by their present owner, Thatcher, and others. It is, perhaps, one of the most anomalous facts in American Indian history that the natives of America could have been divested of the ownership of their own political documents which on their face show what must have been the property in common of a confederacy, no member of which could have had the legal right to convey a title thereto in the absence of conclusive proof that he came by these documents with authority to sell the same in the common interest.

day, never imagining that the Iroquois eventually would lose both their land and their constitution, it is impossible to believe their conception was that of a stone-age intellect. It must be remembered too that the Hodeno-saunnee was founded on the very eve of the massacre of Coligny and the Protestants in France, while William the Silent was uniting the Dutch against Spain, while Philip Sidney was trying to unite the Protestant princes of Germany to oppose Philip, and while Martin Loyola, the nephew of the Grand Inquisitor, as Captain General of Chile, was engaged in trying to unite the hostile Indian tribes under the dominion of Spain for exactly the same reason that the Jesuits in Canada had in mind.*

In truth, the foundation of the League of the Five Nations was but the first step in the far-flung Jesuit scheme of Indian federation. Soon followed either the foundation or the extension, just as among the Iroquois, of the League of The Three Fires in Michigan and Wisconsin constituted by the Ojibway Ottawas, Chippewas and Potawatomis, and the Dakota or League of Friends among the seven Sioux tribes. The first of these was to form the basis of the so-called conspiracy of Pontiac in 1762; the other was to ally itself like the Three Fires with the French in 1753 and later with Pontiac. In California the Spanish Jesuits formed a large league and in Texas the Caddo or Commanche Confederacy. Of the small tribes in Maryland and Virginia the Powhatan or "Tidewater" Confederacy was formed in 1590, while soon were formed the Cherokee Confederacy in Georgia, and the Muskogean league among the Creeks, Chickasaws and Choctaws, the latter, no doubt, with the aid of Jesuits in Florida. This left only two regions unorganized—New England and that embracing New Jersey and Pennsylvania. The difficulties here were very great. As shown, in New England the Norsemen had so completely broken up the Indian tribes that they amounted to no more than widely dispersed villages. There the task of confederation remained to be carried out by King Philip late in the next century, undoubtedly with the aid of both the Jesuits of Canada and the Dutch. In the other section several obstacles existed. First, it was occupied in part by the Iroquoian Susquehannas and Conestogas as far east as the Delaware River. They were vassals of the Five Nations who did not wish them independently organized, so that they would not admit the Jesuits to work among them. Moreover, it was occupied in part by Algonquins between whom and the Iroquois there was a traditional feud. The drawing of these enemies together required a statesmanship which the Jesuits could not supply from their distant positions, one that only the great Tamenend could furnish early in the next century when eventually, after visiting the Jesuits in Mexico, he finally succeeded

* Loyola was killed by the Indians in 1583.

against the opposition of the Five Nations in founding the Lenni-Lenape or Delaware Confederacy.

In the organization which the Jesuits had given the red race by 1600, there is no necessary reflection upon the Catholic Church. Certainly nothing that was done among the tribes under the orders of the Vatican was more unpardonable than what is known of the Protestants in their struggle for empire in North America. In 1775 the Continental Congress actually undertook to engage the tribes to war not only upon their coreligionists but their fellow countrymen who were opposed to independence, while the British government employed them usefully against the colonists and both used them in the war with France.

7

Raleigh and the First English Colony

AMONG THE HALF DOZEN FRENCHMEN who had escaped from Fort Caroline was Lemoine, an artist who with the other survivors had been picked up while wandering along the coast of South Carolina and brought to London whence he made his way to the almost regal palace of Coligny. What a company gathers there to hear him!

Beside the Grand Admiral of France sits his son-in-law, William the Silent, whose voice still rings down the ages. Then there is young Walter Raleigh, scion of a noble Devon family of seamen. Possessed of a hundred graces, the younger half-brother of the great Sir Humphrey Gilbert is on his way to Flanders to fight under William. They do not know as they sit there quivering with emotion while the voice of Lemoine mingles with the Angelus coming to them sweetly through casements thrown open to the balmy air of spring, that death is even then lurking in the dusky streets below.

The speaker's tale has been told with the art which only the deep emotion of an actor in the tragedy could inspire. Not one of his hearers but has already entered into a soulful covenant to persist in the effort to make of the New World a Utopia where Catholics, Protestants, the despised Jews, even the savage heathens who claim it may worship as they please.

A few months pass. Hawkins is treacherously assailed by the Spaniards for trespassing in Mexican waters. A battle follows. From the neighborhood of Tampico, David Ingram and two companions left ashore actually march overland to Nova Scotia, often sheltered and fed by the natives.

The Englishmen who seek asylum in Mexico City fall into the hands of the Inquisition now established there, and are burned at the stake by their fellow Christians! Now, too, Alva is goading the Dutch to resistance. In 1570 the Pope declares Elizabeth deposed. Five years later on St. Bartholomew's Eve, while the carillons are summoning the devout to worship, Coligny lies a mangled corpse in the gutters of Paris!

With his imaginative being fired by the grand conception of Coligny, young Raleigh now abandons the army in Flanders. While Sidney is in Germany trying to organize the Protestant princes in aid of Elizabeth and William the Silent, Raleigh sets out to execute Coligny's will through the agency of England. It is no easy task to accomplish this, for still England is a poor country while the power of Spain—the mistress of the seas—is unbroken. Moreover, Elizabeth is a born autocrat who fears free worship, free governments, republics and the like almost as much as Rome.

In 1573 Drake gazes upon the Pacific at Darien and four years later sets out upon his immortal voyage with the one purpose of weakening the empire of Philip II in South America by invading the Eldorado there. By this time Raleigh has introduced Gilbert into his scheme. If only she will permit it, Gilbert writes the Queen a month after Drake's departure, her captains will deliver to her the West Indies with all their gold as well as the fisheries of Newfoundland. Sidney and Cecil plead with her to grant Gilbert's request, but it is not until the Anglo-Dutch alliance of 1578, while Drake is ravaging the coasts of Chile, that she grants Gilbert a patent authorizing him to "hold by homage remote heathen and barbarous lands, not actually possessed by any Christian prince, nor inhabited by any Christian prince, nor inhabited by Christian people, which he might discover within the next six years."

From this it is apparent that Raleigh's idealism did not extend to the natives of America. Yet, the patent was broad enough to embrace the communistic scheme of More and Coligny for the relief of English labor. Hakluyt's *Discourse of Western Planting* written largely by Raleigh shows this idea had fixed a strong hold upon the latter's mind. Thus, the colonies planted by Gilbert were to have "all the privileges of free denizens and persons native of England, in such ample manner as if they were born and personally resided in our said realm of England." No law to the contrary was to have any effect. Moreover, the colonists were to be free to govern themselves according to such statutes as they might choose to establish for themselves provided that they conformed "as near as conveniently may be with those of England, and do not oppose the Christian faith, or anyway withdraw the people of those lands from our allegiance." Thus, as declared by Fiske, Gilbert and Raleigh demanded, and Elizabeth granted in principle, just what Patrick Henry and Samuel Adams demanded and George III refused to concede.

Yet, it is exceedingly doubtful if Elizabeth knew that under this patent More's communistic theories could be tested out legally by Raleigh on a large scale, for the reason that communism already existed in the practice of several small communities in England which were recognized by the law of England. According to English jurisprudence communism is that economic system, or theory, which rests upon the total or partial abolition of the right of private property and which ascribes to the community as a whole, or to the state. In other words Raleigh had cleverly contrived to make it possible to adopt for Anglo-Americans the same economic system common among the aborigines themselves. This, he believed, had received no fair test under the fanatical rule of the preachers at Fort Caroline.

That Elizabeth was insincere in the limitation imposed by her of the right to settle in lands not actually possessed by a Christian prince is shown by the arguments which induced the granting of the patent and the point selected for settlement. Immediately Gilbert and Raleigh set sail with seven ships for Norumbega, where the French were known to be established, and would have gone there had the watchful Spaniards not driven them back at once.

Four years passed during which Raleigh pleaded in vain to be allowed to resume his efforts. Now, fascinated by him and thoroughly alarmed lest harm come to one in whose person she was far more interested than in American colonization, Elizabeth refused her assent. Therefore, in 1583 Gilbert proceeded alone to Newfoundland while the Queen's favorite, captain of her bodyguard and Lord Warden of the Stannaries, remained at court against his will soon to learn of Gilbert's death by drowning on the way home from an unsuccessful venture.

Meantime, however, in order to enable Raleigh to carry on his enterprise, Elizabeth had presented him with all the confiscated estates of traitors in England and Ireland, so that in 1584 after Gilbert's patent had been renewed in his name, he despatched two ships to find a suitable place of settlement on the coast of the present North Carolina which was sighted July 4, 1584—a significant date in American history.* After visiting the friendly Roanokes or "northern people," Amedas and Barlow returned to report in favor of Roanoke Island.

But what is this? In January, 1585, William the Silent, as Stadtholder, proclaims the independence of the Dutch states in terms which two centuries later Jefferson is to appropriate for the republic of free worship which More, Coligny, William and Raleigh had striven to found in America. How fitting that the following spring Raleigh should send his cousin, Sir Richard Grenville, to convoy Lane and about a hundred colonists to

---

* The Declaration of Independence; the surrender of Vicksburg; the defeat of Lee at Gettysburg; California admitted to the Union; the death of Thomas Jefferson and John Adams within an hour of each other, all occurred on this day.

Roanoke Island in the realm he had named Virginia at the Queen's suggestion!

But unfortunately for the whole enterprise, despite the friendliness of the Indians, when one of them smuggled off a silver cup which pleased his fancy as a toy that of a child, the great hero Grenville, with utter lack of forethought, set fire to the standing corn of the native hosts. After thus sowing the germs of discord he departed to join in the war on Spain which had been declared by England in July. It was the spring of 1587 when a reenforcement of 150 colonists, including seventeen women, under John White, reached Roanoke Island. Meantime, the original party had almost starved after ceaseless troubles with the Roanokes. On August 18, 1587, Virginia Dare was born, the first Anglo-American.

In 1588, Drake, Hawkins, Winter, Howard, Raleigh and Grenville destroyed the great Armada sent from Spain to crush England three years after Raleigh's colony had been sent to Virginia. Much impaired in fortune, in 1589 Raleigh assigned his colonial rights to Hakluyt and others, and it was not until three years more had elapsed that relief could be sent to the colonists.

It had been agreed that in the event they should move they would carve on a tree the name of the place to which they were going. Upon reaching the island the settlement had been abandoned. On a tree was carved the word "Croatan"—the name of a neighboring island. After a brief search the relief party returned to England. Five attempts, extending down to 1602, were made by Raleigh to locate possible survivors, but the lips of the natives remained sealed until 1607. Then it was learned that the colony had mingled with the natives and lived with them amicably for a number of years until "certain medicine men" had caused to be killed all but four men, two boys, and a young woman who were spared by request of the chief.

Who were these "certain medicine men"? Is it not obvious that in all probability they were agents of the Hodenosaunnee from the neighboring Tuscaroras who, at the instance of the French, demanded that the Croatans elect between executing the English and war with the powerful Iroquois? Was this not the decree which the shamans sent forth from the Indian oracle?

Nothing could be more likely. Under irresistible duress the friendly chief obeyed, and apparently to hasten his action was allowed to save those who were probably his best friends—so few that they constituted no real danger. The "young woman" was probably Virginia Dare so that no doubt some of the Croatans who in recent years had been declared by a North Carolina statute to be the descendants of the lost colony of Roanoke, are descended from her.

In 1592 after his marriage Raleigh fell into disfavor with the Queen and was imprisoned, but soon ransomed himself with a vast capture of Spanish gold made by one of his captains—Christopher Newport. Two years later he was in South America searching for the Spanish Eldorado in the Orinoco country. This gave added offense to Rome. Next, just after the death of Hawkins and of Drake, whom the Spanish now called the dragon, in 1596 with Essex and Howard, Raleigh destroyed the Spanish fleet at Cadiz. Assuredly now he was doomed.

Elizabeth died in 1602 and was succeeded by James I. In the Anglo-Spanish treaty of peace with which the Stuart reign began, it was stipulated that no English encroachments upon Spanish territory in America were to be made, but soon Raleigh was involved in all manner of schemes to outwit Rome and free the hand of James I. Although imprisoned in 1602 in the Tower of London, there he organized an expedition which under one of his old captains, Bartholomew Gosnold, visited the Chesapeake and also New England, where he tarried a while on Cuttyhunk Island. "I shall yet live to see Virginia an English nation," wrote Raleigh to Robert Cecil at this time. Moreover, he continued to appeal, through Hakluyt, to public sentiment in favor of his communistic scheme of western planting as a means of relieving the poor. But it was not the relief of human suffering that interested the leaders in England now, any more than in the time of More and Coligny. What they wanted was gold. Accordingly, with the Spanish fleet out of the way, in 1603 a group of merchants sent Martin Pring to search for the old Eldorado of Norumbega which was now firmly believed to be somewhere in New England.

Fully advised of what was taking place, the Grand Monarch of France, who had employed Vattel to revise the Law of Nations to suit his plan of empire, instantly despatched Champlain with a large number of Frenchmen, including more priests, to reenforce the Canadians. Nor was he to allow the British under any circumstances to settle upon the great river leading from the strategic harbor in the land of the Manhattans into the realm of the Hodenosaunnee and the great lakes in the Mohawk domain. In other words, at all costs he was to split what had come to be known as Virginia wide apart and thus make it possible for the Iroquois and the French in union to crush any English colony that might be founded in either section.

Mace, who was sent to reconnoiter in the southern section, like Pring came home to report on the movements of the French. What they said evidently encouraged Raleigh, who now called the dramatists to his aid.

"I tell thee," said Seagull to Scapethrift in one of Ben Jonson's supposed stage comedies, "it is more plentiful than copper with us; and for as much red copper as I can bring I'll have thrice the weight in gold. Why, man, all the dripping-pans . . . are pure gold," etc.

Such was the clever propaganda Raleigh was dictating from his dungeon to arouse all the cupidity of a commercial race in support of his continuing purpose, knowing that in the end Rome would be thwarted by the overruling of James I. Nor was he disappointed. While Johnson's play was stirring England to fever heat, the Earl of Southampton and others, including Sir Ferdinando Gorges, Governor of the garrisons of Plymouth, sent a party under Weymouth to the Kennebec River ostensibly in search of the Eldorado.* After remaining there a month the explorers returned with no gold but five captive Indians! Unquestionably they had been instructed to seize them to be employed for propaganda purposes in overcoming the fear of the natives engendered by the supposed massacre of the Roanoke colonists. Now England was regaled with its first "Wild West Show" in which the helpless captives unwittingly played the part designed for them. A more harmless looking people the delighted Londoners could not imagine.

Almost coincidently Raleigh sent his nephew Bartholomew, the son of Sir Humphrey Gilbert, to Virginia to locate the site of San Miguel, whither, way back in 1585, he had planned to remove the Roanoke colony for better protection against the Spanish. He was not fortunate, however, in the misunderstanding of the natives by his relatives. First Grenville had done him immeasurable harm. Now the son of the man who had lost his life trying to colonize Norumbega, was to lay the foundation for certain trouble in the Chesapeake country. It seems certain that Gilbert tried to extort from the Indians by punishment information about the site of San Miguel and the whereabouts of the Roanoke colonists. The result was a conflict in which many Indians were killed as well as Gilbert and two other Englishmen.

The upshot of the propaganda conducted by Raleigh through Hakluyt and others was the formation of two Virginia companies, one in Plymouth and one in London, which included many of the most powerful personages in England. Moreover, they completely dominated Parliament, against which James I was powerless to contend. Willing enough to share in the gain to be had so long as he could avoid trouble with Rome, to hide his real purpose he let Popham, Lord Chief Justice of England, a director of the newly formed Plymouth Company, deliberately lie to Zúñiga, the Spanish Ambassador. On March 16, 1606, the latter wrote Philip III that when reminded that the proposed enterprise was an encroachment upon Spanish territory and a violation of the existing treaty, "this astute judge" whom he described as a "terrible Puritan," had told him they were only trying "to clear England of thieves and get them drowned in the sea." Then he went on to say: "I have not yet complained of this to the King; but I shall do so."

* Upon this, claims of priority of settlement have been advanced in New England.

Long and loud were the protests of Zúñiga. But they were useless. Raleigh, Sidney, Bacon, Jonson, Shakespeare, Drayton, and others, had long since clothed the colonization of Virginia with the aspect of a patriotic duty on the part of Englishmen, all the while adroitly appealing to their lust for gold. Thus, in the end the despicable James was compelled by a highly organized public sentiment to act. Therefore, on April 10, 1606, the charter authorizing the colonization of Virginia by the two companies was issued by him. Virginia was declared to include the region from the present Vermont-Canadian boundary to the Cape Fear River in North Carolina. The exploring parties, however, had long since discovered the power of the Hodenosaunnee. Thus, in order to avoid antagonizing the Iroquois, their realm was not to be touched for the present. Accordingly only the part of Virginia north and east of the Hudson, or New England, was set apart to the Plymouth, and the southern portion, or that below the Potomac, to the London Company. After England's hold had been firmly fixed on these regions the crushing of the Iroquois and their French allies might be possible. Nevertheless, he now granted away two territories, known to be inhabited, together larger than England, purposefully with no provision whatever for their "heathen and barbarous" denizens, specimens of whom were actually being exhibited in London at the time in order to convince investors that they were a friendly people.

It was on December 19, 1606, when the expedition quickly organized by the London Company finally embarked at Blackwall on the *Black-eyed Susan,* Raleigh's old captain, Newport, commanding, and January 1, 1607, when it sailed from the Downs for the Chesapeake. Having profited by the experience of the Roanoke venture, the directors of the company did not fail to anticipate the difficulties in their way. In the "instructions" it was urged that in all their "passages" the colonists have great care not to offend the "naturals," if they could "eschew" it, and to employ "some few" of the company to trade with the Indians for corn and all other "lasting victuals" if they had any. This they were to do before the natives perceived a planting among them was intended. The colony was to be established at the site of the vanished San Miguel, while explicit provisions were made for its communal organization. At last Raleigh had outwitted both Rome and James I.

In vain the King explained to Zúñiga that the almost unlimited power which Elizabeth had conferred upon Raleigh was not to be employed by these new plantations. Rome was not deceived. It was obvious what was the real purpose behind the expedition. The problem of Rome now was to prevent the colonization of America by England without actually coming to blows. Therefore, while Newport was on his way, Zúñiga was directed by Philip III to report in detail just what the English were doing so that steps could be taken to block them. Was there to be another Fort Caroline, another Roanoke Island tragedy? If students of American history bear in mind all that had occurred among the red race prior to 1607, the predisposed attitude of that race, and the further fact that Englishmen were to come to America with consciences uncharged with any sense of responsibility for it, they should be able to understand better much that followed.

The long-heralded prophet had already come among the Indians. In 1570 Hiawatha and Dekanawida had founded the Iroquois Confederacy of the Five Nations, or the Hodenosaunnee, and under the influence of their teachings the Powhatan or "Tidewater" Confederacy of the Chesapeake, or "land of the big river," had been formed about 1590. Over this league its founder, Wahunsonacock, was the reigning chieftain. It consisted of ten or eleven tribes in eastern Virginia, including the inhabitants of the little peninsula between the bay and the ocean known as the Kingdome of Accowmacke, all Algonquins. Its principal seat was at Kighoutan (Hampton) nearly opposite Cape Henry, while the favorite residence of its venerable ruler was at Powhatan, or "the place where the two waters meet." This was at the falls of the James near the present site of Richmond.

According to a tradition extant in early Virginia, the ruler of the Powhatans had been driven there from the West Indies. This, of course, is not true. He was a pure Algonquin. The idea was probably based on statements by the Indians that their chief had been ousted of his lands by the Spaniards from the West Indies under de Allyon. But although unpleasant traditions must have remained of the Norse and possibly the Welsh visitations, and of those of Verrazano and de Allyon, it is by no means certain that the Indians had destroyed San Miguel as commonly assumed. From all we know to the contrary it was just such a transitory settlement as that made by Gudrid and Thornfinn on Bird Island in 1007, and by Weymouth on the Kennebec in 1605. De Allyon did not come to plant a colony. He was merely looking for gold, and when Verrazano's map appeared in 1529 indicating that the Eldorado had been located in New England, the Spanish settlers no doubt "pulled up stakes" and moved at once. Moreover, it has been shown that when Rome adopted Cartier's plan of French colonization the Spaniards officially abandoned the region north of Florida to France so that San Miguel had to be abandoned in 1540 even if still in existence.

No. The hostility which the colonists under Newport encountered was due to the savage assault by Gilbert upon the natives just two years before. Naturally when the visitors landed at Cape Henry on April 26, 1607, they were received by a shower of arrows.

From there they passed on up the James River searching for the site of San Miguel which Gilbert's party had no doubt located, and here, the present site of Jamestown, with befitting ceremonies on May 14, 1607, was planted the standard of James I at what they called James' Fort in honor of their king.

The communal organization which their instructions enforced upon the colonists was a handicap from the first. Fortunately for the settlers, among them was one Captain John Smith, who had gained much experience with other than white peoples while campaigning in "Tartary." Typical of the great company of empire builders who, beginning about this time, Britain sent forth from her shores, this remarkable man, whose history of Virginia was long unjustly relegated to the realm of romance, by reason of sheer ability was soon to find himself the admitted leader of his associates. It seems certain that without him at their head they would have perished immediately, since no other manifested the slightest notion of how to deal with the natives.

Learning that there was a principal sachem of the country who held sway over many vassal chieftains, at once he made his fearless way with Newport to the Indian village of "Werowocomoco," on the York River, about fifteen miles from Jamestown, and there found Wahunsonacock, whom he styled "emperor," surrounded by his vassal "kings" and "princes." Immediately he set out to win the old chief by flattery and the other arts known to such an one as Smith. It was now, too, that the visitors learned of the fate of the Roanoke colonists and with reasonable certainty of the site of San Miguel. Of the disappearance of the latter the Indians had no explanation, which indicates two things—that they had no part in it and, therefore, felt no necessity for explaining it, and second, that it had occurred even before the memory of their oldest living men like their chief. This fixes its abandonment just about 1529, or when Wahunsonacock was two or three years old.

With no understanding whatever of the Indians, and basing their accounts solely on the unintelligent gossip of the first settlers who endowed everything with an inexplicable strangeness, the historians have given no insight whatever into what followed. The truth is, while the old chief and his brother Opitchapan, second in rank, were dissembling and detaining Smith, a younger brother, Opechancanough, the soldier and more virile leader of the Confederacy, attacked Fort James with a large force of natives. Though the assailants were driven off with heavy loss, the defenders also suffered casualties, including one man killed and eleven wounded. Suspecting something, Smith and Newport managed to get back to the fort and during the next fortnight of ceaseless strife with the good captaincy of the former the colonists were saved.

At the end of these futile efforts to destroy the visitors, the mission of Smith to the "Emperor" bore good fruit. Having learned that he was no ordinary warrior, that the whites under him had come to stay, Wahunsonacock dictated a new course. Disclaiming all control over those who had attacked the English, he now sent emissaries with presents of food to disarm those whom his people had already twice failed to drive off. Instead of showing malice, Smith took advantage of the truce to go in search of a supply of food for the winter.

Again the purpose of the chieftains appears. While searching the country Smith was seized in White Oak swamp by Opechancanough at the head of a large force of Chickahominies and taken to the seat of the Confederacy for trial. This, instead of his murder in the woods by a subordinate chieftain, shows a very high respect on the part of the Powhatans for law. After a long absence he returned to the fort to find that only thirty-eight of the settlers had survived the hopeless communism to which they were subject.

As to what had happened to the settlers we know from their contemporaneous accounts. As to what had happened to Smith we only know from an account published by him in 1624 after the death of all the principal actors except Opechancanough. According to him he was tried in the regular and orderly way prescribed by the laws of the Confederacy and condemned to death. When the sentence was about to be legally executed in the presence of the council, the old chief's daughter—Matoaka, called Pocahontés or Pocahontas, "the joyous one," a child then not over twelve years of age, intervened by throwing herself between him and the club of the executioner. Around this incident an amazing mass of misunderstanding and sentiment has gathered. Over and over Smith has been branded as the Munchausen of his age. But just as in the case of Herodotus and of Raleigh in connection with his explorations of the Orinoco, time has tended to vindicate Smith's veracity.

Certain it is Smith won the friendship of Pocahontas and possibly the enduring affection of the child through a display of natural gratitude on his part. Now, at the instance of Smith and Pocahontas, the Powhatans came to the rescue of the whites with a bountiful supply of food, and to cement their alliance Smith and the old chieftain exchanged hostages. The latter adopted as his own son the little cabin boy, Thomas, who had come with the first expedition.*

Is it not plain what had happened? The clever Smith had done what he did not dare disclose in his book for fear of bringing down upon himself the wrath of Rome. Undoubtedly he had assured the Powhatans as the basis of this treaty of amity that the English had come to protect them

* This boy was thereafter called Thomas Savage, possibly because adopted by the native chieftain. He is the only member of the first expedition from whom descendants, the Savages of Acomack County, Virginia, whither he was later sent by The Powhatan for safety, may be traced. It is odd that the oldest white family in America should be that of Savage.

against the Pale Faces in the North and the dreadful Iroquois who for centuries had ravaged the Algonquins. This was just the diplomacy best calculated to embolden them to resist the demands of the Iroquois agents.

The result was, in 1608, attended by less than a dozen others, Smith was able to explore unmolested the entire Chesapeake country, being everywhere received by the Indians in the most kindly way.

Two events of the utmost importance now occurred. Having made a settlement in 1608 at the old Norse Stadcona, which the French called Quebec, Champlain set out to fulfill home-made instructions which utterly ignored the power of the Hodenosaunnee. In other words, as between the danger of it and the English, the French government saw no choice. Accordingly, Champlain was compelled to take advantage of an assault by the Mohawks upon the hated Wendats as an excuse for invading the domain of the former and seizing the great lake which took his name. A fatal step, it was enough to make Cartier and Roberval turn over in their graves. The result was instant. From being the logical, long-prepared basis of French empire in America, the Hodenosaunnee at once became the rock on which it was ultimately wrecked.

Hardly had this terrible blunder occurred when Heindrick Hudson, in the employ of the Dutch East Indian Company, looking for a northern passage to Asia and also, no doubt, for the far-famed Eldorado of Norumbega, arrived at the great river in the land of the Manhattans, Mohegans and Mohawks which later took his name. Learning of the Mohawk war that was going on, in order to ingratiate the Dutch with the Iroquois, at once he furnished them with firearms and gave them their first instruction in musketry. Though not perceived at the moment, this spelt the doom of French empire in America, for once armed with modern weapons the Hodenosaunnee became even more powerful. Henceforth no Jesuitical blandishments could win them back to France, for the simple reason that to condone the first act of aggression upon their territory meant their doom.

Unfortunately for Virginia, after saving the colony more than once and establishing it upon an orderly basis, despite the handicap of a personnel uninured to labor and its communistic organization, Smith now departed from James Town, no doubt under secret orders to make before returning to England a reconnaisance of the situation in the north as the basis for the colonization of New England.* Thereupon the colony quickly fell into complete chaos and was on the point of starvation as well as destruction by the Powhatans who, of course, were being threatened by the Hodenosaunnee. To save themselves the colonists now resorted to a clever move.

---

* Such a reconnaisance he actually made.

Argal was sent to lure Pocahontas upon a vessel. Thus entrapped, she was carried to James Town and held hostage there against the danger of her uncles. In this the colonists saw no lack of gratitude to her since Smith had not yet confessed to her saving him. Obviously he had not wished her to pay the penalty to the Hodenosaunnee. Nevertheless, in the Indian mind it was but another instance of English perfidy to be scored up in a growing account.

Meantime Raleigh's communism had failed utterly so that steps were taken in London looking to the complete reorganization of the colony. In 1609 a new charter with a new governing personnel was obtained from the king and the following year, the noble Thomas West, Lord Delaware, Raleigh's cousin, whose name was given by the English to the Iroquois tribes of the Lenni-Lenape north of the Potomac to distinguish them from their Iroquoian cousins, was appointed Governor. His instructions were to reorganize the whole scheme of things in Virginia. Thus ended the third pitiful attempt to establish a European community in America. About its possibilities Raleigh retained no illusions. Just as the colonists were about to leave Virginia, the news of Newport's arrival with new settlers arrived.

The charter of 1609 was soon replaced by that of March 12, 1611, which like the original one was silent with respect to the natives. Nevertheless, Thomas Dale, Delaware's deputy, a rugged though practical soldier, soon reestablished order. Moreover, like Smith, he displayed great genius in dealing with the Indians. Believing that the marriage of Pocahontas to an Englishman would bind the Powhatans to the English and prevent their succumbing to fear of the Iroquois, the authorities at home had directed him to spare no effort to educate her and convince her that she could best serve her own people by keeping them out of the clutches of the French. On the other hand, her marriage was adroitly designed as the first step in taking advantage of the growing hostility between the Iroquois and the French to court an alliance with the former. Gradually won over by Dale, in 1613 Pocahontas was baptized by the Anglican parson Whittaker, and in April of the following year, when nineteen years of age, was married to John Rolfe, Gent., the first tobacco planter in Virginia. In this purely political marriage Rolfe, of course, felt fully justified by reason of the great service he was rendering his country. Nor is there any doubt that he was sincerely fond of "the joyous one" before he consented to the union.

The marriage took place in the presence of Opitchapan and other kin of Pocahontas, though Powhatan dared not attend for fear of the Iroquois, while Opechancanough held sullenly aloof. Among the English there was great rejoicing. Now a veritable "love feast" ensued between the Powhatans and the colonists. Even the Chickahomonies, "a lustie and daring people who have a long time lived free from Powhatan's subjection," came at once to tender voluntarily their allegiance to the King and to urge Dale to be their chief; "and ever since," wrote Hamor, "we have had friendly commerce and trade not only with Powhatan himself, but also with his subjects round about us, so as now I see no reason why the Collonie should not thrive apace."

Dale, however, was not lulled by such things into a sense of security. He knew that the Powhatan chief had been compelled to commit Thomas Savage to the guardianship of his feudatory, Kiptopeke, or the "Laughing King" of Accawmacke, to remove him from reach of Opechancanough, who like the Iroquois, had looked askance on the alliance of Pocahontas with Rolfe. When, therefore, the old chief refused to give another daugh-ter in marriage to the English, Dale was quick to take advantage of the advances of the Chickahominies to bind them by a treaty of peace. In consideration he promised to defend them against all their enemies which, of course, meant the Iroquois as well as Opechancanough.

Soon the deep-laid scheme behind the marriage of Rolfe and Poca-hontas became apparent. Not only must the Iroquois be led to believe that the English would back them in their fight with the French, but France and Rome must be shown that their own designs were useless. On the other hand, the evidence of an Anglo-Indian union was expected to start a flow of much needed colonists of a better sort to Virginia. Accordingly, in 1616 Dale was ordered to bring Rolfe, "the Lady Rebecca," her brother-in-law, and several other Indians to England where with great éclat Poca-hontas was received at court as a princess and the daughter of the "Great American Emperor Powhatan, ally of Great Britain." This same year, having been sentenced to death after completing his *History of the World*, Raleigh was released from the Tower in order that he might return to the Orinoco to find the Eldorado there for his royal master. In his absence Pocahontas died of smallpox in 1617 on the ship which was intended to bear her back to Virginia with Rolfe and her son Thomas who had been born in England. After her burial at Gravesend the child was left to be reared by his uncle, Henry Rolfe, and his father returned to Virginia as Secretary of the Colony.*

What a year was 1618! The venerable Wahunsonacock died and was succeeded by Opitchapan as the nominal ruler of the Powhatan Confed-eracy, while Raleigh, who had been reimprisoned upon the demand of Rome on returning from an unsuccessful venture in South America, after much dilly-dallying back and forth by the King, was finally ordered to be executed under the old sentence. No more craven act is to be found in the reign of the cowardly James I.

---

* From Thomas Rolfe many Virginians are descended.

Raleigh's *History of the World* showed that he knew his prediction of 1602 was about to be realized. The very year after he went to the gallows, representative government was established in Virginia and an English Republic became a fact. Oddly enough three events of the utmost importance concurred with this. The first English women arrived at James City, the first Negro slaves were imported, and the English Pilgrims in Leyden, having obtained permission to come to James City, set out from Holland to settle in the Utopia of the New World.

The new laws, however, had not overcome the innate selfishness of Englishmen. True, the governing authorities professed the most humane motives under the guidance of the noble Delaware. Moreover, provision had been made for the building of the first college in America at Henricopolis, or the curls of the James, which was to be devoted to the education and Christianization of the Indians. But despite all this, the colonists generally had but one thought with respect to the aboriginal tenants of the soil. They must yield their land when it was desired. Consequently the amity was but apparent.

The truth is, the Dutch had proved smarter than either the English or French. Taking advantage of the gap which the former had left between New England and Virginia and of the French-Iroquois feud, their agents had been secretly at work among the Iroquois ever since Hudson's visit in 1609. Therefore, when in 1620 the Pilgrims arrived in New England whence the control of the Hudson and St. Lawrence River fur trades could be threatened, under the promptings of the Dutch the Hodenosaunnee became more and more restless under English encroachments. Now again, therefore, threatened by the Tuscaroras to the south and the Senecas in the north, the Powhatans had to choose between their own destruction by the Iroquois and that of the English by them. At last, in 1622, just as the first Dutch colony arrived at the mouth of the Hudson, the Indian College which was about to be opened, like many of the outlying settlements, fell victim to the devastating flames of the general attack which Opechancanough had contrived. Only the devotion of the Indian lad Chanco saved the colony from complete destruction.

It was a sad sequel to the alleged union of the white and red races in Virginia. Yet, all that is necessary to explain it is a comparison between the methods being employed by the Virginians and those of the Dutch and of the French since Champlain's mistake. Both were courting the Hodenosaunnee by every known means.

Nevertheless, and despite the immunity afforded by the New England tribes to both the Dutch and the godly Pilgrims because of their justice in dealing with the Indians, the Virginians learned little from the calamity of 1622. In the hostility of Opechancanough they only saw the implacable hatred of a barbarous savage. To him they could ascribe no such motive as the desire to save his people from the destructive effects of contact with a civilization which he knew had everywhere vitiated the character of the red man and in the end either destroyed him, or reduced him to virtual slavery. They never imagined that the new ruler of the natives in Virginia might have reasoned from a knowledge of what had occurred elsewhere, and from what he saw only too plainly was taking place in Virginia, that now was the time for a real patriot to strike a blow for the salvation of his race. And there was no Smith or Dale to deal with him. Thus, while they made haste to put the colony in a better state of defense, it never occurred to them that the reactionary methods which they adopted in dealing with the Indians merely served to keep alive among them a smoldering hatred during the next two decades.

The English attitude of mind in 1632 is reflected in the charter of Charles I granting Maryland to Lord Baltimore. In it "savages having no knowledge of the Divine Being" and occupying a region "hitherto uncultivated," were mentioned as among the enemies who might be encountered. Against these "barbarians" the colonists were authorized to wage war, and "to pursue them even beyond the limits of their province," and "if God shall grant it, to vanquish and capitvate them; and the captives to put to death, or according to their discretion, to save." Thus it is manifest the British Government still entertained no solicitude for the welfare of the Indians who were left to be dealt with by the colonists as might prove most expedient.* Opechancanough and his people were forced to sign a treaty fixing a boundary line beyond which they might not settle, while within a few years large grants within the domain set apart to them by the treaty were made to individuals by Governor Harvey, necessitating another treaty about 1642.

By this time the Virginians and Marylanders had engaged in a war over the ownership of Kent Island in the Chesapeake on the boundary line of the two colonies, while the Dutch of New Netherland and the English in New England were also embroiled with each other. Over the native wire Opechancanough learned that the tribes of Massachusetts were preparing to rise in concert against the English, and no doubt he, too, was encouraged by Dutch emissaries. At any rate, he decided to make one more effort to drive out the usurpers of his country, who were plainly divided. Therefore, while the Puritans were warring on the Pequots in Massachusetts, with the encouragement of the Hodenosaunnee he contrived another general massacre for the year 1644. Although it nearly succeeded in destroying the colony, this only served to seal the fate of his race. Pursued

* For a comparison of the Indian policies of Spain, France, England, Holland, and the thirteen colonies, and all the laws of each relating to the Indians, see Bulletin 30, Bureau of American Ethnology, Smithsonian Institute, 1896-7, Part 2, introduction by Cyrus Thomas.

and captured, he was shot by one of his guards while a prisoner. Laws were now enacted authorizing and encouraging the destruction of Indian villages and the killing of Indians at sight, the war of extermination upon them ending concurrently with the Pequot War only upon their execution of a treaty in October, 1646, in which they agreed to dissolve the Powhatan Confederacy. But from now on "a dead Indian was a good Indian," if for no other reason than the land he yielded; so that when in 1656 the Susquehannocks and other Iroquois armed with Dutch muskets began to arrive in Virginia from over the mountains, the ten tribes of native Indians had been almost wiped out. Nevertheless, by legislative acts in 1655, 1657, 1658, 1660, 1661 and 1665, restraints were put upon the alienation of Indian lands, not out of consideration for the owners, but to avoid the necessity of having to allot them new areas.

Over in Accawmacke, which had become the County of Northampton, no hostile hand had been raised against the "brothers of Thomas Savage." Nevertheless, Colonel Edmund Scarburgh, Surveyor General of the colony, with three saracen heads in his shield, and known as "The Conjurer" by reason of his methods of dealing with the natives, undertook with energy to destroy the remnant of the race. Inviting the few who remained to assemble as his guests, a barrage of culverins was opened on them while they were feasting! Summary and effective were such methods! By 1666 it only remained for him to marshal a force of six hundred men and lead them against the Assateague Indians to the north. These he quickly drove across the Maryland boundary with infinite gain in acres for the Virginians.

On the western shore the "good work" of extermination progressed rapidly. When in 1676 Governor Berkeley refused to lead the planters against the Indians of the west, taking up arms in a revolt against him under the leadership of Nathaniel Bacon, whose plantation in the west was threatened, they drove the local Indians to bay at the present site of Richmond. There, in what was no less than a massacre, which gave to Bacon's Quarter Branch the name of Bloody Bun, the power of the Indians of Virginia was broken.

The following year there occurred what is conclusive—that the Indians of Virginia were innocent victims in Bacon's Rebellion. In the treaty into which the ten tribes formerly subject to Powhatan entered in 1676 through the Queen of the Pamunkey's, they were promised sufficient land for their subsistence and that the whites should not settle within three miles of them, while they were to have full justice in the future. Moreover, they were armed against the encroaching Iroquois, obviously for the benefit of the whites. Upon the conclusion of this treaty, Colonel Herbert Jeffreys, the acting governor, expressed the fond hope they would cut each other's throats!

The Indian problem in Virginia having been solved in the manner shown, it was only when Spotiswood became governor, after the natives had been despoiled of all the colonists coveted, and were no longer able to offer serious opposition to the whites, that any attention whatsoever was devoted to the miserable remnant of those who for the better part of a century had been mercilessly ravaged.*

* For Indian Policy of Virginia see Indian Land Cessions in the United States, Introduction by Cyrus Thomas, 18th Annual Report, Bureau of American Ethnology, Smithsonian Institute, 1896-97, Part 2, H. R. Doc. No. 736, 56th Cong., 1st Sess.

8

New England–the Iroquois and New Netherlands

THE "GREAT PATENT OF NEW ENGLAND," like the Virginia charters, made no reference to the natives. From the proviso that the grant to the Plymouth Company was not to include any lands "actually possessed or inhabited by any other Christian prince or state," it is obvious it imposed no obligation to respect native rights.

The little company of Pilgrims who, setting out from Holland for Virginia, arrived by accident of the sea in New England in 1620, like Newport's Expedition of 1607 were by no means the first Europeans of whom their Indian hosts had had experience. As shown, Gosnold had visited Cuttyhunk Island in 1602 and a temporary English settlement had been made at the mouth of the Kennebec in 1605, while Smith had only recently explored the country and the French who had long been visiting the St. Lawrence and coasting southward as far as Mount Desert had been on Lake Champlain since 1608.

Just as in Virginia, the Pilgrims were to find friends who saved them from their own ignorance by sharing during the winters of 1620 and 1621 their meager reserve of maize. First Samoset, then Massasoit, or Yellow Feather, taught them corn culture, which alone enabled them to survive. Nor did the "godly, gentle Pilgrims" in any way seek to take advantage of their hosts.

Two years after their arrival in Massachusetts, the first colonists sent out by the Dutch West India Company began to arrive at the Hudson River. Imbued as the Dutch were with the philosophy of Vitoria through their contacts with Spain and the teachings of Grotius, Peter Minuit, the directing genius of the enterprise, had instructed them to treat with the natives before erecting any buildings, and this they did even before building Fort Orange on the North, and Fort Nassau on the South River in 1624, while in 1626 they purchased "11,000 morgens" constituting Manhattan Island for 60 guilders as the site for Fort Amsterdam.

The disparity between such a price—about $11.50—and the value of the land acquired by the New Amsterdam colonists in 1626, is often cited by the English to impugn the moral sincerity of the Dutch when, as a matter of fact, the relatively small prices paid by them are not of primary importance to the moral question that was involved in their transactions. The Indians had not the least conception of the ownership of land in the European sense, and, therefore, none of its value to those who purchased it. But they did have a very highly developed idea of their proprietorship over it, so that what for the Dutch was a purchase price was for them something symbolizing the recognition of their exclusive right of occupancy. It was not then to insure the receipt of what was fair compensation for Indian lands measured according to their value to the Dutch, but to satisfy the native mind and thereby avoid conflict with the Indians, that in the "New Prospect of Freedoms and Exemptions," probably drawn up in 1629, the patroons were required to purchase from the sachems such lands as the natives were willing to surrender, while ordinary individuals were forbidden to take Indian lands under any circumstances.

The Puritans who began to arrive in Massachusetts in 1630 were of an entirely different type from the Pilgrims. They had no idea of adopting the Dutch policy. While perhaps more prone to biblical expression, they were no more "godly" than the Virginians, and were equally intent on taking the lands of the Indians. With no thought of creating a college for their Christianization, like the early Spaniards, they did not hesitate to enslave them when they could do so with advantage. Soon they had executed an accused Indian upon the sentence of a "drum-head court martial."

Warned by the late massacre in Virginia, and aware of the harmony that prevailed between the Dutch and Indians in New Netherland, the authorities of the Plymouth Company urged upon these new colonists the adoption of the Dutch policy. But their purpose was plain from the first. Thus in a paper bearing the title, "General Considerations for the Plantation in New England, with an Answer to Several Objections," written by John Winthrop, of Massachusetts, about 1631, it was said:

That which is common to all is proper to none. This savage people ruleth over many lands without title or property; for they inclose no ground, neither have they cattell to maintayne it, but remove their dwellings as they have occasion, or as they can prevail against their neighbors. And why may not Christians have liberty to go and dwell amongst them in their waste lands and woods (leaving them such places as they have manured for their corne) as lawfully as Abraham did among the Sodomites? For God hath given to the sons of men a two-fould right to the earth; there is a natural right and a civil right. The first right was naturall when men held the earth in common, every man sowing and feeding where he pleased; Then, as men and cattell increased, they appropriated some parcells of ground by enclosing and peculiar manurance, and this in tyme got them a civil right. Such was the right which Ephron the Hittite had to the field of Machpelah, wherein Abraham could not bury a dead corpse without leave, though for the out parts of the country which lay common, he dwelt upon them and tooke the fruite of them at his pleasure. This appears also in Jacob and his sons, who fedd their flocks as bouldly in the Canaanites land, for he is said to be lord of the country; and at Gothem and all other places men accounted nothing their owne, but that which they had appropriated by their own industry, as appears plainly by Abimelech's servants, who in their owne country did often contend with Isaac's servants about wells which they had digged; but never about the lands which they occupied. So likewise betweene Jacob and Laban; he would not take a kidd of Laban's without special contract; but he makes no bargaine with him for the land where he fedd. And it is probable that if the country had not been as free for Jacob as for Laban, that covetous wretch would have made his advantage of him and upbraided Jacob with it as he did with the rest. 2dly, There is more than enough for them and us. 3dly, God hath consumed the natives with a miraculous plague, whereby the greater part of the countrey is left voide of inhabitants 4thly, We shall come in with the good leave of the natives.

Winthrop tells how, finding their sanction in the Old Testament, the Puritans "smote" the Indians "from behind" and were rewarded with much fine land. Told without any humbuggery, ignoring the new international testament of Grotius, they took what land they wanted, kicked the Indian tenants off *vis a tergo*, and like the Jews of old would have taken their cattle without a twinge of conscience had the native livestock consisted of more than wild beasts of the forest.

It was not because the Puritans were unfamiliar with the Indian communal system that they were bent on pursuing this policy. Indeed, lands were long held by the Pilgrims of New England under the communal system. Moreover, a very considerable portion of New England was originally held under title acquired by grants made by the colonial legislatures to a township or other large tracts of lands, or to a number of proprietors who, though they acquired and exercised in regard to their property the general power of corporate bodies, administered the same in accordance with certain communistic principles.

The sophistry of such arguments as those of Winthrop was at once manifest to Roger Williams, who arrived in Massachusetts from Holland in 1631. When the following year the Dutch of New Netherland purchased large tracts on both banks of the Connecticut River from the

Indians, he insisted that this was the proper policy, pointing out that the taking of the lands of the Indians was a direct blow at their very existence and could have but one result. Furthermore, following Vitoria, he contended that the title to the soil, being in the tribes, could not be granted away even by the King without their consent. Fearlessly he declared that the taking of Indian lands without compensation not only found no sanction in Scripture, but was contrary to the teachings of Christ.

But in Massachusetts, as in Virginia, there was no place for a Christianity that breathed tolerance for an inferior race. Immediately, therefore, the humane Williams was damned and cursed from the pulpits for giving utterance to "outrageous" doctrines of foreign origin. Having made a direct attack upon those whose views were but expressed by Winthrop, he was expelled from the colony.

When in the spring of 1636, with twelve companions, he succeeded in passing beyond the boundary of the Plymouth Company, he found himself in the country of the Narragansetts, where the simple story of their unhappy state so excited the pity of Canonicus, the chief, that he gave them at once a large tract of land. Here, as Williams observed to his weary companions, "The Providence of God found out a place for them among the savages, where they might peaceably worship God according to their consciences; a privilege which had been denied them in all the Christian countries they had ever been in."

Thus, it is seen that the little colony of Rhode Island which Roger Williams now founded and where, with the lasting friendship of Canonicus, he undertook to put into practice among the English the Indian policy of the Dutch, owes its origin not only to a spirit of justice which did not characterize the English colonists, but the friendship of an Indian. "I have never suffered any wrong to be offered to the English since they landed," said Canonicus in his old age, "nor ever will." And it was recognized by both Williams and Winthrop that he was a powerful influence for peace.

Between the Indians of New England and Virginia there was the closest connection just as between the whites of those regions who from the earliest times conducted a thriving trade with each other. In fact, the New England Indians and those of Eastern Virginia were both of the Algonquin family and even spoke the same dialect. Thus, the little peninsula between Chesapeake Bay and the Atlantic Ocean and Plymouth were both called "Accawmacke," meaning peninsula in the native tongue. Undoubtedly there was a connection between the Pequot disturbances of 1637-45 and the Virginia massacre of 1644.

It has frequently been said that Hooker's New England Confederacy of 1639, which was the first attempt at a political union in America other than the Indian leagues, grew out of the necessity for a better common defense against the Pequots and the Dutch. In a measure this is true. The principle that "in union there is strength" has ever been a guiding one in political development. But the truth is, the fundamental purpose of the Confederacy of 1639, like the one of 1643, which included most of the towns in New England outside of Rhode Island, was to give the whites the power to preempt Indian lands without fear of the consequences.

After the Pequots in New England and the Indians in Eastern Virginia had been crushed, for a full generation they remained supinely submissive. But that they were to be permitted to retain such lands as remained to them was a vain hope on their part. By 1675 the population in New England had increased to such an extent that the Indians were everywhere being hard pressed again, just as in Virginia. Now it was that Pokanoket, or King Philip, the son of Massasoit, who had died in 1662, saw a last stand must be made. Having organized with consummate skill the remnants of about twenty New England tribes into a confederacy under his able leadership, when all seemed ready he launched what was designed to be a defensive war conducted along lines of offensive strategy. Of the ninety towns in New England, fifty-two were attacked and twelve completely destroyed.

The Indians fought with desperate bravery. In all probability, treachery among them alone saved the colonists from complete extinction. Due to the discords among the tribes, Philip's force was decisively defeated on August 12, 1676, in a night attack, and the great leader himself killed. "He almost made himself a king by his marvelous energy and statecraft put forth among the New England tribes. Had the opposing power been a little weaker, he might have founded a temporary kingdom on the ashes of the colonies."*

Thus, concurrently Indian resistance was crushed finally in New England and Virginia. But who, let history say, was the greater patriot—Massasoit's heroic son, or Bacon?

Harsh, indeed, have been the judgments passed upon the people of New England.

They deemed themselves commissioned, like Joshua of old, to a work of blood; and they sought an excuse for their uniform harshness to the Indians in those dreadful tragedies which were enacted, far back in primeval ages, on the shores of the Red Sea and the fertile plains of Palestine, and in which Almighty Wisdom saw fit to make the descendants of Israel the instruments of His wrath. So early as 1632, the Indians "began to quarrel with the English about the bounds of their land"; for the Puritan Pilgrims, maintaining that "the whole earth is the Lord's garden," and, therefore, the peculiar property of

---

* *History of King Philip's War*, Church, Garret Press, 1970; *Eulogy on King Philip*, Apes; *Civilization and Barbarism*, Freeman; *Narrative of King Philip's War*, Markham.

his saints, admitted the natural right of the aborigines to so much soil only as they could occupy and improve. In 1633, this principle was made to assume the shape of law; and, "for settling the Indians' title to lands in the jurisdiction," the general court ordered, that "what lands any of the Indians have possessed and improved, by subduing the same, they have just right unto, according to that in Genesis, ch. I, 28, and ch. LX, 1." Thus the argument used was *vacuum domicilium cedit occupanti*: and, by an application of the customs of civilization to the wilderness, it was held that all land not occupied by the Indians as agriculturists "lay open to any that could or would improve it."

It has been the fashion of late to assert for the Puritans that they regarded European rights, resting on discovery, to be a Popish doctrine, derived from Alexander VI, and that they recognized the justice of the Indian claims. But this position cannot be maintained. The rude garden, which surrounded the savage wigwam, was alone considered as savage property. The boundless landscape, with its forests, fields and waters, he was despoiled of, on the harsh plea of Christian right. In this way, Charlestown, Boston, Dorchester, Salem, Hingham, and other places, were intruded into by the Puritan Pilgrims, without condescending to any inquiry concerning the Indian title. They were seized and settled, because they were not waving with fields of yellow corn duly fenced in with square-cut hawthorne.*

An even more dreadful indictment may be quoted:

Champions of liberty, but merciless and unprincipled tyrants, fugitives from persecution, but the most senseless and reckless persecutors; claimants of an enlightened religion, but the last upholders of the cruel and ignorant creed of the witch doctors; whining over the ferocity of the Indian, yet outdoing that ferocity a hundredfold; complaining of his treachery, yet, as their descendants have been to this day, treacherous, with a deliberate indifference to plighted faith, such as the Indians have seldom shown—the ancestors of the heroes of the Revolutionary and the Civil Wars might be held as examples of the power of a Calvinistic religion and a bigoted republicanism to demoralize fair average specimens of a race which, under better influences, has shown itself the least cruel, least treacherous, least tyrannical of the master races of the world.†

Intemperate as such criticism may be, it is only too true, as declared by Lowell, that the Puritans fell first upon their own knees and then upon the aborigines. Consequently, the Indians of the seventeenth century could only judge European civilization by what they saw of it, and with that civilization they naturally confused Christianity, unknowing that the religion of the ecclesiastics formed no real part of the philosophy of Christ. Indeed, the wonder is the Indians were able to accept the outward forms of Christianity as readily as they did when it must have been apparent to those who understood the real character of Jesus that their own ethical

* *The Puritan Commonwealth*, Peter Oliver.
† The Saturday Review (London), January 28, 1881.

practices conformed more nearly to His precepts than did the conduct of the self-professed Christians.

Such a philosophy it was utterly impossible for the Indian mind to grasp. Until tutored in the true precepts of Christ, it was not unnatural for it to hold the Great Spirit it conceived to be the creative force of nature, superior to the God of the Europeans—a God that adjudged him the "Spawn of Hell" and sanctioned his destruction.

The point was well illustrated by Voltaire in a novelette in which personified principles are the heroes and superstitions the villians:

An Indian comes to France with some returning explorers. The first problem to which he gives rise is that of making him a Christian. An abbé gives him a copy of the New Testament, which he likes so much he soon offers himself not only for baptism but for circumcision as well. "For," he says, "I do not find in the book that was put into my hands a single person who was not circumcised. It is, therefore, evident that I must make a sacrifice to the Hebrew custom, and the sooner the better." Hardly has this difficulty been smoothed over when he has trouble over confession; he asks where in the Gospel this is commanded, and is directed to a passage in the Epistle of St. James: "Confess your sins to one another." He confesses; but when he has done he drags the abbé from the confessional chair, places himself in the seat, and bids the abbé confess in turn. "Come, my friend; it is said, 'We must confess our sins to one another'; I have related my sins to you, and you shall not stir till you recount yours." He falls in love with a French girl, but is told that he cannot marry her because she has acted as godmother at his baptism; he is very angry at this little trick of the fates, and threatens to get unbaptized. Having received permission to marry her, he is surprised to find that for marriage "notaries, priests, witnesses, contracts and dispensations are absolutely necessary. 'You are then very great rogues, since so many precautions are required.' " And so, as the story passes on from incident to incident, the contradictions between primitive religion and ecclesiastical Christianity are forced upon the stage.

What must have been the mental attitude of more than one Indian toward Christianity is also illustrated by the following story:

On his way to church a white missionary spied the bronze form of an old Indian, hoeing the weeds between the rustling green ribbons of his corn patch.

"You are breaking the Sabbath," shouted the stern Christian. "I have spent many years preaching to your people. How many times must I tell you that the seventh day is God's day? Put away your hoe along with these pagan ways and come to worship in God's holy temple."

The Indian stopped his labor, looking placidly a while upon the angry countenance of the speaker before replying.

"You address me, an old Indian, born and bred in the ancient faith that is dear to the heart of my people. You enter my peaceful cornfield to drive me to the great lodge where on the seventh day you worship your God. I have been taught that the Great Spirit that watches over my people is always present, everywhere. Amid this corn my heart sings with joy, for here I feel the living presence of the Great Spirit. My labor helps the corn to grow. Helping it I am helping my God express himself. I do no sin. I would not, could not live six days without Him. In such aloofness my soul would perish. Go, then, your way to worship in your church. I seek not to take it from you. Leave me in my garden with the Great Spirit—the only God I know."

What answer could the missionary make to one for whom every day was a Sabbath? What of brightness could the intolerance of this stranger who had invaded the sanctity of the Indian's spiritual communings add to the latter's life? What, in truth, did a religious philosophy manifesting more of the spirit of a jealous Jehovah than of Christ offer the aborigines of America but to silence the voice of a God that spoke to them in every wind?

Upon the seizure of New Netherland, the British colonists had come face to face with a situation the like of which they had not encountered in Virginia and New England.

In the first place, the greater part of New York was inhabited by the Iroquois, who, having been driven from Canada by Champlain, were still the most potent and advanced Indians in America. In the region of the upper St. Lawrence and Hudson lay the domain of the Mohawks, including the Mohawk Valley. Westward lay the Oneidas, Onondagas, Cayugas in the order mentioned, with the Senecas occupying the march of the Confederacy along the Niagara River and Lake Erie facing the Twitchwees or Miamis. Eminently aware of their power, they were wont to illustrate it by binding five arrows together into an unbreakable bundle and then snapping each one singly.

The Dutch had found the Iroquois still carrying on the feud which had persisted since the thirteenth century with the Wendats, or the Indians along the St. Lawrence. When Champlain and the French took up the cause of the latter in his attack on the Mohawks, the Iroquois naturally assumed the whites of Canada were the allies of their traditional enemies, hatred of whom had persisted throughout four centuries. Therefore, when the Dutch furnished them with firearms in order that they might defend the sources of the profitable fur trade conducted from the Hudson, they had become at once the firm allies of the Dutch. Moreover, according to Macaulay, between 1616 and 1620 some Dutchmen belonging to the East Indian Company had visited Ohnowalagantle (Schenectady), the Mohawk capital, and established the precedent of paying them for land.

While in any event the Iroquois would not have surrendered their lands lightly, this, coupled with their enhanced ability to defend them, had only served to make them more tenacious of their holdings.

It was 1615 when the Recollect fathers took up their work among the Hurons. Fathers LeCaron, Poulain, Viel, and Daillion, labored earnestly among them for many years. But in 1625 Father Brebeuf with an able corps of Jesuits had assumed exclusive control of affairs by reason of the political blunders during the régime of Champlain. Upon their demand Richelieu himself had taken a hand in the direction of Canadian affairs and in 1627 had organized the Company of New France. Seven years later Champlain had sent Jean Nicolet, an intrepid *coureur des bois*, westward along the Great Lakes to make treaties with the Ojibways. Seeing that the Jesuits, of whom twenty-five were working among the Hurons, now intended to break down the Hodenosaunnee (since they could not make it accept the Huron claims to Iroquois territory, when in 1642 the Jesuits established themselves at Hochelaga which they named Montreal), the Five Nations had begun to prepare themselves for a war to the death with the French. By this time they had over 400 muskets in their possession and at least 1500 in 1648 when the blow fell. During the next two years the Huron Confederacy, though fully supported by the French, had been so completely dispersed that it was no longer a political or military factor. The Jesuits saw now that they had blundered again. And what was this? In 1653 Father Le Moine had found more than a thousand Christian Hurons among the Onondagas alone! Dutch agents had been actually winning the Catholic converts over to the Iroquois who had adopted a policy of recruiting their strength by incorporating defeated enemies. This, of course, was fatal to the French. So widespread had become the defection of the Hurons that in 1656 the Iroquois had incorporated most of the three major tribes. Only the Cord People and the Neuter Nation, henceforth called Wyandots, remained in Canada.

This terrible blow having been delivered to the French, in 1656 Iroquois war parties, passing down the Susquehanna, had appeared in Maryland and Virginia in support of their Dutch allies. Everywhere it was then rumored that the Five Nations were going to drive the English out of America. Such was the situation when in 1659 Laval-Montmorency, the titular bishop of Petrea in Arabia, was placed at the head of the Catholic Church in New France and the government virtually turned over to him. Soon a system of paternalism was established which not only destroyed the initiative of the population but did not appeal to the Iroquois.

Having established possession of half the territory north of the Ohio and east of the Mississippi with the aid of the Dutch, the Iroquois, however, had no idea of being made the pawns of the latter. Therefore they did not assail the English during their trade war with the Dutch. Moreover, so powerful were they that when Great Britain seized New Nether-land in 1664 and converted it into the colony of New York, Richard Nicholls, the royal governor, did not dare openly reverse the Dutch policy of paying for Indian lands despite the complaints of the New England and Virginia colonists. Consequently the first patentees of land along the upper Hudson were required in 1665 to pay the Mohawks for it.

Nevertheless, as the demand for Indian land increased with population, the disposition manifested by the British elsewhere began to show itself. Thus, when Governor Fletcher was instructed by the King to use his discretion to purchase "great tracts of land" for "small sums" where occasion presented, he made huge grants of territory, sometimes seventy miles long, to individuals, disposing in this way of a very considerable portion of the Mohawk domain.

The result was inevitable. Again all the Iroquois became restless, now and then sending forth such expeditions against the rear of the colonies as that which penetrated eastern Virginia in 1676, giving rise to Bacon's Rebellion. And the truth is, the Jesuits were still hard at work among the Iroquois, seeing in an alliance with them their last hope of overthrowing the English, who had simply let the Dutch develop New Netherland until they were ready to seize it. So successful were these French agents that they actually induced several bands of Mohawks to settle along the St. Lawrence at Caughnawanda, St. Regis and Oka as open allies of France. Thus, the unity of the Five Nations was threatened by the Jesuits. Appar-ently it was impossible for the Iroquois to dwell in peace. Long and carefully they deliberated upon their policy for the future. Should they join France or England?

In retrospect it appears that in the decision they now made hung the balance of European empire in North America, for at this time, with twenty-four villages in New York and a total population of 16,000 adults, they were capable of putting 2,250 well-armed warriors in the field. In the end they cast their decision against France, for while they despised the British, long since their sachems had detected in the Jesuits a force more to be feared than the stupid English. Therefore, it was French power that they must destroy first, so that in 1680 they outlawed the Mohawk seces-sionists and proposed a union between themselves and the English with an immediate concerted attack on Canada that would put an end to French power forever.

But although the wise Iroquois sachems were statesmen enough to envisage the future, among the British as yet there were none capable of grasping the political situation in America. The Iroquois were simply regarded as savages with no other object than the satisfaction of a native lust for warfare. It was not perceived then, and has not generally been understood since, that a people called by thoughtless writers "the Huns of

America," had anticipated the colonists in the formation of a league to enforce a definite political policy and to insure self-determination on the part of the British colonists as well as themselves. But the more these people are studied, the more closely their aims and acts are scrutinized, the wiser and more civilized they appear to have been, and the greater appears the tremendous influence they exerted upon the American people. Fortunately for the English, just after the Iroquois proposal was rejected, when they were at the very height of their power their unrest was brought to an end for the time being by a dreadful epidemic of smallpox which had soon reduced their strength by half.

9

"Saint Tammany"
and the First Proposal of Union

HIAWATHA, Wahunsonacock, Samoset, Massasoit, Canonicus and King Philip were not the only Indian chieftains who were to exert a profound effect upon early American history.

No puny saint was he, with fasting pale. He climbed the mountains, and swept the vale; rushed through the torrent with unequalled might;—Your ancient saints would tremble at the sight—caught the swift boar, the swifter deer with ease, and worked a thousand miracles like these. To public views, he added private ends, and loved his country most, and next his friends.

So in anonymous verse was Tamenend, the great chief of the Delawares, eulogized by one of his white contemporaries.

While it is impossible at this time to disentangle truth from the pure fiction concerning him, enough of substance remains to give him a true historical position. So great was his wisdom, so famed his statesmanship, it was contemporaneously declared that he counselled the reigning Manco Capac of Peru at the latter's request upon political matters. At any rate, after a long absence during which he is supposed to have gone to Mexico for the purpose mentioned, he reappeared among the Lenape to institute among them the great reforms of which he had conceived and to complete the federation of the native tribes by forming the great confederacy which took Lord Delaware's name.

At this time the people whom the Canadians called wolves had fallen into a state of moral and physical decadence. It is said that, finding them disorganized and rotting with disease, including syphilis, after teaching them to cure themselves of their physical ills, Tamenend organized them into thirteen district groups, each with a sachem subordinate to him as Grand Sachem, and to each group gave an animal fetish symbolic of a virtue to be emulated. Thus, the agility of the panther, the industry of the beaver, the frugality of the squirrel, the benevolent patience of the tortoise, the courage of the tiger, were accentuated as those virtues which in combination would enable his people to survive.* Moreover, Tamenend seems to have believed, like Confucius, that government found justification only in service to the governed. Now only did he preach democracy, but he spared no effort to enforce the observance among his people of democratic principles. Plainly he was but the disciple of Dekanawida and Hiawatha, as well as an anti-Jesuit.

Such was the Indian statesman with whom William Penn came immediately in contact upon arriving in America in October, 1682, to assume personal control over the territory granted him by the King in a charter similar in its provisions concerning the Indians to that of Lord Baltimore. He was authorized to destroy them at his own discretion.

While most of the tribes along the North Atlantic seaboard had been uprooted from their immemorial domains, the Lenape of Pennsylvania, New Jersey and Delaware had been little disturbed by the whites. But Penn was not ignorant of colonial history. Moreover, like Roger Williams, through intellectual contacts with Holland he, too, was imbued with the philosophy of Vitoria. Even the King, he believed, had not the right to grant away Indian lands without their consent. They must be dealt with by treaties. The Dutch policy of Roger Williams was to be his policy. Nor did he fail to recognize in Tamenend more than a chieftain to be placated because his friendship would be useful. He was a statesman whose counsel it would be unintelligent not to seek. Therefore, with the governor he dispatched to America in 1682, he sent a pledge to the Indians.

The influence of Tamenend quickly asserted itself. Almost immediately after Penn's own arrival, with the former's aid the Delawares were assembled at their great council fire at Shackamaxon on the banks of the Delaware, where Penn proposed to found his capital—The City of Brotherly Love. There under a spreading elm, as depicted by Benjamin West in a great painting that has become an heirloom of the American people, he first met those whom he came to rule.†

The details of what followed have been fully recorded. Contrary to the common belief, Penn did not demand lands of them at this time.

---

* Although the tiger fetish is undoubtedly the origin of the term Tammany Tiger, for the statement that the Lenni-Lenape were organized in thirteen subgroups, there is no authority in the literature of the Bureau of American Ethnology.

† The elm was blown down in 1810, and is commemorated by a little riverside park at the foot of Shackamaxon Street near Cramp's shipyard.

Tamenend was too wise a counsellor to permit him to do this. The Delawares knew only too well how their race was being treated elsewhere. They must be reassured, made to see that this new white overlord who had come to assert authority over them was both friendly and just. Accordingly, in the understanding which was arrived at, of which there is no written evidence, but which is known to history as the treaty of Shackamaxson, no more was attempted than to establish a system of relations between the overlord and his Indian fiefs, to arrive at what diplomatists called a *modus vivendi*, and to provide methods of adjusting peaceably any disputes that might occur thereafter between them.

This, indeed, was an auspicious event. The elm of Shackamaxson does not fail to suggest the charter oak, for beneath it the inherent rights of the red race were recognized, and by this simple bit of justice the foundations laid of an enduring friendship between Penn and the Indians which was to prove of the utmost value to the colonies as a whole.

It was easy now for Penn to obtain from them such land as he required. On June 23 and July 14, 1683, deeds were recorded stipulating that he should have as much land beyond the mouth of Neshaminy Creek and between the Schuylkill and Chester Creek, "as a man could walk in three days." Having thus provided for a site for his capital, in November Penn and a party of Indians paced off the ceded tract. "In a day and a half they arrived at a point about thirty miles distant at the mouth of a creek they called 'Baker's' (from the name of the man who first reached it). Here they marked a spruce tree; and Governor Penn decided that this was as much land as would be immediately wanted for settlement, and walked no farther. They walked at leisure, the Indians sitting down sometimes to smoke their pipes, and the white men to eat a biscuit and cheese and drink a bottle of wine. A line was afterwards run from the spruce tree to Neshaminy and marked, the remainder was left to be walked out when wanted for settlement."*

Inasmuch as the consideration which passed to the Indians for the three hundred square miles thus released by them to Penn consisted of a mere handful of trinkets, Penn's critics find nothing of idealism in the transaction. In their view he was merely a shrewd trader, governed solely by expediency and not by principle. Such criticism, of course, ignores Indian nature as well as the political sagacity of Penn. In the eyes of Tamenend and his people the material things which passed to them were but the symbol of the fealty which they were pledging to an acknowledged overlord in exchange for that protection which for them had become a matter of life and death.

During the autumn and winter of 1683 the Great Proprietor made an

* Hegard's Register, Vol. VI.

exploration of his domains, penetrating far into the wilderness. Everywhere he was hospitably received by the natives, who were constantly at his heels, bringing him fish, game, and fruits of all sorts. At the end of his journey he was actually able to communicate with them both by sign and spoken words. The description of the country and its people which he now wrote is one of the most interesting of the time, showing only too plainly that in all fundamental respects Indian nature in Pennsylvania was similar to that noted by Columbus. "In liberality they excel. Nothing is too good for their friend," he wrote. "Some kings have sold, others presented me with several parcels of land." The details of their communal system of ownership were clearly described by him and much there was among them that led him to believe they were the descendants of the Ten Lost Tribes of Israel.

Nor did the statesmanship of Tamenend fail to envisage the future. Accordingly, over and over he warned Penn of the impending danger of the "Friendly Ones," whom he knew were forming alliances with the western tribes of which they proposed to make full use in the extension of French empire, while even among the Iroquois, still they were actively at work endeavoring to seduce them from their attachment to the British. Only lately Père Marquette and Père Joliet had been among the Illinois and Father Hennepin with the Chippewas and Sioux, while La Salle had traversed the entire Mississippi Valley and set up the Cross at the mouth of the "Father of Waters," claiming it for the Grand Monarch. Aware of all this, Penn saw only too plainly that a fuller measure of justice must be done the Indians by the British colonies unless all the tribes were to be driven into the arms of the French, and it was largely at his suggestion that the Five Nations were summoned to meet with commissioners from Massachusetts, New York, Maryland and Virginia at Albany in 1684 for the purpose of reaching an accord similar to that he had established with the Delawares. So effective did this council prove that ten years later another was held at which Massachusetts, New York, Connecticut and New Jersey were represented, resulting in the treaty of Albany of 1694, in which the Five Nations pledged their allegiance to the King in consideration of his promise to protect them against the French.

Nevertheless, upon Penn's head anathemas continued to be heaped because of his pacific Indian policy, and it seems certain now that many of his troubles found their origin in the hostility which that policy had engendered among neighbors who had no idea of recognizing Indian rights as the best means of defending against the French. At any rate, compelled to go to England to defend himself before the King even against a charge of treason growing out of his pacific policy, he was there when the situation in America became urgent by reason of the threatening alliance between the Five Nations and the French. Something had to be done.

It was at this time that Penn proposed the union of all the colonies for the purpose of defense, the regulation of commerce, and particularly for the establishment of that uniform Indian policy which he declared alone would prevent the tribes from joining the French. His plan contemplated a Congress composed of two representatives from each colony, which was to be presided over by a commissioner appointed by the King. While this plan was not adopted, at his insistence on the date of August 31, 1697, the Earl of Bellomont directed the Governor of New York to call before him the Five Nations, and assure them that if they would stand fast in their allegiance to the British King he would protect them as his subjects from the French, which was a clear recognition of their right of occupancy.

This having been accomplished, and released at last from the toils in which he had become involved, Penn returned to Pennsylvania in 1699, and at once resumed his efforts to insure the security of the colonies, negotiating treaties in 1700 and 1701 with the Susquehannocks and Conestogas, binding them not to make war upon the English nor to trade outside his province. Meantime, pursuant to Lord Bellomont's instructions, the Five Nations had been summoned by the Governor of New York to a great council at Albany and afforded an opportunity to express all their grievances.

Since the coming of the English they had found themselves between two hostile forces and, despite the treaty of 1679, their rights had not been respected. Yet, their sachems were diplomats. During the past several decades they had overrun a vast territory in the northwest beyond New York approximately eight hundred by four hundred miles in extent, which they called the "Beaver Hunting Grounds," their claim to which was constantly being disputed by the French and its Indian inhabitants. King William was still desirous of acquiring large tracts. Since the Iroquois could not themselves hold the region mentioned, it was a good idea to transfer their claim to it to the King and let him assume the responsibility for protecting the game for them. Therefore, on July 19, 1701, they executed a deed at Albany, conveying to the King and his heirs forever their title in this vast region, retaining therein a hunting right. Thus, while reaffirming their allegiance to King William, they converted into crown lands about 320,000 acres beyond the organized limits of the colonies.

To complete his work, in 1700 Penn caused to be enacted by Pennsylvania a statute providing "that if any persons presumed to buy any land of the natives within the limits of this Province and Territories, without leave from the Proprietary therof, every such bargain or purchase shall be void and of no effect," this being the model for similar statutes subsequently enacted in almost every colony. Thus, there came into being, as a result of his treaties and the Iroquois deed, a vast Indian barrier state or crown domain at the back of the colonies which subsequently was to make possible their westward expansion.

With Penn's political demise, his wise proposal of union among the Colonies was cast into the discard, there to remain until revived first in 1731 and then by Franklin in 1754. Nor is anything more heard of Tamenend as an actor upon the stage of events after 1700, down to which time his relations with Penn continued. Yet, the memory of him was to survive.

About 1772 a benevolent and fraternal order was founded in Philadelphia which took his name to commemorate the democratic tenets of his philosophy, and which was accustomed to hold annual parades in May.* Out of this society grew the present Improved Order of Red Men, a semi-secret society represented in every state, with a membership of nearly half a million.

Tammany societies having become very popular, after the Revolution William Mooney, a former soldier, founded in New York City a patriotic association known as the fraternal and benevolent "Society of Saint Tammany, or The Columbian Order," as an agency to insure the adoption of the Constitution. The names of few men in American history are, therefore, better known than that of Tammany. Writing of him nearly a century after his time, a biographer said:

Let Asia extoll her Zamolies, Confucius, and Zoroaster; let Africa be proud of her Dido, Ptolemy, and Barbarossa; let Europe applaud her numberless worthies, who, from Romulus to Charlemagne, and from Charlemagne down to the present day, have founded, conquered, inherited, or governed states; and where, among them all, will you find coercion so tempered by gentleness, influence so cooperative with legal authority, and speculation so happily connected with practice, as in the institutions of Tammany? Avaunt, then, ye boasters! Cease, too, your prating about your St. Patrick, St. George and St. Louis; and be silent concerning your St. Andrew and St. David. Tammany, though no Saint, was, you see, as valorous, intrepid and heroic as the best of them; and, besides that, did a thousand times more good.†

While the idea of union put forward by Penn is not accredited to Tamenend in this florid appreciation, it is significant that his name was connected with that of the discoverer of America by the founders of the society designed to secure the adoption of the Federal Constitution and the more perfect union of the thirteen states. And more significant still is it that this society was organized on the basis of thirteen tribes, each with a sachem, all ruled by a Grand Sachem, exactly after the political system instituted by Tamenend among the Lenni-Lenape.

Knowing how loath is tradition to enshrine the human name in the

* Philadelphia Chronicle, May 4, 1772.
† The Life Exploits and Precepts of Tammany, the Famous Indian Chief; Being the anniversary Oration Pronounced Before the Tammany Society or Columbian Order (1795) by Dr. Samuel L. Mitchell, Professor of Chemistry, Natural History and Agriculture in the College of New York.

memory of man, the historian can but conclude from the evidence given that the wisdom, patriotism and moral nobility of "Saint Tammany" exerted a powerful influence upon American history. And it is particularly interesting to contemplate the fact that the idea of a union among the American people, as well as certain constitutional forms at the basis of the union eventually formed by them, had its inception in the aboriginal mind.

According to tradition Tamenend died about 1740 and was buried in Bucks County, Pennsylvania, but if we should stop here with the ordinary historian we would have no complete understanding of this remarkable man. He was far more than the disciple of Hiawatha. The mystery surrounding his visit to Mexico is due to the fact that the Manco Capac with whom he is supposed to have conferred was nothing more than the Jesuits in Mexico. Cut off from the Delaware country by the Five Nations, Montmorency and the Jesuits in Canada were unable to work among the Lenni-Lenape. Therefore, after the Hodenosaunnee had turned against France it was necessary for Tamenend to be schooled in Mexico in preparation for his task of turning the Lenni-Lenape against the Five Nations. But when, after this had been achieved, Tamenend was able to commit Penn to a policy of justice toward the red race, his part in the Jesuit conspiracy had ended at once. Thus, having regenerated the Lenni-Lenape and liberated them from the oppression of the Hodenosaunnee, being a true patriot he was unwilling to be a cat's paw of Rome.

Although the scheme devised by him for the regeneration of his people unquestionably had a common origin with that of Hiawatha, just like that of King Philip, in many respects Tamenend was the most farseeing statesman of his age in America, if not in Europe, Penn not excepted. Having learned from the Jesuits in Mexico what was their ultimate purpose, it was he that first saw the one possible means of defeating their design. The union of the tribes under the direction of the French and Spanish Jesuits and their ultimate joint uprising against the British could only be prevented by the union of the British colonies and their alliance with the intervening tribes under an equitable system of governance for the latter. Therefore, Tamenend marked a climax in the two great politico-religious conspiracies which had been going on in Europe, since the Society of Jesus and Sir Thomas More had undertaken to seize the North American Continent for Rome and the Protestants, respectively.

The truth is, both sides had outdone themselves. The colonists were determined to have the land free of both British dominion and Indian rights. Already the Hodenosaunnee had proved a boomerang for the Jesuits. In federating the other ethnic groups they had merely created a common fear among the British colonists so that English, Scotch, Irish, Dutch, Germans, Swedes, Swiss and French alike, without regard to religion, were soon to recognize what Tamenend and Penn pointed out to them—the absolute necessity of union among themselves after the manner of the Hodenosaunnee. In the end, just as the Christians emerged from the catacombs, and ascending from the depths of oppression to the Capitoline Hill, governed Rome, these homeseekers in the wilderness of America, following their conquest of Canada in 1760 and the final overthrow of the Jesuits with the crushing of Pontiac, would come to rule America!

Naturally when Tamenend disclosed the designs of the Jesuits to Penn, he was looked upon by Montmorency and Rome as a traitor. Nor is there any doubt that many of Penn's troubles, like those of Raleigh, traced to the Vatican where it was seen that Penn and Tamenend were working in union to defeat the designs of Montmorency. Then, the same old charge of treason was lodged against him that sent Raleigh to the Tower and ultimately to his death.

10

Sir William Johnson and the French War

THE IDEALISM OF PENN did not extend far beyond his time. By 1708 one-quarter of the slaves in South Carolina were Indians. In his history of Queen Anne's reign Oldmixon tells how five Iroquois chieftains—one of each of the Confederated Nations—were brought to England in 1710 and paraded about London in the royal coaches as evidence of the success of Britain's colonial policy. Entertained at Court as vassal kings, a review of the Life Guards was held for them, obviously to impress them with British might. Boyer declares that Lord Cotterel and the Duke of Ormond outdid themselves to make political capital of the event while Smith in his history of New York tells how the bewildered chieftains were followed about by the English mobs who bought cuts of them.

Nevertheless, the old processes common to Virginia and New England went on apace in the south, resulting finally in 1717 in the uprising of the Tuscaroras, a people of Iroquois blood who were the only tribe left in North Carolina deemed worthy of serious consideration except the Cherokees in the far west of the state. With the aid of South Carolina this revolt was repressed with such appalling harshness that all the Tuscaroras, with the exception of a few who were not involved, migrated to New York upon the invitation of their kinsmen to be admitted about 1712 as a sixth nation in the Iroquois League of the Long House. Having thus recruited their strength, about 1720 the Six Nations overwhelmed and disarmed the Delawares, reduced them to a state of virtual vassalage, which caused them for a time to be known as "women-men" among the Algonquins, and in this ignominious state they were to remain for forty years.

Fearful again of the intentions of the increasingly powerful Iroquois, in 1731 the colonists revived Penn's proposal of union, though nothing came of this. When, however, the following year a charter to found the colony of Georgia was granted General Oglethorpe, he was quick to institute Penn's lenient policy. Thus, with the aid of the venerable chief Temo-chi-chi, treaties of purchase were at once made by him with the Lower Creeks and Uchees, while in 1739 a treaty was negotiated with the Cherokees, Creeks and Chickasaws in which their rights in the lands reserved by them were clearly defined. Soon Penn's law, forbidding the purchase by individuals of Indian lands and invalidating any made, was also adopted.

Meantime, there had begun to arrive in Pennsylvania itself a type of colonists to whom Penn's Indian policy was entirely unacceptable. These were the Scotch-Irish from the north of Ireland, who came to America to seek new homes in the wilderness beyond the control of the hated British government, and little thought they had of recognizing any restrictions upon their new-found liberty. From Vermont to Tennessee they pressed forward through the older settlements, dirk in hand, prepared to fight for "freedom." Woe betide those who stood in their way, for now it was clansman against tribesman! In 1729 alone over five thousand of these hardy, relentless homeseekers and implacable foes of the red man arrived in America, and from then on about 12,000 a year up to 1750.*

Now, when the Pennsylvanians desired to acquire land, they took it. Even Thomas Penn forgot the treaty of Shackamaxson. Thus, in 1733, when he desired to acquire land under deeds worded like those of 1682, he employed a sprinter to pace it off, who covered eighty miles in a day and a half! Such methods the Indians, of course, resisted.

Humanity blushes at the events of this period of colonial history. General Jeffrey Amherst, representing the Imperial Government in America, but reflected the popular attitude when in 1732 he wrote one of his subordinates: "You will do well to try to inoculate the Indians by means of blankets in which smallpox patients have slept, as well as by every other method that can serve to extirpate this execrable race. I should be very glad if your scheme of hunting them down by dogs could take effect."†

With a representative of the government advocating practices going one better than those employed by the Spaniards in Haiti, no wonder that, expressing the general view of the Indians, the Iroquois chieftain, Canassatego, should have said at Lancaster in 1744:

Some of the young men of the English would every now and then tell us that we should have perished had they not come into the country and furnished us

---

* In 1775 one-third of the Colonial population was of Scottish origin, for the most part occupying the advancing frontier.

† Quoted in *Story of America*, Van Loon, Hart Publishing Co., Inc., New York, p. 42.

with strouds (coarse woolen goods) and hatchets and guns and other things, necessary for the support of life. But we always gave them to understand that we lived before they came amongst us; and as well, or better, if we may believe what our forefathers have told us. We had then room enough, and plenty of deer, which were easily caught, and though we had not knives, we had hatchets of stone and bows and arrows, and those served our uses as well as English things do now.*

Such was the Indian attitude of mind when, in 1738, there arrived in the colony of New York the moral successor of Roger Williams and William Penn in the person of a young Irishman named William Johnson, who is entitled to high rank in the roll of Britain's empire builders.†

Just twenty-three years of age when he assumed the personal management of the vast estate of his uncle, Sir Peter Warren, embracing the present Mohawk Valley, for Indian diplomacy he manifested from the first great aptitude. Studying the native dialects, mastering Indian lore and customs, marrying the sister of the powerful Mohawk chieftain, Joseph Brant, his ascendancy over the Iroquois was soon complete, with the result that in 1744 he was appointed Colonel of the Six Nations by the governor of the colony, and two years later succeeded Colonel Schuyler as Commissioner of New York for Indian Affairs. His letters dealing with Indian affairs are among the most illuminating in American history.

Johnson was no less far seeing than Penn. Now in possession of the Mississippi Valley from Canada to Louisiana, the French had begun to push their trade into the interior, protecting it with a line of advancing military posts. Their obvious intention was to dispute the claim of Great Britain to the basin of the Ohio River and the region of the Great Lakes. Basing their claims upon the possession of settlements at the mouth of the St. Lawrence and the Mississippi rivers, it was plain they would soon deny the validity of the British King's colonial grants "from sea to sea," and seek to limit his jurisdiction to the narrow strip between the Ohio and the coast. Knowing the power of the Indian tribes, Johnson, like Tamenend and Penn, did not fail to see the wisdom of cementing their allegiance before the test came, of welding them more firmly into a vast barrier state along the frontier as a first line of defense. Upon this the French military power in America would have to expend much of its force.

The Iroquois League alone was now capable of putting 10,000 warriors in the field. French agents had continued to work among the Six Nations, trying to undo the harm done by Champlain. In the north the Iroquois, in the south the Cherokees, Choctaws, Chickasaws, Creeks and Seminoles, along the midfrontier the Shawnees, Miamis, Wyandots, and associated

---

* *History of The Five Indian Nations*, Colden, Cornell Univ. Press, Ithaca.

† Johnson was now to do in effect just what Dekanawida, perhaps Roberval, had done—make himself more Indian than white man.

tribes must, he declared, be held as military vassals of the British. But in vain Johnson urged his plan. The day of judgment was drawing near. A foolish Imperial Government and thirteen selfish colonies would not see his wisdom until too late.

Next to William Johnson—the "Indian Guardian" of New York—in 1750 Major George Washington, of Virginia, probably knew more about Indians than any man of eminence in America. A surveyor by profession, he had visited the Kanawha country and, in fact, had interests in some of the royal grants in that region. It was not unnatural, therefore, that in the autumn of 1753 Governor Dinwiddie of Virginia should send him on a mission to the French outposts along the Allegheny to protest against the further advance of the French.

His mission yielded no fruit. The following spring a force of Virginians over which he had been placed was assailed at Fort Pitt by the French and their Indians allies, so that he was compelled to retire from the disputed territory. These were the hostilities which ushered in the French War. The Seven Years War, or the European War of the Austrian Succession, had at last and inevitably involved the American colonies and the Indians!

Now the British government perceived the wisdom of the broad policy which had long since been proposed by Penn and Johnson, but all too late. During the summer of 1754 a council of all the colonies was called to assemble at Albany to consider means of cooperation and the enlisting of the frontier tribes against the French. Meantime, Benjamin Franklin, who has been called the "first civilized native-born American," had removed from New England and succeeded to the leadership of Penn in Pennsylvania. From the first the policy of Roger Williams and Penn, with respect to the Indians, had strongly recommended itself to him. Going to the Albany Council as the delegate of his colony, he there collaborated with Johnson, and not only urged the union of the colonies, which Penn had long since proposed, but the adoption of Johnson's Indian policy.

Among other things, his plan contemplated a president-general to be appointed and supported by the Crown, and a council to be chosen by the colonial legislatures. Together they were to have power to make treaties with the Indians, to regulate Indian trade, purchases of land from them and all settlements in their country. But although it was approved by all the Commissioners except those from Connecticut, and was sent to England in the hope it would be put into effect by act of Parliament, because it was thought to give too much power to the colonial representatives no action was taken on it.

Nevertheless, another great council was now held with the Iroquois, who, notwithstanding Johnson's influence, had become lukewarm in their attachment to the British. One of the leading sachems unbraided the English with great boldness for their neglect of the Indians, the invasion of their lands, and the dilatory conduct with regard to the French, who, as the speaker averred, had behaved like men and warriors.* So great was Johnson's personal influence, however, that a treaty of alliance was finally concluded with the Iroquois deputies. With astonishing perverseness the agents of the proprietary government of Pennsylvania now took advantage of the council to procure from the assembled sachems the grant of an extensive tract including the lands of their subordinate tribes which the French were at the moment endeavoring to seduce. When the Delawares and Shawnees heard of their betrayal, convinced that there was no limit to English greed and Iroquois injustice, even the children of Tamenend, or the faithful allies of Penn, went over to the enemy, thus freeing themselves from the servitude which the Iroquois had imposed upon them.

In June, 1755, General Braddock set out for Fort du Quesne at the head of a force of British regulars, accompanied by Colonel Washington and a small body of Virginians. As he approached his destination, where the Chippewas, Ottawas, Potawatomis, Hurons, Abnakis, Caughnawagas and Delawares were encamped, at first the French endeavored in vain to lead the Indians against him. Finally, in July, led by the great Ottawa chieftain, Pontiac, the tribes were induced to take to the warpath and, accompanied by Beaujeu and about 250 French and Canadians, moved out to the attack.

The slaughter lasted three hours. The Indians did not pursue beyond the Monongahela. Braddock and the remnants of his force fled to Philadelphia, their flight covered to some extent by Colonel Washington's rangers. Everywhere the frontier tribes, with the exception of the Six Nations, now joined the French.

During the first two years of the war, Johnson, who had been commissioned a Major-General, was alone successful. One disaster after another fell to the British arms in quarters other than New York. Loyally supported by the Six Nations, Johnson demonstrated beyond a doubt the value of the Iroquois alliance. Under the leadership of the French the western tribes had also proved their value as military allies, so that when in 1756, undoubtedly at the instance of Johnson, the Lords of Trade called on Governor Hardy of New York to report what should be the proper and general system for the management of Indian affairs, he replied that with respect to the Six Nations, the Governor of the Province should have the chief direction of their affairs, and that no steps should be taken with them without consulting him, as he had always directed the transactions with them. He suggested, however, that "some proper person under this direction should have the management and conduct of Indian affairs," recommending for this task Sir William Johnson, who had previ-

* Minutes of Conference at Albany, 1754.

ously been commissioned for the same purpose by General Braddock and who, meantime, with extraordinary promptness, had been knighted and given great grants of land as a reward for his outstanding services.

Nor was the confidence now reposed in Johnson misplaced, for in 1757 he succeeded in ways known to him in winning away from the French a part of the Delawares, while the following year all the Ohio tribes were induced to join the Iroquois as allies of the English. Thus deprived of the aid of many of the western tribes, French strength began to fail so that Montcalm's forces were driven back into Canada. Finally, in September, 1760, after Louisburg, Quebec and Montreal had fallen, Canada with all its dependencies was surrendered to the British.

Still, however, it was necessary for the British to take possession of Detroit and the other western posts, which was not possible without resorting to diplomacy with the western Indians.

At this time, Pontiac, chief of the Ottawas, and until now a firm ally of the French, was, perhaps, the most influential representative of his race. About fifty years of age, he was at the height of his prestige. Not long before the French war broke out, he had saved the garrison of Detroit from imminent peril of an attack from some of the discontented tribes of the north, and during the war he had commanded the Ottawas with such gallantry and ability at the defeat of Braddock as to win high praise from Montcalm himself. Preeminently endowed with courage, resolution, address and eloquence—the sure passports to distinction among Indians— over the Ottawas, Chippewas, and Potawatomis, or the Ojibwas Confederacy of the Three Fires, his authority was almost despotic, extending far beyond the limits of the three tribes mentioned. Throughout the farthest boundaries of the widespread Algonquin race his name was known and respected. Naturally his fame was not unknown to Sir William Johnson, who selected the celebrated Major Rogers to proceed at the head of his rangers to take over the French posts, instructing him how to deal with Pontiac.

While Rogers' little command was toiling along Lake Erie, a party of Indian chiefs and warriors entered their camp. Declaring themselves an embassy from Pontiac, ruler of all that country, they directed, in his name, that the British should advance no farther until they had had an interview with the great chief, who was already close at hand. In truth, before the day closed Pontiac himself appeared; and it is here, for the first time, that this remarkable man stands forth distinctly on the page of history.

Pontiac greeted Rogers with the haughty demand, what was his business in that country, and how he dared enter it without his permission. Rogers informed him that the French were defeated, that Canada had surrendered, that he was on his way to take possession of Detroit and restore a general peace to white men and Indians alike.

Undoubtedly the messages of Johnson, who was revered by all Indians, sufficed to gain Pontiac's attention, for, listening with keen interest, after enquiring if Rogers and his command were in need of anything his country afforded, he withdrew with his chiefs at nightfall to hold council with them. "I will stand in your way until morning," he declared upon parting.

Suspecting treachery, Rogers stood well on his guard, but in the morning, true to his promise, Pontiac returned with his attendant chiefs to make reply to Rogers' speech of the previous day. He was willing, he said, to live at peace with the English and suffer them to remain in his country as long as they treated him with due respect and deference, whereupon the Indian chiefs and provincial officers smoked the calumet together. Perfect harmony seemed established between them.

In this result the matchless Indian diplomacy of Johnson is manifest. What was the sign he sent Pontiac through Rogers, history will never know, but at least it had its effect.

A cold storm of rain set in, so that the rangers were detained several days in their encampment. During this time Rogers had several interviews with Pontiac, and was constrained to admire the native vigor of his intellect, no less than the singular control which he exercised over those around him.

On the twelfth of November the detachment was again in motion and within a few days they had reached the western end of Lake Erie. Here they heard that the Indians of Detroit were in arms against them, and that four hundred warriors lay in ambush at the entrance of the river to cut them off. But the powerful influence of Pontiac was exerted in behalf of his new friends. The Indians abandoned their design, and the rangers continued their progress towards Detroit, now within a short distance.

In the meantime, Lieutenant Brehm had been sent forward with a letter to Captain Beletre, the commandant at Detroit, informing him that Canada had capitulated, that his garrison was included in the capitulation, and that an English detachment was approaching to relieve it. The Frenchman, in great wrath at the tidings, disregarded the message as an informal communication, and resolved to maintain a hostile attitude to the last, doing his best to rouse the fury of the Indians. Among other devices, he displayed upon a pole, before the yelling multitude, the effigy of a crow pecking a man's head; the crow representing himself, and the head, observes Rogers, "being meant for my own." Thoroughly under the influence of Pontiac, the Indians refused to respond to the appeals of the Frenchman, who stormed and railed at them in vain, calling them ingrates, traitors and the like.

Rogers had now entered the mouth of the River Detroit, whence he sent forward Captain Campbell, with a copy of the capitulation, and a letter from the Marquis de Vaudreuil, directing that the place should be given up in accordance with the terms agreed upon between him and General Amherst. Beletre was forced to yield, and, with a very ill grace, declared himself and his garrison at the disposal of the English commander.

The whaleboats of the rangers moved slowly upwards between the low banks of the Detroit, until at length the green uniformity of marsh and forest was relieved by the Canadian houses, which began to appear on each bank—the outskirts of the secluded and isolated settlement. Before them, on the right side, they could see the village of the Wyandots, and on the left the clustered lodges of the Potawatomis; while, a little beyond, the flag of France was flying for the last time above the bark roofs and weatherbeaten palisades of the little fortified town.

The rangers landed on the opposite bank, and pitched their tents upon a meadow, while two officers, with a small detachment, went across the river to take possession of the place. In obedience to their summons, the French garrison laid down their arms. The Fleur-de-Lis was lowered from the flagstaff, and the cross of St. George rose aloft in its place, while seven hundred Indian warriors, lately the active allies of France, greeted the sight with a burst of triumphant yells. The Canadian militia were next called together and disarmed. The Indians looked on with amazement at their obsequious behavior, quite at a loss to understand why so many men should humble themselves before so few. Nothing is more effective in gaining the respect, or even the attachment, of Indians than a display of power. From all this the spectators conceived the loftiest idea of the powers of the English or "The Long Knives," who seemed to be far superior to the "Friendly Ones."

It was on the twenty-ninth of November, 1760, that Detroit fell into the hands of the English. The garrison was sent as prisoners down the lake, but the Canadian inhabitants were allowed to retain their farms and houses, on condition of swearing allegiance to the British Crown. An officer was sent southward to take possession of forts Miami and Ouatanon, which guarded the communication between Lake Erie and the Ohio; while Rogers himself, with a small party, proceeded northward to relieve the French garrison of Michillimackinac. The storms and gathering ice of Lake Huron forced him back without accomplishing his object; and Michillimackinac, with the three remoter posts of St. Marie, Green Bay and St. Joseph, remained for the time in the hands of the French. During the next season, however, a detachment of the 60th Regiment, then called the Royal Americans, took possession of them; and nothing now remained within the power of the French except the few posts and settlements on the Mississippi and the Wabash, not included in the capitulation of Montreal.

The work of conquest was finished. The fertile wilderness beyond the Alleghenies, over which France had claimed sovereignty—that boundless forest, with its tracery of interlacing streams, like veins and arteries giving it life and nourishment—had passed into the hands of her rival. And as the heroic Wolfe recognized when he fell at Quebec, it was the Iroquois who had won an empire for the British Crown.

11

The Revolt of Pontiac
and a New British Policy

It was a brilliant opportunity with which the diplomacy of Sir William Johnson and Pontiac presented the British government in 1760. Yet hardly had hostilities between the British and French ceased when smothered murmurs of discontent with the English began to be audible among the tribes. From the head of the Potomac to Lake Superior, and from the Alleghenies to the Mississippi, in every wigwam and hamlet of the forest, a deep-rooted hatred of them increased with rapid growth: Nor is this surprising.

With sagacious policy the French had labored long to ingratiate themselves with the Indians. The recent slaughter of the Monongahela, the horrible devastation of the western frontier, the outrages perpetrated at Oswego, and the massacre at Fort William Henry, bore witness to the success of their efforts.

Under these circumstances, it behooved the English to use the utmost care in their conduct toward the tribes. But even when the conflict with France was impending, and the alliance with the Indians was of the utmost importance, they had treated them with indifference and neglect, driving even the Delawares into alliance with the French. They were not likely to adopt a different course now that their friendship seemed a matter of no consequence. In the zeal for retrenchment which prevailed after the close of hostilities, the presents which it had always been customary to give the Indians at stated intervals, were either withheld altogether, or doled out with a niggardly and reluctant hand; while, to make matters worse, the agents and officers of the government often appropriated the presents to themselves, and afterwards sold them at an exorbitant price to the Indians. When the French had possession of the remote forts, they were accustomed, with a wise liberality, to supply the surrounding Indians with guns, ammunition and clothing, until the latter had forgotten the weapons and garments of their forefathers, and depended on the white man for support. The sudden withholding of these supplies was, therefore, a grievous calamity. Want, suffering and death were the consequences; and this cause alone would have been enough to produce general discontent. But, unhappily, other grievances were superadded.

The English fur trade had never been well regulated, and it was now in a worse condition than ever. Many of the traders, and those in their employ, were ruffians of the coarsest stamp, who vied with each other in rapacity, violence and profligacy. They cheated, cursed and plundered the Indians and outraged their families, offering, when compared with the French traders, who were under better regulation, a most unfavorable example of the character of their nation.

The officers and soldiers of the garrisons did their full part in exciting the general resentment. Formerly, when the warriors came to the forts, they had been welcomed by the French with attention and respect. The inconvenience which their presence occasioned had been disregarded, and their peculiarities overlooked. But now they were received after the English manner with cold looks and harsh words from the officers, and with oaths, menaces, and sometimes blows, from the reckless and brutal soldiers. When the Indians were lounging everywhere about the fort, or reclining in the shadow of the walls, they were met with muttered ejaculations of impatience, or abrupt orders to be gone, enforced, perhaps, by a touch from the butt of a soldier's musket. These marks of contempt were unspeakably galling to their spirit.

But what most contributed to the growing discontent of the tribes was the intrusion of settlers upon their lands, at all times a fruitful source of Indian hostility. Its effects, it is true, could only be felt by those whose country bordered upon the English settlements; but among these were the most powerful and influential of the tribes. The Delawares and Shawnees, in particular, were roused to the highest pitch of exasperation. Their best lands had been invaded, and all remonstrance had been fruitless. They viewed with wrath and fear the steady progress of the white man, whose settlements had passed the Susquehanna and were fast extending to the Alleghenies, eating away the forest like a spreading canker. The anger of the Delawares was abundantly shared by their late masters, the Six Nations, to whom the threatened occupation of Wyoming Valley in Pennsylvania by settlers from Connecticut gave great umbrage.*

The Senecas were more especially incensed at English intrusion, since, from their position, they were farthest removed from the soothing influ-

* Minutes of Conference with the Six Nations at Hartford, 1763, MS. Letter—Hamilton to Amherst, May 10, 1761.

ence of Sir William Johnson, and most exposed to the seductions of the French; while the Mohawks, another member of the Confederacy, were justly alarmed at seeing the better part of their lands patented out without their consent. Some Christian Indians of the Oneida tribe, in the simplicity of their hearts, sent an earnest petition to Sir William Johnson that the English forts within the limits of the Six Nations might be removed, or, as the petition expresses it, kicked out of the way.*

The discontent of the Indians gave great satisfaction to the French, who saw in it an assurance of safe and bloody vengeance on their conquerors. Canada, it is true, was gone beyond hope of recovery; but they still might hope to revenge its loss. Interest, moreover, as well as passion, prompted them to inflame the resentment of the Indians; for most of the inhabitants of the French settlements upon the lakes and the Mississippi were engaged in the fur trade, and, fearing the English as formidable rivals, they would gladly have seen them driven out of the country. Traders, habitants, *coureurs de bois*, and all classes of this singular population accordingly dispersed themselves among the villages of the Indians, or held councils with them in the secret places of the woods, urging them to take up arms against the English. They exhibited the conduct of the latter in its worst light, and spared neither misrepresentation nor falsehood. They told their excited hearers that the English had formed a deliberate scheme to root out the whole Indian race, and, with that design, had already begun to hem them in with settlements on the one hand and a chain of forts on the other. Among other atrocious plans for their destruction, they had instigated the Cherokees to attack and destroy the tribes of the Ohio Valley. These groundless calumnies found ready belief. The French declared, in addition, that the King of France had of late years fallen asleep; that, during his slumbers, the English had seized upon Canada, but that he was now awake again, and that his armies were advancing up the St. Lawrence and the Mississippi, to drive out the intruders from the country of his red children. To these fabrications was added the more substantial encouragement of arms, ammunition, clothing and provisions, which the French trading companies, if not the officers of the crown, distributed with a liberal hand.

Pontiac was not the mere "intractable, scheming, ambitious savage" he has been represented to be even by Parkman. Upon the things described, as one who loved his race he looked with bitterness in his heart.

Believing that French power had been broken, he had shown himself both statesman and diplomat enough to accept at once the British victory

* "We are now left in Peace, and have nothing to do but to plant our corn, hunt the wild Beasts, smoke our Pipes, and mind Religion. But as these Forts, which are built among us, disturb our Peace, and are a great hurt to Religion, because some of our Warriors are foolish, and some of our Brother Soldiers don't fear God, we therefore desire that these Forts may be pull'd down, and kick'd out of the way."

as a fact with all it implied, so it cannot be said he was governed by motives of personal hostility alone. Seeing that he was bitterly disappointed, the French irreconcilables spared no effort to inflame his wrath. He had been very stupid, they declared, in abandoning them, in putting his faith in the British. Now that the King of France was to recover his lost possessions, it would be better for him to aid rather than oppose the French.

Perhaps, after all, he had been too quick to abandon his old allies. What could have been more natural than that now, in the light of the knowledge which was his, again he should have looked to the welfare of his race? Certainly no European statesman would have done less.

Long the Ottawas, Chippewas, and Potawatomis had been united in a loose confederacy. Recalling the teachings of Tamenend and knowing the power the Iroquois and the Lenni-Lenape had developed through the union of their tribes, he did not fail to see the wonderful opportunity that was his to unite the western tribes, to upbuild upon the existing foundation of those immediately subject to him a greater Indian confederacy than that which Hiawatha had created. Unlike many of his people, he was not, however, under the delusion that the Indians by their unaided strength could expel the British. This must be accomplished, if at all, in concert with the French.

Looking at him from the point of view of his own race, it is, therefore, utterly absurd to speak of one as a conspirator who adopted the only course consistent with reason—the better organization of his people in order that they might cooperate with the French.

In view of Indian nature, such a task involved enormous difficulties. In all that followed Pontiac was to demonstrate a capacity as a leader equal to his statesmanship, which clearly marked him as the moral successor of King Philip.

It is characteristic of people that one of the manifestations of their inward stirrings is the appearance among them of a prophet. Lacking other means of effecting cooperation, such a one is employed by the chieftains to preach the common cause. Pontiac was to have his prophet in the person of one of great ability and character, plainly familiar with the morals of "Saint Tammany."

The Kingdom which Pontiac's prophet had in mind was not limited, however, to the spiritual world. "Lay aside the weapons and clothing which you have received from the white man," he enjoined. "Return to the primitive life of your ancestors. Thus, and by strictly observing my precepts, the tribes will be restored to their ancient greatness and power, and be able to drive out the Swannok!"*

* Never was the Red Pipe Stone Quarry Shrine used more effectively by the Jesuits.

The prophet soon had many followers. Indians came from far and near, gathering together in large encampments to listen to his exhortations.* His fame spread even to the nations of the northern lakes; but though his disciples followed most of his injunctions, flinging away flint and steel, and making copious use of emetics, with other observances equally troublesome, yet the admonition to abandon the use of firearms was too inconvenient to be complied with.

The English were not long left in doubt as to what was astir. Addressing the Wyandots and Ottawas at Detroit in July, 1761, a Seneca sachem said:

The English treat us with much disrespect, and we have the greatest reason to believe, by their behavior, they intend to cut us off entirely; They have possessed themselves of our country, it is now in our power to dispossess them and recover it, if we will but Embrace the opportunity before they have time to assemble together, and fortify themselves, there is no time to be lost, let us strike immediately.

Learning of these inflammatory words, Captain Campbell reported them to Sir Jeffrey Amherst at once. Upon investigation it was discovered that a plot was on foot to destroy Niagara, Fort Pitt and other posts, all of which were at once put on guard. At the same time a council of the Iroquois was called to meet in Philadelphia in August.

"We, your brethren of the Six Nations, are penned up like hogs. There are forts all around us, and therefore we are apprehensive that death is coming upon us," complained a Seneca sachem at this council.

Fully informed by Sir William Johnson of what might be expected, the British government saw at last that something must be done, since, after all, the Indians were human beings whom Johnson alone could not control. Thus, in a report of the Lords of Trade which was read before the King and Council at the Court of St. James, November 23, 1761, the following appears:

That it is as unnecessary as it would be tedious to enter into a Detail of all the Causes of Complaint which our Indian Allies had against us at the commencement of the troubles in America, and which not only induced them tho reluctantly to take up the Hatchet against us and desolate the Settlement on the Frontiers but encouraged our enemies to pursue those Measures which have involved us in a dangerous and critical war, it will be sufficient for the present purpose to observe that the primary cause of that discontent which produced these fatal Effects was the Cruelty and Injustice with which they had been treated with respect to their hunting grounds, in open violation of those solemn compacts by which they had yielded to us the Dominion, but not the property of those Lands. It was happy for us that we were early awakened to a

* Pontiac's prophet undoubtedly was tutored by some Jesuit or shrewd Frenchman and this Johnson knew.

proper sense of the Injustice and bad Policy of such a Conduct towards the Indians, and no sooner were those measures pursued which indicated a Disposition to do them all possible justice upon this head of Complaint than those hostilities which had produced such horrid scenes of devastation ceased and the Six Nations and their Dependents became at once from the most inveterate Enemies our fast and faithful Friends.

That their steady and intrepid Conduct upon the Expedition under General Amherst for the Reduction of Canada is a striking example of this truth, and they now, trusting to our good Faith, impatiently wait for that event which by putting an End to the War shall not only ascertain the British Empire in America but enable Your Majesty to renew those Compacts by which their property in their Lands shall be ascertained and such a system of Reformation introduced with respect to our Interests and Commerce with them as shall at the same time that it redresses their Complaints and establishes their Right give equal Security and Stability to the rights and interests of all Your Majesty's American Subjects.

That under these Circumstances and in this situation the granting Lands hitherto unsettled and establishing Colonies upon the Frontiers before the claims of the Indians are ascertained appears to be a measure of the most dangerous tendency, and is more particularly so in the present case, as these settlements now proposed to be made, especially those upon the Mohawk River are in that part of the Country of the Possession of which the Indians are the most jealous having at different times expressed in the strongest terms their Resolution to oppose all settlements therein as a manifest violation of their Rights.*

The King was evidently much impressed by this statement, for on December 2, 1761, the Lords of Trade submitted to him the following:

Draft of an Instruction for the Governors of Nova Scotia, New Hampshire, New York, Virginia, North Carolina, and Georgia, forbidding them to Grant Lands or make Settlements which may interfere with the Indians bordering on these Colonies.

Whereas the peace and security of Our Colonies and Plantations upon the Continent of North America does greatly depend upon the Amity and Alliance of the several Nations or Tribes of Indians bordering upon the said Colonies and upon a just and faithful Observance of those Treaties and Compacts which have been heretofore solemnly entered into with the said Indians by Our Royall Predecessors Kings & Queens of this Realm. And whereas notwithstanding the repeated Instructions which have been from time to time given by Our Royal Grandfather to the Governor of Our several Colonies upon this head the said Indians have made and do still continue to make great complaints that Settlements have been made and possession taken of Lands, the property of which they have by Treaties reserved to themselves by persons claiming the said lands under pretense of deeds of Sale and Conveyance illegally fraudulently and surreptitiously obtained of the said Indians; and Whereas it has

* *New York Colonial Documents*, Vol. VI, p. 473.

likewise been represented unto Us that some of Our Governors or other Chief Officers of Our said Colonies have countenanced such unjust claims and pretensions by passing Grants of the Lands so pretended to have been purchased of the Indians We therefor taking this matter into Our Royal Consideration, as also the fatal Effects which would attend a discontent amongst the Indians in the present situation of affairs, and being determined upon all occasions to support and protect the said Indians in their just Rights and Possessions and to keep inviolable the Treaties and Compacts which have been entered into with them, Do hereby strictly enjoyn and demand that neither yourself nor any Lieutenant Governor, President of the Council or Commander in Chief of

Our said $\begin{cases} \text{Colony} \\ \text{Province of} \end{cases}$ do upon any pretence what-

ever upon pain of Our highest Displeasure and of being forthwith removed from your or his office, pass any Grant or Grants to any persons whatever of any lands within or adjacent to the Territories possessed or occupied by the said Indians or the Property Possession of which has at any time been reserved to or claimed by them. And it is Our further Will and Pleasure that you do publish a proclamation in Our Name strictly enjoining and requiring all persons whatever who may either wilfully or inadvertently have seated themselves on any Lands so reserved to or claimed by the said Indians without any lawful Authority for so doing forthwith to remove therefrom And in case you shall find upon strict enquiry to be made for that purpose that any person or persons do claim to hold or possess any lands within Our said

$\begin{cases} \text{Province} \\ \text{Colony} \end{cases}$ upon pretence of purchases made of the said Indians

without a proper license first had and obtained either from Us or any of Our Royal Predecessors or any person acting under Our or their Authority you are forthwith to cause a prosecution to be carried on against such person or persons who shall have made such fraudulent purchases to the end that the land may be recovered by due Course of Law and whereas the wholesome Laws that have at different times been passed in several of Our said Colonies and the instructions which have been given by Our Royal Predecessors for restraining persons from purchasing lands of the Indians without a License for that purpose and for regulating the proceedings upon such purchases have not been duly observed, It is therefore Our express Will and Pleasure that when any application shall be made to you for licence to purchase lands of the Indians you do forbear to grant such licence untill you shall have first transmitted to Us by Our Commissioners for Trade and Plantations the particulars of such applications as well as in respect to the situation as the extent of the lands so proposed to be purchased and shall have received Our further directions therein; and it is Our further Will and Pleasure that you do forthwith cause this Our Instruction to you to be

made Publick not only within all parts of your said $\begin{cases} \text{Province} \\ \text{Colony} \end{cases}$

inhabited by Our Subjects, but also amongst the several Tribes of Indians living within the same to the end, that Our Royal Will and Pleasure in the Premises may be known and that the Indians may be apprized of Our determ's

Resolution to support them in their just Rights, and inviolably to observe Our Engagements with them.*

Two Imperial Indian agents were now appointed, one for the northern district, or that embracing the northern colonies and the unorganized territory of the West, and one for the southern district.

These agents were, of course, powerless to enforce upon the colonial governments the new policy outlined on the instructions of the King, which was by no means acceptable to the colonists. Already they were chafing over the Navigation Acts which interfered with the freedom of their trade, and the Forestry Act of 1729, which in its conservative nature tended to limit the freedom of the frontier population. Now the Imperial Government was seeking to deny to the very people who had left the British Isles in search of homes in the American wilderness freedom to usurp Indian lands. This was not to be tolerated. Nor were the colonial governments strong enough to carry out the King's instructions, to which little heed was paid.

During the summer of 1762 still further evidence was discovered indicating only too clearly that the Indians were contemplating a general uprising, and the truth is Pontiac's plans were about matured, for no one better than he knew that the English colonies at this time were ill-fitted to bear the brunt of an Indian war. The army which had conquered Canada was broken up and dissolved; the provincials were disbanded, and most of the regulars sent home. A few fragments of regiments, miserably wasted by war and sickness, had just arrived from the West Indies; and of these, several were already ordered to England to be disbanded. There remained barely enough troops to furnish feeble garrisons for the various posts on the frontier and in the Indian country. Moreover, at the head of this dilapidated army was Sir Jeffrey Amherst, who, though he had achieved the reduction of Canada after the death of Wolfe, inspired not the least fear in Pontiac. Of the professional English soldier of his type, whose weakness lay in contempt for all adversaries, he had already taken the measure on the Monongahela in the person of Braddock.

Finally assured by the French that they were staunch of purpose, to ward the close of the year 1762 Pontiac sent forth his ambassadors. Visiting the country of the Ohio and its tributaries, they passed northward to the region of the upper lakes and the borders of the River Ottawa, and far southward towards the mouth of the Mississippi. Bearing with them the war-belt of wampum, broad and long, as the importance of the message demanded, and the tomahawk stained red, in token of war, they went from camp to camp, and village to village. Wherever they appeared, the sachems and old men assembled to hear the words of the great Pontiac.

* *New York Colonial Documents*, Vol. VII, pp. 478-9.

Then the chief of the embassy flung down the tomahawk on the ground before them, and, holding the war-belt in his hand, delivered, with vehement gesture, word for word, the speech with which he was charged. It was heard everywhere with approval; the belt was accepted, the hatchet snatched up, and the assembled chiefs stood pledged to Pontiac. Thus, although their inveterate hatred of the French and Johnson's influence over the Iroquois kept all but the Senecas from joining the Confederacy, almost the entire Algonquin family was united and with it was joined several tribes of the lower Mississippi.

The blow was to be struck at a certain time in the month of May following, to be indicated by the change of the moon. The confederated tribes were to rise together, each destroying the British garrison in its neighborhood, and then, with a general rush, the whole were to turn against the settlements of the frontier. Such was the Confederacy of Pontiac, such his plan of attack.

With the approach of spring, the Indians, coming in from their wintering grounds, began to appear in small parties about the various forts; but now they seldom entered them, encamping at a little distance in the woods. They were fast pushing their preparations for the meditated blow, waiting with stifled eagerness for the appointed hours. While thus on the very eve of an outbreak they concealed their designs; the warriors still lounged about the forts, with calm impenetrable expressions, begging, as usual, for tobacco, gun-powder and whisky. Now and then, some slight intimation of danger would startle the garrisons from their security. An English trader, coming in from the Indian villages, would report that, from their manner and behavior, he suspected them of brooding mischief; or some scoundrel half-breed would be heard boasting in his cups that before next summer he would have English hair to fringe his hunting frock. On one occasion, their purpose was nearly discovered. Early in March, 1763, Ensign Holmes, commanding at Fort Miami, was told by a friendly Indian that the warriors in the neighboring village had lately received a war-belt, with a message urging them to destroy him and his garrison, and that this they were preparing to do. Holmes called the Indians together and boldly charged them with their design. They did as Indians on such occasions have often done, confessed their fault with much apparent contrition, laid the blame on a neighboring tribe, and professed eternal friendship to their brethren—the English. Holmes wrote to report his discovery to Major Gladwyn, who, in his turn, sent the information to Sir Jeffrey Amherst, expressing his opinion that there had been a general irritation among the Indians, but that the affair would soon blow over, and that, in the neighborhood of his own post, the savages were perfectly tranquil. Within cannon shot of the deluded officer's palisades was the village of Pontiac himself.

Meantime, on February 10, 1763, the King of France had signed the Treaty of Paris, finally establishing peace and ceding to Great Britain all his possessions east of the Mississippi River and south of the Great Lakes. But of this the Indians, like the French in America, were still ignorant when, exactly in accordance with Pontiac's instructions, delegations from all the tribes assembled on April 27th for the final council at the preappointed rendezvous on the banks of the little River Ecorces near Detroit. Thither they came from all sides, pitching their tepees in a far-flung meadow green with the verdure of spring.

Having arrived well ahead of the others, Pontiac lost no time. After the pipes were passed, the great chieftain rose and walked forward into the midst of the council. According to tradition, he was not above the middle height, though his muscular figure was cast in a mould of remarkable symmetry and vigor. His complexion was darker than is usual with his race, and his features, though by no means regular, had a cold and stern expression; while his habitual bearing was imperious and peremptory, like that of a man accustomed to sweep away all opposition by the force of his impetuous will.

With loud, impassioned voice he inveighed against the arrogance, rapacity, and injustice of the English, and contrasted them with the French, whom they had driven from the soil. He declared that the British commandant had treated him with neglect and contempt; that the soldiers of the garrison had abused the Indians; and that one of them had struck a follower of his own. He represented the danger that would arise from the supremacy of the English. They had expelled the French, and now they only waited for a pretext to turn upon the Indians and destroy them. Then, holding out a broad belt of wampum, he told the council that he had received it from their great father the King of France, in token that he had heard the voice of his red children; that his sleep was at an end; and that his great war canoes would soon sail up the St. Lawrence to win back Canada and wreak vengeance on his enemies. The Indians and their French brethren would fight once more side by side, as they had always fought; they would strike the English as they had struck them many moons ago, when their great army marched down the Monongahela, and they had shot them from their ambush, like a flock of pigeons in the woods.

Having roused his warlike listeners, he next addressed himself to their superstition, and told the following tale. Its precise origin is not easy to determine. It is possible that the Delaware prophet may have had some part in it; or it might have been the offspring of Pontiac's heated imagination during his period of fasting and dreaming. That he deliberately invented it for the sake of the effect it would produce is the least probable conclusion of all; for it evidently proceeds from the superstitious mind of an Indian, brooding upon the evil days in which his lot was cast, and turning for relief to the mysterious Author of his being. It is, at all events, a characteristic specimen of the Indian legendary tales, and, like many of them, bears an allegoric significance.

A Delaware Indian conceived an eager desire to learn wisdom from the Master of Life; but, being ignorant where to find him, he had recourse to fasting, dreaming and magical incantations. By these means it was revealed to him that, by moving forward in a straight, undeviating course, he would reach the abode of the Great Spirit. He told his purpose to no one, and, having provided the equipments of a hunter—gun, powder-horn, ammunition, and a kettle for preparing his food—he set out on his errand. For some time he journeyed on in high hope and confidence. On the evening of the eighth day he stopped by the side of a brook at the edge of a meadow, where he began to make ready his evening meal, when, looking up, he saw three large openings in the woods before him, and three well-beaten paths which entered them. He was much surprised; but his wonder increased, when, after it had grown dark, the three paths were more clearly visible than ever. Remembering the important object of his journey, he could neither rest nor sleep, and, leaving his fire, he crossed the meadow and entered the largest of the three openings. He had advanced but a short distance into the forest, when a bright flame sprang out of the ground before him, and arrested his steps. In great amazement he turned back and entered the second path, where the same wonderful phenomenon again encountered him; and now, in terror and bewilderment, yet still resolved to persevere, he took the last of the three paths. On this he journeyed a whole day without interruption, when at length, emerging from the forest, he saw before him a vast mountain of dazzling whiteness. So precipitous was the ascent, that the Indian thought it hopeless to go farther, and looked around him in despair; at that moment he saw, seated at some distance above, the figure of a beautiful woman arrayed in white, who arose as he looked upon her, and thus accosted him: "How can you hope, encumbered as you are, to succeed in your design? Go down to the foot of the mountain, throw away your gun, your ammunition, your provisions, and your clothing; wash yourself in the stream which flows there, and you will then be prepared to stand before the Master of Life." The Indian obeyed and again began to ascend among the rocks, while the woman, seeing him still discouraged, laughed at his faintness of heart, and told him that, if he wished for success, he must climb by the aid of one hand and one foot only. After great toil and suffering, he at length found himself at the summit. The woman had disappeared, and he was left alone. A rich and beautiful plain lay before him, and at a little distance he saw three great villages, far superior to the squalid wigwams of the Delawares. As he approached the largest and stood hesitating whether he should enter, a man gorgeously attired stepped forth, and, taking him by the hand, welcomed him to the celestial abode. He then conducted him into the presence of the Great Spirit, where the Indian stood confounded at the unspeakable splendor which surrounded him. The Great Spirit bade him be seated, and thus addressed him:—

"I am the Maker of heaven and earth, the trees, lakes, rivers, and all things else. I am the Maker of mankind; and because I love you, you must do my will. The land on which you live I have made for you and not for others. Why do you suffer the white man to dwell among you? My children, you have forgotten the customs and traditions of your forefathers. Why do you not clothe yourselves in skins, as they did, and use the bows and arrows, and the stone-pointed lances which they used? You have bought guns, knives, kettles and blankets from the white men, until you can no longer do without them; and, what is worse, you have drunk the poison fire-water, which turns you into fools. Fling all these things away; live as your wise forefathers lived before you. And as for these English—these dogs dressed in red, who have come to rob you of your hunting-grounds, and drive away the game—you must lift the hatchet against them. Wipe them from the face of the earth, and then you will win my favor back again and once more be happy and prosperous. The children of your great father, the King of France, are not like the English. Never forget that they are your brethren. They are very dear to me, for they love the red men, and understand the true mode of worshipping me."

The Great Spirit next gave his hearer various precepts of morality and religion, such as the prohibition to marry more than one wife; and a warning against the practice of magic, which is worshipping the devil. A prayer, embodying the substance of all that he had heard, was then presented to the Delaware. It was cut in hieroglyphics upon a wooden stick, after the custom of his people; and he was directed to send copies of it to all the Indian villages.

The adventurer now departed, and, returning to the earth, reported all the wonders he had seen in the celestial regions.

Such was the tale told by Pontiac to the Council. Plainly the French missionaries had already exerted a powerful effect upon Indian religion which, even among the more remote tribes of the American wilderness, had passed far beyond totemism. How utterly absurd it is to try to make of a man who could couch his appeal to his people in such noble language a faithless savage. No European ever made a higher moral appeal. The Philippics of Demosthenes do not surpass it. Pontiac was not only appealing to the patriotism of his race but, as Tecumseh did half a century later, for its moral self-regeneration.

Almost coincidently arrived news of the peace of which in vain the French notified Pontiac. When the Indian confederates heard that they and their lands had been abandoned to the British by the King of France without their being consulted, with bitterness in their hearts against the French, they were determined to strike for their freedom even if they had to do so unaided.

Yet, it is one of the most tragic facts in their history that on May 5, 1763, the day before their general attack was launched, Lord Egremont addressed the Lords of Trade as follows:

The second question which relates to the security of North America, seems to include two objects to be provided for; the first is the security of the whole against any European Power; the next is the preservation of the internal peace and tranquility of the Country against any Indian disturbances. Of these two objects the latter appears to call more immediately for such Regulations and Precautions as your Lordships shall think proper to suggest & ca.

Tho in order to succeed effectually in this point it may become necessary to erect some Forts in the Indian Country with their consent, yet his Majesty's Justice and Moderation inclines him to adopt the more eligible Method of conciliating the minds of the Indians by the mildness of His Government, by protecting their persons and property, and securing to them all the possessions rights and Privileges they have hitherto enjoyed and are entitled to be most cautiously guarded against any Invasion or Occupation of their hunting Lands, the possession of which is to be acquired by fair purchase only, and it has been thought so highly expedient to give the earliest and most convincing proofs of his Majesty's gracious and friendly Intentions on this head, that I have already received and transmitted the King's commands to this purpose to the Governors of Virginia, the two Carolinas and Georgia, and to the Agent for Indian Affairs in the Southern Department, as your Lordships will see fully in the enclosed copy of my circular letter to them on this subject.*

When the outbreak occurred there were fourteen British posts between Pennsylvania and Lake Superior which had been taken from the French. The skill with which Pontiac had organized the attack is clearly indicated by the fact that all of them except Niagara, Fort Pitt, Legonier and Detroit were captured at once, the last escaping only through the loyalty of an Indian girl. A sortie of the garrison was, however, defeated in July, at Bloody Run by Pontiac, who again proved himself a superb general.

Apparently uninformed of the outbreak, the Lords of Trade, who were now ignoring Amherst, addressed a communication to Sir William Johnson in August in which they said they had proposed to His Majesty that a proclamation should be issued declaratory of His Majesty's final determination to permit no grants of lands nor any settlement to be made within certain fixed bounds under pretense of purchase, or any pretext whatever, leaving all the territory within these bounds free for the hunting grounds of the Indian Nations, and for the free trade of all his subjects.

Nevertheless, the King, who had procrastinated all along, fearing the political effect upon the colonies of asserting jurisdiction over the tribes within their boundaries, was still hesitating when advices were received of what was occurring in America. Convinced at last nothing constructive was to be expected of the colonies, and that the Indian problem must be handled by the British government, finally on October 7, 1763, the following royal proclamation was published:

And whereas, it is just and reasonable, and essential to our interest and the security of our colonies, that the several nations or tribes of Indians with

* *New York Colonial Documents*, pp. 520-521.

whom we are connected, and who live under our protection, should not be molested or disturbed in the possession of such parts of our dominions and territories, as, not having been ceded to, or purchased by us, are reserved to them, or any of them, as their hunting grounds; we do, therefore, with the advice of our privy council, declare it to be our royal will and pleasure, that no Governor or commander in chief, in any of our colonies of Quebec, East Florida, and West Florida, do presume, upon any pretence whatever, to grant warrants of survey or pass any patents for lands beyond the bounds of their respective governments, as described in their commissions; as, also, that no Governor or commander in chief of our other colonies or plantations in America, do presume for the present, and until our further pleasure be known, to grant warrants of survey, or pass patents for any lands beyond the heads or sources of any of the rivers which fall into the Atlantic ocean from the West or Northwest; or upon any lands whatever, which, not having been ceded to, or purchased by, us, as aforesaid, are reserved to the said Indians or any of them.

And we do further declare it to be our royal will and pleasure, for the present, as aforesaid, to reserve under our sovereignty, protection, and dominion, for the use of the said Indians, all the land and territories not included within the limits of our said three new Governments, or within the limits of the territory granted to the Hudson's Bay Company; as also all the lands and territories lying to the Westward of the sources of the rivers which fall into the sea from the West and Northwest as aforesaid; and we do hereby strictly forbid, on pain of our displeasure, all our loving subjects from making any purchases or settlements whatever, or taking possession of any of the lands above reserved, without our special leave and license for that purpose first obtained.

And we do further strictly enjoin and require all persons whatever, who have either wilfully or inadvertently seated themselves upon any lands within the countries above described, or upon any other lands, which, not having been ceded to, or purchased by us, are still reserved to the said Indians as aforesaid, forthwith to remove themselves from such settlements.

And whereas great frauds and abuses have been committed in the purchasing lands of the Indians, to the great prejudice of our interests, and to the great dissatisfaction of the said Indians; in order, therefore to prevent such irregularities for the future, and to the end that the Indians may be convinced of our justice and determined resolution to remove all reasonable cause of discontent, we do, with the advice of our privy council, strictly enjoin and require that no private person do presume to make any purchase from the said Indians, of any lands reserved to the said Indians, within those parts of our colonies where we have thought proper to allow settlement; but that, if, at any time, any of the said Indians should be inclined to dispose of the said lands, the same shall be purchased only for us, in our name, at some public meeting or assembly of the said Indians, to be held for that purpose, by the Governor or Commander-in-chief of our Colony, respectively, within the limits of any proprietaries, conformable to such directions and instruction as we or they shall think proper to give for that purpose.

Although primarily relating to the colonies of Quebec, East Florida and West Florida, it is evident from the distinct statements therein that this proclamation was intended to be of general application.*

Upon receiving notice of this epochal though belated act for which he was primarily responsible, Sir William Johnson lost no time in notifying the Senecas, and through them the other confederates, of what had occurred, in order to make them see the futility of further fighting. But this was not enough. The Six Nations must be reassured that not only their rights in their immemorial domain but the deed of 1701 was to be respected. Therefore, on October 13, 1763, he wrote the Lords of Trade as follows:

My Lords:

"In obedience to your Lordships' commands of the 5th of August last, I am now to lay before you the claims of the Nations mentioned in the State of the Confederacies. The Five Nations have in the last century subdued the Shawanese, Delawares, Twighties, and Western Indians, so far as Lakes Michigan and Superior, received them into an alliance, allowed them the possession of the lands they occupied, and have ever since been in peace with the greatest part of them; and such was the prowess of the Five Nations' Confederacy, that had they been properly supported by us, they would have long since put a period to the Colony of Canada, which alone they were near effecting in the year 1688. Since that time, they have admitted the Tuscaroras from the Southward, beyond Oneida, and they have ever since formed a part of that Confederacy.

As original proprietors, this Confederacy claim the country of their residence, south of Lake Ontario to the great Ridge of the Blue Mountains, with all the West Part of the Province of New York towards Hudson River, west of the Catskill, thence to Lake Champlain, and from Regioghne, a Rock at the East side of said Lake, to Oswegatche or La Gallette, on the River St. Lawrence, (having long since ceded their claim north of said line in favor of the Canada Indians, as Hunting-ground,) thence up the River St. Lawrence and along the South side of Lake Ontario to Niagara.

In right of conquest, they claim all the country (comprehending the Ohio) along the great Ridge of Blue Mountains at the back of Virginia, thence to the head of Kentucky River, and down the same to the Ohio above the Rifts, thence Northerly to the South end of Lake Michigan, then along the Eastern shore of said lake to Michillimackinac, thence Easterly across the North end of Lake Huron to the Great Ottawa River, (including the Chippewa or Mississagey County,) and down the said River to the Island of Montreal. However, these more distant claims being possessed by many powerful nations, the Inhabitants have long begun to render themselves independent, by the assistance of the French, and the great decrease of the Six Nations; but their claim to the Ohio, and thence to the Lakes, is not in the least disputed by the Shawnees, Delawares, etc., who never transacted any sales of land or other

* Letter of Lord Colden to Earl of Halifax, Dec. 8, 1763.

matters without their consent, and who sent Deputies to the grand Council at Onondaga on all important occasions.

The effect of Johnson's efforts was instant. After the Indian forces besieging Fort Pitt and Ligonier were defeated by Bouquet at Bushy Run, and British reenforcements succeeded in entering Detroit, desertions increased so rapidly with the approach of winter that Pontiac was compelled to lift the siege of Detroit, which he had maintained five months. The following year a second successful expedition was conducted by Bouquet into Ohio, whereupon, now fully informed by Johnson of the royal proclamation of 1763, Pontiac, with the same statesmanship he had hitherto displayed, entered into a formal treaty of peace with the English at Detroit on August 17, 1765.

Thus came to an end the great Indian revolt commonly and erroneously called the Conspiracy of Pontiac.

The success of a man is to be judged by his achievements. From a strictly military point of view, Pontiac suffered defeat. In truth, however, he won a great victory, enforcing as he did upon the British government an Indian policy designed to do justice to his people, one which clearly recognized the doctrine of Roger Williams that even the King could not grant away the lands of the Indians without their consent. Moreover, he gave to his people a doctrine of Indian nationality of which Little Turtle, Tecumseh, Osceola, Black Hawk and Sitting Bull were each in turn to become exponents.

The punishment of Pontiac was almost instant. A short time after he had smoked the peace pipe with the English at Oswego, and while maintaining the pleasantest relations with Sir William Johnson, with whom he was now cooperating, he was foully murdered by an English trader.

In 1768 Sir William Johnson was to realize in the treaty of Fort Stanwix with the Iroquois the dream of his life. Seeing that the colony of New York was bent on taking their lands sooner or later, the Six Nations now did just what in 1701 they had done with respect to their "Beaver Hunting Grounds." Conveying the fee in their immemorial domain to the King and reserving a perpetual right of occupancy therein, the treaty clearly defined the territory and expressly declared it to be at the back of the colony of New York and Pennsylvania, and "no part of any colony."* Thus, the Iroquois domain was clearly converted into crown land.

The Virginians, however, had no idea of being denied the rich lands of the West. Thus, one year after Pontiac had been overthrown, great numbers of immigrants began to cross the Cumberland and Allegheny Mountains. In 1770 the Legislature was compelled to obtain from the Cherokees the treaty of Lochaber, relinquishing all their claims to the territory

* *New York Colonial Documents,* VII, p. 13.

south of the Ohio basin and east of the Cumberland Mountains. The way having thus been cleared, in 1773 Daniel Boone led his first party into Kentucky.

There was no longer a Pontiac to hold the western Indians in leash. The situation had long since passed beyond the control of Johnson. Boone's party was promptly attacked by the Shawnees under Cornstalk and partly destroyed. When, the following year, a large party under John Floyd, with surveyors, crossed the Alleghenies, the Shawnees, Wyandots, Ottawas and Ojibwas again combined to resist further encroachments upon their lands.

The same year the whites in the Ohio country perpetrated an unprovoked massacre of the Shawnees at Yellow Creek in which the brother and sister of Logan, a famous Mingo chief, were killed while guests of their murderers. Frenzied by this outrageous deed the Indians rose *en masse*.

So commenced Dunmore's War in which the Shawnees under Cornstalk were decisively defeated at Point Pleasant on October 10, 1774, by the force of Virginians which Governor Dunmore sent against them. Though they fought desperately to the last against overwhelming odds, their spirit was so broken by this defeat and the killing of their chief, that they soon entered into a treaty with Virginia, in which they relinquished all their lands south of the Ohio. This treaty they kept faithfully until they were compelled by circumstances and constant importunings to choose between the contestants in the Revolution.

12

Indians and the American Revolution

THE HISTORIAN ordinarily dwells upon the Navigation Acts and taxation as the cause of the American Revolution. He ignores utterly the fact that the very year Patrick Henry proclaimed his incendiary anti-British doctrine, the new Indian policy of the Crown was put into effect, reserving to the Indian tribes exclusively the domains occupied by them. To repeat, the Scotch-Irish people of the frontier who had left the British Isles out of open disloyalty to the British government, and come to the New World in search of homes, had no idea of being denied free access to the Indian lands, nor of tolerating control by the Imperial Government. Not as loyal to that government as the Indians themselves, and equally fierce as the red denizens of the forest, they made up their minds to fight for their independence long before the English people in the settled portions of the colonies. They were willing enough to have the latter engage for ten years in a constitutional debate with the Imperial Government, but from the first they themselves, now one-third of the entire colonial population, had had but one purpose. Compared with this, many other contributing causes of the Revolution sink into insignificance.

That the Imperial Government was under no illusions about the influence of its Indian policy upon the rising spirit of independence in America is shown by the fact that when the Virginians assembled in Williamsburg in 1769 to protest against the home government, Governor Botetourt, before hearing their professed grievances, was quick to lay before them a new treaty with the Cherokees opening up additional lands to the entry of the whites.

As armed revolution became more and more imminent, it was a question with which side would the Indian tribes ally themselves. Unfortunately for the colonies, in 1774 Sir William Johnson, who had espoused their cause, died at the age of fifty-nine so that his influence was removed as a determining factor.*

At this time, Thayendanegea, or Col. Joseph Brant, chief of the Mohawks, was undoubtedly the foremost representative of his race. He was born in 1742, the son of a full-blood Mohawk father of the Wolf gens, and possibly a half-breed mother, while his parents were on a hunting expedition in Ohio. His father died while he was young. He took the name of Brant from his stepfather. As a young warrior he knew Pontiac well and had imbibed his political principles. His Indian name was significant —"He places two bets"—for his life was devoted to harmonizing Indian rights with the sovereignty asserted over the Iroquois first by the French, then the Dutch, English and Americans in turn.

Over his brother-in-law, the late Sir William Johnson, who in 1755 sent him to school in Connecticut, unquestionably he exerted a powerful influence, serving under him with great distinction both against the French and Pontiac. In 1765 he married the daughter of an Oneida chief and settled at Canojaharie Castle in the Mohawk Valley. Like most men of his race Brant had little faith in the colonies. Therefore, immediately upon the death of Johnson he proceeded to England, no doubt to discover exactly what was the purpose of the government and to what extent the tribes could rely upon its support.

In the struggle between the English and French both had courted the military aid of the Indians, nor had the British colonists disapproved of the employment of the red warrior against the French. In this conflict the Indians had been taught to regard themselves not only as legitimate but as desirable allies of the whites. Certainly they would not now be permitted to observe the neutrality which the aborigines had long known in intertribal strife with which they were not concerned, even had it been to their interest to do so, for they were being ceaselessly importuned by partisans of both sides.

From the time when Champlain on one of his western expeditions had joined a party of Canadian Indians in an attack upon the Iroquois, bringing down upon France their undying enmity, they had in the main, been the staunch military allies of Great Britain, and with the exception of a few isolated groups of Mohawks and Onondagas whom astute French missionaries had seduced from their tribal associations, had valiantly and effectively fought with the British to the close of the French regime. So, too, the Cherokees had been ardent allies of the British against the

* Succeeded by Butler as Royal Commissioner of Indian Affairs.

French. Now there was even more reason than prior to 1763 why the Indians should support the British. No longer ground between the French and British millstones, it was from the colonists alone that they felt the pressure of continued encroachment upon their lands. The absent government of Great Britian which they could only have experience of through skillful agents, seemed far more desirable as a friend than the peoples at whose hands they were suffering directly. They well knew that the colonial frontiersmen regarded a dead Indian as the only good Indian, for the plain reason that every dead Indian yielded up with his ghost another bit of coveted land. Then, too, the colonial governments had done nothing since 1763 to win the confidence of the Indian, while Sir William Johnson, by the wise and lenient administration of the office of "Indian Guardian," had caused them to look upon the "Great Father Across the Waters" as their protector against the colonies.

But affection for the British government was by no means the determining factor. In the result of the recent war there seemed to be convincing evidence that the strength of Great Britain was superior to that of the colonies. It would be foolheardly, they naturally argued, to support these relatively weak children against the "Great Father" who had overthrown the French, and besides, it was the King and not the colonies who would secure them in the possession of their lands.

Nor did the British in Canada after the death of Johnson spare any reasonable effort to foster the natural predisposition of these military vassals, and to render them effective with proper leadership, arms and equipment. Long before hostilities commenced, the Ohio tribes, encouraged by the recollection of Pontiac's success, smarting under that of his foul murder, the Yellow Creek and other massacres, and Dunmore's War, were tugging at the leash. Again the Indian problem was of vital importance. Contemplating the Indians as certain allies of the British, many a frontiersman must have trembled as he saw hostilities approaching.

The colonial governments were not unaware of the dreadful menace that was threatening their common cause. Together, the Iroquois and Cherokees were able to put not less than fifteen thousand warriors upon the warpath on the flanks of the colonies; the Ohio tribes certainly half that number. The Indians alone, therefore, could neutralize much of the colonial man-power by making recruiting in the frontier regions for an organized army virtually impossible. When, therefore, the Continental Congress was assembled in 1774 as the political agency through which the thirteen revolting colonies were to cooperate, a Committee of Indian Affairs was empowered to continue the British colonial policy of dealing with the Indian tribes through treaties designed to secure their allegiance. So, too, even before hostilities commenced, Massachusetts undertook to negotiate a treaty of alliance with the Oneidas, and actually recruited a military company among the Stockbridge tribe. Therefore, two months after the Battle of Lexington, that is, on June 30, 1775, the Continental Congress resolved "that the Committee for Indian Affairs do prepare proper talks to the several tribes of Indians." At the same time commissioners empowered "to treat with the Indians" were appointed and directed to seek the assistance of persons of influence among the Indians. Inasmuch as any Indians that might be won over to the cause of the colonies would require arms, ammunition and clothing, the sum of 40,000 pounds sterling was appropriated for their equipment and subsistence. To provide for the governance of those already within the jurisdiction of the colonies, and for such tribes as might join their cause, a month later the Department of Indian Affairs was created to succeed that which had existed under the British government. Straightway Lewis Morris, who was specifically directed by Congress "to secure the cooperation of the Western Indians," proceeded to Fort Pitt with that end in view, while innumerable other agents of the Continental Congress sought to enlist their active aid or to insure their neutrality. Thus is seen the absolute absurdity of charging the British with the crime of employing the Indians in their effort to save their colonies.

To counteract the influence of Guy Johnson and Butler, the Indian leaders who were bending every effort to break the neutrality of the Six Nations, Eleazor Wheelock, the President of Dartmouth College, an institution founded for the education of the Indians where many of the Iroquois youth had been hospitably received, sent among the Iroquois the young and humane preacher, James Deane, who was much beloved by them. Coincidently, to the Mohawks and Oneidas, Massachusetts deputed Samuel Kirkland who had lived among them many years, while delegations from the friendly Stockbridge Indians of Massachusetts were sent to the Six Nations to urge them to remain neutral, but Johnson was soon able to expel these agents.

Nevertheless, again the Iroquois sachems were to display the wisdom for which they had ever been noted. With true diplomacy they decided to watch awhile the course of events before following their natural inclinations. Thus, at a Great Council held at the German Flats in 1775, the Mohawk chief, Little Abraham, counselled neutrality for the time being. Certainly, they should not join the Revolutionists, he declared, for "if you conquer you will pull us off altogether," meaning, of course, that even should the colonists establish their independence they would take the Indian lands.

As was to be expected, meantime, the mission of Morris and of Colonel Morgan to the western tribes also came to naught, so that while La Corne and Hamilton, with headquarters at Detroit, busied themselves organizing and equipping the western tribes, British agents in the South quickly won over the Cherokees who agreed to furnish 3,000 warriors for the purpose of ravaging the southern colonies in concert with the Chickasaws. Although the Choctaws, Creeks and Seminoles refused to enter into formal engagements with the British, it was plain that they could be counted upon not to assist the colonies, and that a goodly number of their warriors would take to the warpath upon a favorable opportunity.

It was now that Brant returned to Canada from England. Having been thoroughly committed to the cause of the King, he was commissioned a colonel in the British Army by Governor Carlton. Being a highly educated man and a patriotic Indian, his ascendancy over the Iroquois as head councillor of their league was so complete there can be little surprise at what followed.

So soon as it was certain that the Indians were not to be seduced from their old allegiance, Congress, willing enough until now to utilize them, resorted to a bit of patent hypocrisy in the hope that public sentiment in England might prevent their employment for military purposes. Thus, on July 12, 1775, it declared it to be the intention of the United Colonies "to seek only a neutrality of the Indian nations unless the ministerial agents should engage them in hostilities or in an offensive alliance." But the British had been fully informed of the mission of Morris and the efforts of the other colonial agents. The belated protestations of Congress, therefore, deceived no one.

The attempt of the Continental Congress to deal with the Indians through its Committee of Indian Affairs and the Indian Department created by it, at once gave rise to the most perplexing questions of jurisdiction. At every turn it was confronted by the claims of the revolting colonies that Congress was treading upon their rights. Hardly had they declared their independence when each of them began to extend their jurisdiction over the Indians.

It had not taken them long to discover the impossibility of conducting a war effectively through the agency of the Continental Congress, so that in July, 1775, Franklin submitted the Articles of Confederation looking to a more perfect union among them. Closely following the Iroquois system, his plan formed the basis of that reported to Congress by a committee appointed for the purpose.

Meantime Arnold had employed the Abnakis with such good results in his Canadian campaign that on May 25, 1776, less than six weeks before the Declaration of Independence, having failed to accomplish the political object of its attempted deception, Congress resolved "that it was highly expedient to engage the Indians in the service of the United Colonies." The enlistment of two thousand was then authorized. Thereafter, not less than 1,000 Indians served in the American army.

Amazed though the Iroquois were by the expulsion of the British from

Boston, the most that even Washington, whose influence among them was very great, could do was to keep them from the warpath a little longer. Finally, on May 31, 1776, the Six Nations entered into an engagement, not as Bancroft declares, to observe a strict neutrality, but merely permitting each tribe to exercise its own discretion. It was but a short time before, succumbing to the tremendous pressure brought to bear upon them, all the Iroquois except the Oneidas and about half of the Tuscaroras cast in their lot with the British.

Franklin's proposal was debated and amended until November 17, 1777, when it was agreed to by Congress and reported to the various States. At this time there were both Indian tribes and individual Indians in all the states and there were many tribes inhabiting the unorganized territory whose domains had been declared by the royal Proclamation of 1762 not to be within the jurisdiction of any colony. Since no one State could possibly assert jurisdiction over all of the latter, Para. 4 of Article 7 of the Articles of Confederation provided that "The United States in Congress assembled shall have the sole and exclusive right and power of regulating the trade and managing all affairs with the Indians not members of any of the States, provided that the legislative right of any State within its own limits be not infringed or violated." Thus, in the Constitution adopted by the sovereign State of New York in 1777, jurisdiction was asserted by it over the Indians within its borders in the following terms:

And whereas, it is of great importance to the safety of this State, that peace and unity with the Indians within the same be at all times supported and maintained; and whereas, the frauds too often practiced towards said Indians, in contracts made for their lands, have in divers instances, been productive of dangerous discontents and animosities:

Be it ordained, That no purchase or contracts for the sale of lands made since the fourteenth day of October, in the year of our Lord, one thousand seven hundred and seventy-five, and which may hereafter be made with any of the said Indians, within the limits of this State, shall be binding on the said Indians, or deemed valid, unless made under the authority, and with the consent of the Legislature of this State.

The question remained, however, how could the United States deal with the Indians who were not members of any state? All such Indians were organized in tribes. Without exception the tribes not within the jurisdiction of a state were the belligerent vassals of Great Britain, having the right under the Law of Nations to abandon their domains even should they be conquered by the United States, and continue as dependent British communities. On the other hand, if a tribe were willing to foreswear allegiance to the King and acknowledge the sovereignty of the United States, not being a part of the body politic of any state, it could have no other status until it had assumed the new allegiance except that of an independent community. Therefore, since there was no way for Congress to deal with such a people other than by mutual agreement, the Secretary of War proceeded at once to invoke the treaty-making power of Congress to negotiate a treaty with the Delawares, which was executed September 17, 1778. This was the first Indian treaty of the United States.

At this time, however, the states of Massachusetts, Connecticut, New York and Virginia were all laying claims based on their colonial charters to the unorganized territory north of the Ohio and east of the Mississippi rivers, an imperial region consisting of nearly 266,000 square miles. This being so, were the Delawares members of these states? Would New York claim jurisdiction over their territory just as she had over the domain of the Six Nations? Moreover, Rhode Island resolutely refused to ratify the Articles of Confederation until the States mentioned agreed to abandon their claims, so that the Delaware treaty was not ratified.

Assured by the British agents of protection, and well armed, the hostile Iroquois, reenforced by the Wyandots and other western tribes, waged a bloody war in New York and Pennsylvania under the direction of Brant and skillful British leaders in conjunction with Tory contingents, while the Cherokees ravaged Tennessee and marauded along the frontiers of the Carolinas and Virginia. Along the Ohio and in Kentucky the Indians were also active, the smoking ruins of many a lonely cabin sending up to the heavens an unanswered signal of distress. But there the warfare that was waged differed little in its general aspects from that which the pioneers had but lately carried on against the Indians even in the preceding so-called days of peace.

So great were the activities of the Cherokees that the powerful support they were affording the British could not long be ignored. Gradually the frontiersmen of Georgia, the Carolinas and Virginia were organized in units of militia rangers, and under the leadership of such men as Isaac and Evan Shelby, Williamson, and Robertson, succeeded in laying waste the Indian country, compelling the more southern Cherokees to welcome a peace in May, 1777. Those in the north fought on with unabated zeal. Appalling inhumanities occurred on both sides.

Brant and his warriors were particularly conspicuous at the battle of Oriskany in August, 1779. There can be no doubt he was in large measure responsible for the Massacre of Cherry Valley in a retaliatory raid into Orange County. But at the Wyoming Valley Massacre, where British regular troops lost control of the Indians, he was not present as commonly supposed.

Such things were, of course, utterly futile since they could have but one result. Even Washington saw the necessity of crushing the Iroquois who looked in vain to the British for the promised protection. As General Sullivan moved upon them in July, 1779, their chiefs implored for aid.

"Why does not the great King, our father, assist us? Our villages will be cut off, and we can no longer fight his battles." When the advance up the Tioga into the heart of the Seneca country commenced, Little David, a Mohawk chief, made a last appeal for help to Haldimand, then Governor of Canada. "Brother," said he, "for these three years past the Six Nations have been running a race against fresh enemies, and are almost out of breath. Now we shall see whether you are our loving, strong brother, or whether you deceive us. Brother! We are still strong for the King of England, if you will show us that he is a man of his word, and that he will not abandon his brothers, the Six Nations."

In this brief address, characterized by that simple directness common to the Indians, is contained not only the justification of the Iroquois people, but a dreadful condemnation of the white man's duplicity in dealing with them. Where, let it be asked, is there to be found in the history of the Revolution evidence of so great a spirit of loyalty on the part of the colonists and their government; of resolution, of willingness to fight on against adversity?

By this time, however, the employment of the Indians as military allies was believed by the King, the Ministry, and Parliament to be a mistake, for they had been officially informed by the British military authorities that it was practically impossible to manage and control them, and that they had become more and more unreasonable. No longer finding it expedient to employ them, every obligation even to try to protect them was ignored by the British government.

In vain Cornplanter undertook upon the urgent demands of Brant to make a stand in 1779 on the shores of Lake Canandaigua. There he was abandoned by the Senecas under Red Jacket, enabling their country to be destroyed by Sullivan and Clinton almost unopposed. Thus were the Iroquois like the Cherokees to pay dearly for their loyalty to the King. From the punishment now inflicted upon them again they were to learn how little of faith was to be reposed in the promises of the white man. When the Seneca warriors returned to their lodges and told of the killing of their chiefs, the sorrow with which the old men and women wept was not unmingled with scorn for those who had betrayed them. Yet, though abandoned by the British government, the Iroquois and Cherokees for the most part continued on the warpath until the end of the war, knowing only too well the further penalty that would be exacted of them should the Revolutionists succeed.

The Georgians and Carolinians had wreaked a terrible punishment upon the southern Cherokees. To them, Man Killer, a Cherokee chieftain, pointing out with fine understanding the true responsibility for the acts of his people, said: "You have destroyed our home, but it is not my eldest brother's fault; it is the fault of my father over the water."

Thus, without malice did the Indians accept their punishment. Man Killer was right. The Indians were not to be blamed for waging the only kind of warfare known to them. The fault lay with those who employed them.

In the use of semi-civilized peoples in war there are certain inevitable dangers inherent to their nature. In the Revolution the Indians were put by the British and colonists to a choice of allegiance by importunities as compelling as they were varied and ceaseless. If in the exercise of that choice the red man elected to cast his lot with the one or the other it was only natural that he should do so. Furthermore, in doing so he was but observing a precedent which previously had been sanctioned by the colonists in the struggle between Britain and France.

Plainly the protest against the British which is contained in the Declaration of Independence was but a piece of diplomatic insincerity designed to affect English public opinion and make the war being waged by a repressive government even more unpopular than it already was among the mass of the British people. It is equally absurd to anathematize the British Indian leaders—Butler, Guy Johnson, Germaine, St. Leger, and the others. They were responsible to their government for the success of military operations and the maintenance of British supremacy in America. Their acts must be judged in the light of their times. Those times sanctioned the employment of savages in war. Surely, too, they should not be blamed for doing that which Washington himself deemed admissible. In the very particular instructions which he gave Sullivan for the total destruction of the Iroquois country in 1779 he added that such "chastisements" might not only cause them to sue for peace, but that "by address, secrecy, and stratagem" they might be engaged to surprise the British garrison of Niagara and the shipping on the lakes, thereby putting these into the possession of the Americans.

Upon a full consideration of the Indians' part in the Revolution, the wonder is that the British government made so little use of them. Properly organized and supported, provided with able leaders, they would have been a powerful military ally. The fact that no well-considered plan was adopted for their utilization is but another evidence of how unimportant the war of American independence was deemed by the King and his Hanoverian advisers in comparison with the wars in which Great Britain was engaged with Holland and France, how little in sympathy Parliament and the British people as a whole were with the oppression of the colonies. The lack of a more general and concerted effort on the part of the Indians is also evidence of the fact that as a people they were not naturally disposed to war, that when not directly affected they were hard to rouse. From the Indians' standpoint Pontiac with his tremendous organizing capacity was to live a generation too early, and Tecumseh a generation too late.

Even if all credit be denied the Indians for a motive of loyalty on their part in the political cause which they elected, and the hope of recovering their lands and thereby profiting out of the war alone be attributed to them as their impelling motive, in their course they were not governed any more by pure expediency than the great majority of the so-called patriots. In all great political revolutions expediency is the dominating motive of the masses. The leaders may be impelled by high ideals. Not so with the great mass of those who follow. Ordinarily their support must be enlisted by arguments holding out to them an apparent material advantage. So it was in the war of American independence. Washington and the leaders associated with him in the Revolution were idealists who were willing to sacrifice all material benefits in the cause of liberty. Among those who supported the Revolution were undoubtedly many equally self-sacrificing individuals, but a great proportion of the revolutionists were propertyless, with no stake in the existing order of things, actuated in the main by the hope of improving their lot. This is evident from the fact that many of the propertied class of colonials in the urban centers were either Tories or remained passive in the struggle, while the relatively poor rural classes and the hard-pressed frontier element of colonial society espoused the cause of independence. Nor must it be overlooked that the latter were without the protection of the British who could more readily defend the population of the principal cities along the seaboard. This being so, the rural population, or the poorer people, were under an additional pressure of expediency to support the revolution.

No. The Indians were no more self-interested in their loyalty to the King than were the masses of the colonists in their varying support of the Revolution.

One aspect of the Revolution involving the Indians which has never been dealt with fully remains to be considered. It has been shown that in 1772 a patriotic organization called the Saint Tammany Society was founded in Philadelphia and that similar societies were founded elsewhere. What was the real reason for this widespread canonization of an Indian who had been dead three quarters of a century? Why was Tamenend picked out instead of Dekanawida, or Hiawatha, or King Philip, or some other great Indian who had also stood for liberty?

We do not have to search far. Tamenend had stood not merely for the independence of the Delawares and of the red race generally, but for the freeing of America from Jesuitical influences. The first Tammany Society of Philadelphia, known as Sons of King Tammany and later as the Sons of Saint Tammany, was then no more than an anti-European society which took this American native with whom Pennsylvanians were familiar to symbolize the idea of its independence of European intrigue. Similar societies having been formed elsewhere, their efforts were reflected in one of the first resolutions passed by the Continental Congress of which Washington, Adams, and Franklin were members. This was a protest against the maintenance by the King of the Catholic Church as a state institution in Canada, for it was to the Jesuits there that the Revolutionists like Tamenend traced most of their troubles with the Indians.

Unquestionably the Jesuits saw in the Revolution a rare opportunity to project the monarchical influence. During the Seven Years War, Louis XV had undertaken to place a French prince on the throne of Poland, employing as his principal agent Count de Broglie, the brother of the Marshal. The results of the Seven Years War had not been such as to leave France without interest in America. Louis could not have failed to note with pleasure the revolt of the British colonies. Why not help him? So much the better, if incidentally they could be consolidated under a French ruler and reclaimed for Rome!

Exactly how far the plot went it is impossible to determine, but certain it is that when Count de Broglie, accompanied by de Kalb, a German in the French service, who had served on the Marshal's staff, visited Silas Deane, the American agent in Paris, in November, 1776, to propose the candidacy of a French soldier of experience for the command of the American Army, it was not without the King's knowledge. On December 6th, Deane wrote the Secret Committee of Congress:

I submit one thought to you—whether you could engage a great general of the highest character in Europe, such for instance, as Prince Ferdinand, Marshal de Broglie, or others of equal rank, to take the lead of your armies, whether such a step would not be politic, as it would give a character and credit to your military and strike perhaps a greater panic in your enemies.

What was afoot is only too plain. The Jesuits in Canada were at work again and of course were supporting the Iroquois in every way possible. With the ultimate decision of the Senecas, Cayugas, Onondagas and Mohawks they had much to do. De Broglie himself was careful to warn Deane against the danger of Congress discovering a design to establish a monarchical form of government for the united colonies. Nor can there be any doubt why de Kalb was despatched to America in 1777 as his agent, accompanied by the youthful Marquis de Lafayette, and ten other officers of the French Army. If the King knew of de Broglie's dealings with Deane, he must have approved de Kalb's mission in furtherance of the plot.

While all this was going on numerous societies reflecting all manner of ideas designed to effect political opinion had been forming, many of them plainly the agencies of monarchical as well as Jesuit interest. Therefore, by this time, in almost every state Saint Tammany Societies had been formed. Anti-Jesuit and anti-foreign, they often lent themselves to the ridicule of the proponents of monarchy.

It is not necessary here to discuss the defects of the Continental Congress as a political body. It was not unnatural that among a people with no experience of a centralized government and traditionally indisposed to make the sacrifices essential to military efficiency, there should have occurred that which was the direct consequence of indiscipline—a conspiracy against Washington which may be traced directly to French influences, with the effect calculated. In 1777 even John Adams, who was responsible for his selection, as well as Samuel Adams, was impatient with his enforced Fabian tactics.

In the army the discontent found its principal expression through General Mifflin, the Quartermaster General, and one Thomas Conway, an Irish adventurer who had long served under the French flag and whose unduly rapid promotion Washington had disapproved. Believing that Conway's preferment by Congress over Washington's stated objection would force the latter's resignation, the conspirators mustered enough support in Congress to accomplish it. Various anonymous letters and forgeries now followed all designed to discredit Washington, while the victories of Gates at Saratoga and elsewhere in the North were compared with Washington's defeats at Brandywine and Germantown to the disadvantage of the latter. Soon, however, Adams had heard of the French plot, though according to his informant, Marshal Mallebois instead of de Broglie was the one who had been selected to be the "Stadtholder of America." Therefore, when in 1779 Deane's letter was published, it aroused the bitterest ire on the part of many of the patriot leaders. "I will be buried in the Ocean," wrote Adams, "or in any other manner sacrificed before I will voluntarily put on the chains of France when I am struggling to throw off those of Great Britain." Consequently, when the Frenchmen under de Kalb arrived to tender their swords to the Continental Congress, they were coldly received as the result of an uproar from the Sons of Saint Tammany and others.

Although it seems certain that when the young Marquis de Lafayette came to America with the other Bourbon agents he had little knowledge of what the Revolution was all about and was ready to serve his own country in the best way he could, inspired by Washington he soon abandoned all thought, if he ever had any, of being made a cat's paw by the Bourbons and Jesuits, nor did Washington and the colonies have a more faithful servitor. He probably did more than any one else to expose Conway under whom he refused to serve on an expedition to Canada after forcing him to drink Washington's health publicly. It was then that Congress awoke to the truth.

Eventually it became apparent to de Kalb that Washington had gained such an ascendancy over the minds of the American people as to preclude the thought of supplanting him with a Frenchman. Consequently, in November, 1778, he wrote Count de Broglie:

They [the Americans] are insultingly vain towards any nation but their own. . . . They have established their sovereignty alone, without help, against the bravest and most powerful of nations. Their General Washington is the first of all heroes, ancient and modern. Alexander, Conde, Broglie, Ferdinand and the King of Prussia are not to be compared to him. . . . It is not only the lower classes:—clever people,—or those passing for such, have the same opinion, and this is said so often that Washington believes it himself.

Although an armistice was signed upon the surrender of Yorktown, to such a pass had things come by the spring of 1782, it is not unlikely that the army would readily have made Washington king had he sanctioned it. So it was, on May 22, 1782, Colonel Nicola, an Irishman by birth who had served under Washington, wrote his old commander a long letter to feel him out on behalf of a large number of Americans who desired him to found a monarchy and accept the crown. Washington's reply has been said to have been "the grandest single thing in his whole career." Nicola was rebuked unmercifully for the bare suggestion.

Lodge, with good reason, says this incident "has been passed over altogether too carelessly by historians and biographers. . . ." It would have been a perfectly feasible thing to have altered the frame of government and to place the successful soldier in possession of supreme power. At this moment the Confederation was demoralized; the army was the one organized body of the country. Upon the departure of the English troops there was nothing that could withstand it. Congress had been forced by a Pennsylvania regiment to leave Philadelphia. The monarchical principle in government was then entirely familiar to all; they had known no other until the experiment in which they were then engaged.

A month after Nicola's letter, among other things, General J. M. Varnum, of the Rhode Island Militia, and a member of Congress, wrote Washington:

The citizens at large are totally destitute of that love of equality which is absolutely requisite to support a democratic republic. . . . Consequently, absolute monarchy or a military state can alone rescue them from all the horrors of subjugation. In this situation every moment augments our danger. . . . The instability of national policy may give place to the sentiment "that we are too young to govern ourselves."

But while the Saint Tammany Societies were to achieve much in supporting the political ideals of Tamenend, strange to say the welfare of the red race was no part of their program. This was, perhaps, only natural.

The colonists thoroughly understood the extent to which the tribes had neutralized the American man-power which was their greatest service, perhaps, in the war. Without their aid the British government would probably have been overthrown in the early days of the Revolution. Certainly the conflict would have been quite different had the tribes not been fighting for their own independence of unjust colonies.

13

The United States of America and the United Indian Nations

NO HISTORIAN has undertaken up to this time to discuss the legal aspects of the transition of the Indian tribes from the status of Crown vassals which they were given by the Proclamation of 1763, to that of political dependents of the United States. Yet, the process is clearly recorded. By the time hostilities ceased in 1781 it was manifest to all the states that some great sacrifice must be made if their union was to be preserved. Accordingly, with great magnanimity Virginia agreed to surrender not only her claims to the territory in the northwest but to Kentucky as well. Upon the promise of Congress that the land would be put to common use, Massachusetts, New York and Connecticut followed suit, so that the Articles of Confederation were finally ratified by the last state, Maryland, in 1781.

Since the Mohawks and Cayugas had been driven by Sullivan over into Canada, they had settled under Brant and Fishcarrier, respectively, on lands granted them in Ontario.* From there Brant still dominated the affairs of the Six Nations as a whole. Nor did he have any idea of permitting the state of New York to continue to appropriate to itself the domain of the Six Nations which since the treaty of Fort Stanwix of 1768 had been crown land.

In this position the Law of Nations was on his side. Until the colonies declared their independence in 1776, they had merely engaged in a civil war and could not have acquired territory, not being international per-

---

* Ultimately the fragmentary tribal groups there were organized by Brant as the Seven Nations of Canada.

sonages. Although during the period of their sovereignty, that is from 1776 to 1781, they possessed the capacity to acquire territory either by cession or conquest, New York had not acquired the domain of the Six Nations by cession since it entered into no treaty with Great Britain, and had not individually conquered it since at no time did the state make war against Great Britain independently of the other states. And exactly the same principles applied to the other states.

Yet, when the British surrender occurred at Yorktown in 1781, it was at once evident that unless the tribes abandoned their allegiance to the British Crown and submitted to the sovereignty of the United States, they would lose their domains since the United States had conquered them.

It has been shown that even prior to the advent of the English the principle of union was well established in the Indian political system. The tribal confederacy in New England had early enforced that system upon the settlements there, while Tamenend had urged it upon Penn, and Johnson upon Franklin. Johnson undoubtedly had evolved his conceptions of a union out of his experience with the Iroquois just as did Pontiac. Nor is it to be doubted that Brant had had much to do with the formulation of Johnson's proposals to Franklin, that all along there had been germinating in the great Mohawk's mind the idea of uniting all the tribes on the basis of the Iroquois league as the means of enforcing upon the colonies and then the states, a recognition of the racial rights. At any rate hardly had the Iroquois been driven from their domains by Sullivan when Brant began contriving a scheme that would deny to the state of New York that jurisdiction over the territory of the Six Nations which in its Constitution of 1777 it had undertaken to assert, a scheme that would also inure to the benefit of the tribes generally in their relations with the union which the states were evolving. Obviously to his mind the way to accomplish this was to organize a far flung confederacy among the tribes from the St. Lawrence to Florida after the Iroquois pattern, or upon the fundamental principle that tribal lands were owned by the constituent tribes in common. Therefore, before hostilities ceased, with the ready aid of the British irreconcilables in Canada, he had drawn together the Six Nations, Chippewas, Wabash, Miamis, Shawnees, Ottawas, Potawatomis, Hurons, Delawares, Wyandots and even the Cherokees, in a union to which he gave the significant name of United Indian Nations—the most formidable confederacy yet formed by the Indians.

His conception was a vast one transcending anything of which Pontiac was capable of putting into execution, and one which recommended itself at once to British diplomacy since the firm establishment of the United Indian Nations would not only throttle the development of the United States by barring westward expansion, but in the event of any future attempt to recover the lost colonies would prove a powerful military ally.

Therefore, Brant was given all the aid he required. Consequently, before the treaty of peace of 1782 was signed, his confederacy was an accomplished fact.

British diplomacy was now to do its part. The treaty of peace of 1782 was not made with the states individually but with them collectively, and the land ceded included that constituting the organized territory of the former colonies, the Crown lands within their boundaries, and the unorganized territory beyond. All this was expressly ceded to the United States so that under no tenable theory of law could it be said that New York and Georgia had acquired title to the Iroquois and Cherokee domains. Moreover, in the preliminary treaty of peace of 1782 and the definitive treaty of 1783, were inserted provisions framed in broad enough language to include the Six Nations, guaranteeing against the confiscation of property as a penalty for former allegiance to the King, and binding the United States to restore any property already confiscated.

Nevertheless, whatever the law may have been, to the ordinary citizen it was unthinkable that the Indian tribes but lately in arms against the United States, had any rights to the land upon which they had been seated prior to the Revolution. The popular view was well expressed by Brackenridge, a frontier editor who wrote in 1782, that "so far from" admitting the Indian title, he conceived that "not having made a better use" of the land for many hundred years "the aborigines had forfeited all pretense to a claim"; that he would "as soon" admit the title of the buffalo as that of the Indian. Since "the animals vulgarly called Indians," were by nature "fierce and cruel," he considered that their "extirpation" would be "useful to the world" while entirely "honorable" to those who would "effect it." And this view at once became the platform of a political party which, though not mentioned in history, was henceforth to dominate the states and often the general government through a congress reflecting the popular will—the Buffalo Party. At any rate, in 1783, seeing that only the friendly Oneidas and about half the Tuscaroras were occupying the domains, the Legislature of New York did not hesitate, despite the treaties of 1782 and 1783, to appropriate to the State the vacant Iroquois lands.

At this Brant and the Indians generally were highly incensed. Again British diplomacy was invoked. As an enlightened state, Great Britain owed its former Indian subjects the enforcement of the treaty guarantees designed to protect them, so that without regard to its ulterior motives it was right on its part that the British government now insisted upon retaining a number of military posts along the Canadian border within the ceded territory, including Oswego and Detroit, until the guarantees of the treaty were fulfilled. Nor was it slow to call attention to the demands of the tribes who, by reason of the confiscation of their lands and fear of violence, still remained in Canada. At the same time, Brant prepared a memorial designed to secure the recognition of the United Indian Nations which in 1783 was presented to Congress.

A conference between representatives of the United States and the Council of the Confederacy was urged for the purpose of arranging the terms of Indian allegiance. The necessity for a definite agreement embodying the terms upon which such Indian land as might be required by the United States was to be obtained was pointed out. By this means alone, it was said, could trouble with the tribes be avoided. Assuming the attitude of a vassal people of Great Britain whose allegiance did not necessarily pass with the transfer to the new sovereignty of their domains, the Confederacy proposed to establish the new status of the tribes by treaties independent of the treaty of peace between the United States and Great Britain. This it was willing enough to do should Indian rights be respected by the United States. Meantime, the tribes reserved to themselves what they deemed a right on their part at international law, that is, to remain in their old allegiance even if they had to abandon their domains which had been ceded to the United States.

The general government was in a dreadful predicament. How could it enforce upon the powerful state of New York a recognition of the Indian rights it had guaranteed? Yet, the violations of the treaty of peace were but affording the British imperialists legal grounds for resuming hostilities in order to recover their lost colonies.

At this juncture Washington intervened. In his mind, the advantages of placating the Indians far outweighed the disadvantages of recognizing their rights. Moreover, with war threatening, it was imperative that the government pacify and win over the tribes. The United States must secure the allegiance of the Iroquois for the effect it would have on the western tribes if the unorganized territory were to be thrown open to settlement as public domain. Accordingly, he was emphatic in urging the prompt recognition of the Indian title; the establishment of a boundary between Indian and public lands; the strict prohibition of trespasses upon the former; and the compensation of the Indians for such land as was acquired from them in order to create the public domain. "Anyone acquainted with the nature of Indian warfare," he declared, "would not hesitate to acknowledge that such was at once the cheapest and the least distressing way of dealing with the Indians."

The Buffalo Party was opposed to Washington's proposals. Although in fact the United Indian Nations had but urged the adoption of Penn's policy, in this proposal it saw the hand of Machiavelli. Once recognize the Confederacy and the principle that the tribes had a claim of right to their domains, irrevocable harm would be done. Therefore, it refused to deal with the United Indian Nations, deciding to break up the Confederacy by dealing with the tribes separately. Accordingly, on September 22, 1783, by proclamation the general government enjoined all persons "from making settlements on lands inhabited or claimed by Indians without the limit or jurisdiction of any particular State, and from purchasing or receiving any gift or cession of such lands or claims without the express authority and direction of the United States in Congress assembled." Thereupon commissioners were dispatched to the Iroquois to explain the position of the United States, and to offer and assure them of its protection. "We are the proper persons to deal with," they declared. "All the country which lies now within the limits of Congress, formerly yours, is still yours, for we do not claim any part of your lands except the posts. If we should have any of your land we mean to pay for it."

Until now the Six Nations had been the vassals of the British Crown, exercising as such the rights of belligerents. While the cession or conquest of territory carries with it sovereignty over its inhabitants, and the United States had undoubtedly acquired sovereignty over all those who, like the Oneidas and part of the Tuscaroras, had elected to remain upon the Crown lands ceded it by the King, the other tribes possessed the right not to reoccupy their domain, and to adhere to their old allegiance, or with the consent of the King abandon that allegiance and as an independent people submit to the sovereignty of the United States. It was the latter course which the British government was urging them to pursue. Therefore, in October representatives of all the tribes assembled at Fort Stanwix prepared to close the book of their differences with "the thirteen fires."

Straightway it developed that in order to secure to themselves a right of occupancy in their domains, the Six Nations must not only submit themselves to the sovereignty of the United States but relinquish their claim to the region deeded to them by the King in 1701. Brant was not present and the others did not see that so soon as such a treaty were executed, treaties extinguishing the Indian rights in the Northwest would at once be thrust upon the tribes inhabiting that region. Therefore, after hesitating a while, contrary to the counsel of Brant and Red Jacket, Cornplanter and the other chieftains were finally cajoled into executing the proffered treaty on October 22, 1784.

"The United States of America give peace to the Senecas, Mohawks, Onondagas, and Cayugas, and receive them into their protection upon the following conditions." So read the preamble of the treaty. But although it was only necessary for the United States to grant peace to the four tribes that had engaged in hostilities against it, these tribes were not signatory parties to the instrument. The treaty was made with the Six Nations and signed by the tribal representatives in the council of the League. Moreover, it was the Six Nations as an entity that was guaranteed a right of occupancy forever in the immemorial domain of the League with the exception of a few small tracts, in consideration of a release by it of all its right, title and interest in the region beyond a line running for a point four miles east of Niagara on the shore of Lake Ontario, called Johnston's landing place, thence to the western boundary of the state of New York, and south to the Ohio River.

The treaty of Fort Stanwix of 1784 unquestionably is one of the great landmarks in American history for the reason that being the first treaty between the former Indian subjects of the Crown and the United States to become effective, it fixed the policy of the sovereign state which succeeded to the sovereignty of Great Britain within the territory ceded by the latter. And it was no more than an adoption of the Crown policy proclaimed in 1763 which conceived of the Indian tribes as protected communities and political dependencies of the sovereign to whom their allegiance ran.

Nevertheless, for long the exact status of the Six Nations to which this treaty gave rise was not understood by the Indians nor by the American people generally, and often not by the courts. Thus they have been spoken of even by the latter as a sovereign nation, as a semi-independent nation or community, and as a domestic community similar to a corporation. Yet, if the treaty of Fort Stanwix be examined it must be evident from its context that in making it the Six Nations were acting as an independent state which had been released by Great Britain from its allegiance to the Crown. Moreover, by voluntarily transferring its allegiance from Great Britain to the United States in consideration of the protection promised and the perpetual right of occupancy granted it in the territory of the United States, in effect it attorned to the sovereignty of the latter. Furthermore, even if the Indian tribes had refused to pledge their allegiance to the United States, or the succeeding sovereignty, those which elected to remain upon its territory would at international law have become subject as aboriginal tribes to its political jurisdiction, occupying the status of dependencies. And with respect to their persons and property the United States would have been charged with a trust which it could not by any agreement with them have avoided.

Be this as it may, the Buffalo Party knew nothing of and cared less for international law. The state of New York, having presumed in its Constitution of 1777 to assert jurisdiction over all the Indians and their lands within its boundaries, despite the treaty of 1784 simply continued to deal with them as subject to its authority. The result was the Mohawks and Cayugas, fearful of returning to their domains, remained in Canada with Brant, much to the embarrassment of the British government which had spared no effort to provide for them by compelling the United States to recognize their right. And now the Senecas, Onondagas, Oneidas and Tuscaroras were threatening to remove to Canada under the pressure of those who, unrestrained by the state were intruding upon their domains with loud threats of punishment for alleged disloyalty during the late war.

Taking the position that this constituted both confiscation and reprisal in violation of the treaty of Paris, Brant called loudly upon the British government for the enforcement of the treaty guarantees with the result that early in November Haldimand reenforced the British garrison at Oswego.

In vain Monroe, the Secretary of War, appealed to the governor of New York. In the desperate situation in which the general government now found itself, he appealed to Madison for an interpretation of the Articles of Confederation in the drafting of which the latter had taken a leading part. Writing from Richmond under date of November 27, 1784, Madison replied:

Your favor of the 15th inst. came to hand by Thursday's post. Mine by the last post acknowledged your preceding one. The umbrage given to the Comsrs. of the U.S. by the negotiations of N. Y. with the Indians was not altogether unknown to me, though I am less acquainted with the circumstances of it than your letter supposes. The idea which I at present have of the affair leads me to say that as far as N. Y. may claim a right of treating with Indians for the purchase of lands within her limits, she has the confederation on her side; as far as she may have exerted that right in contravention of the Genl. Treaty, or even unconfidentially with the Comsrs. of Congs., she has violated both duty & decorum. The federal articles give Congs. the exclusive right of *managing all affairs* with the Indians *not members* of any State, under a proviso, that the *Legislative authority* of the State within its own limits be not violated. By Indians not members of a State, must be meant those, I conceive who do not live within the body of the Society, or whose Persons or property form no objects of its laws. In the case of Indians of this description the only restraint on Congress is imposed by the *Legislative authority* of the State.

If this proviso be taken in its full latitude, it must destroy the authority of Congress altogether, since no act of Congs. within the limits of a state can be conceived which will not in some way or other encroach upon the authority (of the) state. In order then to give some meaning to both parts of the sentence as a known rule of interpretation requires, we must restrain this proviso to some particular view of the parties. What was this view? My answer is that it was to save the states their right of preëmption of lands from the Indians. My reasons are 1. That this was the principle right formerly exerted by the colonies with regard to the Indians. 2. That it was right asserted by the laws as well as the proceedings of all of them, and therefore being most familiar, wd. be most likely to be in contemplation of the parties. 3. That being of most consequence to the states individually, and least inconsistent with the general powers of Congress, it was most likely to be made a ground of Compromise. 4. It has been always said that the proviso came from the Virga. delegates, who wd. naturally be most vigilant over the territorial rights of their Constituents. But whatever may be the true boundary between the authority of Congs. & that of N. Y., or however indiscreet the latter may have been I join entirely with you in thinking that temperance on the part of the former will be the wisest policy. . . .

It was not long before Cornplanter and the other Indians saw they had been outwitted by Congress since the treaty of Fort Stanwix was to prove but the opening wedge in the plan to break up the United Indian Nations by separating the tribes and obtaining from them sufficient territory out of which to create within their country a public domain. This became obvious when by the ordinance of May 20, 1785 the 41° of north latitude was fixed as the boundary line between the Indian country and the contemplated public domain, and provisions were made for the surveying of the latter. Moreover, commissioners were now appointed to negotiate separate treaties with the western tribes as rapidly as possible in order to extinguish their rights up to the designated boundary line. Yet, it was in vain Brant now pleaded with the tribes to resist the artifices of the treaty commissioners. Had not the Six Nations made a separate treaty? Why should the other tribes sacrifice the opportunity to secure their rights?

The Cherokees were especially insistent on this point. These people were the mountaineers of aboriginal America. Occupying the upper valley of the Tennessee River as far west as Muscle Shoals, and the highlands of Carolina, Georgia, and Alabama, their domain was one of the most salubrious and picturesque regions east of the Mississippi. Writing of them Bancroft says:

Their homes were encircled by blue hills rising beyond hills, of which the lofty points kindle with the early light, and the overshadowing ridges, like masses of clouds, envelop the valleys. There the rocky cliffs, towering in naked grandeur, mock the lightning, and send from peak to peak the loudest peals of the thunderstorm; there the gentler slopes are decorated with magnolias and flowering forest-trees and roving climbers and ring with the perpetual note of the whippoorwill; there wholesome water gushes profusely from the earth in transparent springs; snow-white cascades glitter on the hill-sides; and the rivers, shallow but pleasant to the eye, rush through narrow vales which the abundant strawberry crimsons and coppices of rhododendron and flaming azalea adorn. At the fall of the leaf, the ground is thickly strewn with the fruit of the hickory and the chestnut. The fertile soil teems with luxuriant herbage, on which the roebuck fattens; the vivifying breeze is laden with fragrance; and daybreak is ever welcomed by the shrill cries of the social nighthawk and the liquid carols of the mockingbird. Through this lovely region were scattered the villages of the Cherokees, nearly fifty in number, each consisting of but a few cabins, erected where the bend of the mountain stream offered at once a defence and a strip of alluvial soil for culture. Their towns were always by the side of some creek or river and they loved their native land; above all, they loved its rivers, the Keowee, the Tugeloo, the Flint, and the branches of the Tennessee. Running waters, inviting to the bath, tempting to the angler, alluring wild fowl, were necessary to their paradise. The organization of their language has a common character with other Indian languages east of the Mississippi, but etvmology has not been able to discover conclusive analogies between the roots of their words. The 'beloved' people of the Cherokees were a nation by themselves. Who can say for how many centuries, safe in their undiscovered fastnesses, they had decked their war-chiefs with the feathers of the eagle's tail, and listened to the counsels of their 'old beloved men'? Who can tell how often the waves of barbarous migrations may have broken harmlessly against their cliffs?

Finally, in 1785 a treaty known as the Treaty of Hopewell was negotiated with them setting apart to their exclusive use a large tract on which no one but a Cherokee was to have the right to remain and settle, in consideration of their agreeing to bury the hatchet forever, and under this treaty they acquired the right of representation before Congress through a delegate of their own choice. The same year the treaty of Fort MacIntosh was negotiated with the Wyandots, Delawares, Chippenwas and Ottawas, and in 1786 separate treaties with the Shawnees, Choctaws, and Chickasaws.

So soon as these treaties were concluded, to provide for a general supervision of the Indians and in some measure protect their rights in their unceded lands, an ordinance for the regulation of Indian affairs was enacted by Congress establishing under the jurisdiction of the Secretary of War an Indian department which included two Indian districts, the northern embracing all the tribes north of the Ohio and west of the Hudson River, and the southern all those south of the Ohio. A bonded superintendent was provided for each district who was authorized to employ two deputies. Appointed for a term of two years and required to reside within their district, they were charged with the enforcement of such regulations as Congress might establish.

Trading posts within the tribal domains, at first welcomed by the Indians, with the tremendous influx of whites following upon the cessation of hostilities quickly became settlements. Everywhere the Indians were being importuned to dispose of their land, while individual Indians and subgroups who had no right whatever to give away or sell land were being enticed into doing so. Worse, they were being debauched with vice-money and whiskey by the unprincipled desperadoes that appeared among them in the van of civilization. In addition to legitimate traders at every post there were squatters, gamblers, and cut-throats who recognized no right in anyone, much less in the "savages." Then came the surveyors with strange devices—eyes that could see far off—and all manner of government agents. Led by British agitators and others to believe that the government proposed to lay off their reserved lands as well as those which they had already released, to be disposed of to the unruly horde of whites that was descending upon them, the Indians naturally became more and more restless. Now too late they saw the mistake of the separate treaties and were disposed to harken to Brant. The district superintendents could not convince them that the government was trying to protect their country. Nothing, they said, was left them but war.

For this, the British in Canada were quite ready, but before hostilities could be commenced, they must secure the sanction of the home government. No one was better prepared than Brant to explain the situation. Therefore, he was induced to go to England and lay his plans before the government. To his amazement he now discovered that again the Indians were but the pawns of the white man's diplomacy. Pointing out to him the inevitable result should the tribes take to the warpath before the British government was prepared to support them, which Whitehall was not yet ready to do, Sidney and others not only dissuaded him from permitting his people to make what would be a futile sacrifice but urged him to use his influence to preserve the peace. Therefore, upon returning to Canada Brant called a great council of the United Indian Nations which assembled on December 18, 1786, at the principal Huron village near the mouth of the Detroit River, and urged that another appeal be addressed to Congress.

Red Jacket, the Seneca chief, who is said to have opposed the treaty of Fort Stanwix of 1784, notwithstanding his great friendship for Washington, having despaired of obtaining justice by peaceful means, was violently opposed to further relations with the United States. He was sincere in believing that the welfare of his race demanded complete separation and independence. Beyond the reach of persuasion, he contended in a speech which is said to have been a masterpiece of oratory, that contact with white civilization could only destroy the red race which, he declared, was not by nature adaptable to it. To the young men who had received education at the hands of the whites he said: "What have we here? You are neither a white man nor an Indian; for heaven's sake tell us, what are you?" He did not believe in burying the hatchet longer and there is no doubt the majority of the Indians shared his views.

Just as it is useless to call Pontiac an intractable savage, so it is, because of his record in the Revolution, to go on anathematizing Brant because of his really patriotic motives. No man could have taken the higher position than that which he now took in addressing the United States.

Under his compelling influence the council preserved a pacific attitude. If the American people must have more Indian lands, it was willing to do what was necessary provided the government would deal with the Confederacy alone and establish with it a definite understanding as to the terms of future cessions. If the United States would adopt the policy inaugurated by the British Crown in 1763, that policy would have his support and he would cooperate to preserve the peace. These views were set out in a new memorial drafted by Brant, which is one of the most remarkable documents in the political history of the United States. Beginning with the statement that during the peace negotiations in 1782 the King had urged the tribes to remain quiet, it declared that the Indians had been greatly disappointed by not being included in the peace arrangements. Various overtures had been made to them by the states but these had been ignored; instead of dealing with thirteen governments they had very properly dealt with the Federal Government only, and in 1783 had proposed to Congress a plan which would have avoided all difficulties. The Confederacy had been ignored. More than three years had elapsed since the establishment of peace. Although the Confederacy had recognized the sovereignty of the United States, and submitted to the Federal authority, instead of Congress recognizing and dealing with it, it had been ignored. Instead of recognizing the Indian Confederacy Congress had seen fit to negotiate separate treaties with the tribes which were not uniform and could only lead to dissension among the Indians. One great treaty with the Confederacy would have been far better. Yet, despite the harm that had been done the Confederacy believed it was not too late to insure the preservation of peace. The tribes were still willing and anxious to meet Congress halfway. In the opinion of the Council the first step looking to peace was the recognition by Congress that all treaties "should be with the general voice of the whole Confederacy, and carried on in the most open manner, without any restraint on either side." A great council for the following spring was urged, and Congress was besought until then to permit no more people to come upon the Indian lands. The memorial closed with a declaration that should have been sufficient warning that the Confederacy would be powerless to restrain the individual tribes if its advice continued to be ignored. "You kindled your council fires where you thought proper," it recited, "without consulting with us, at which you held separate treaties, and have entirely neglected our plan of having a general conference with the nations of the Confederacy. It shall not be our fault if the plan we have suggested to you should not be carried into execution; the world will pity us when they think of the amicable proposals we now make to prevent the unnecessary effusions of blood."

The sole response this second appeal of the United Indian Nations received was an appropriation by Congress in 1787 of $25,000 for the negotiation of Indian treaties, and the issuance of instructions to General St. Clair to proceed without delay to extinguish by separate treaties the Indian rights in the previously designated area.

This was now no longer possible. The Indians had learned a lesson. While treaties were obtained with the Potawatomis, Mingoes, and Sacs, St. Clair was compelled to report to Congress that the boundaries already fixed in the treaty of January 21, 1785 with the Wyandots, Delawares, Chippewas and Ottawas could not be altered.

Nevertheless, Congress was not deterred in the least. Following upon the ordinance of 1787 accepting the cession from the states, in which it was expressly stipulated the Indian title should be extinguished before it should become public domain, on July 13, 1787 a constitution was adopted for the Northwest Territory in which provision was made for the erection of not less than three nor more than five States so soon as it came to possess a population of 5,000 white persons, while a provisional government consisting of a governor, a secretary, and three judges was appointed to administer the territory until legislatures were called to adopt constitutions for the prospective states. Meantime the territorial government was to have power to adopt, subject to the approval of Congress, such laws as were requisite. Finally, the Northwest Territory was dedicated to freedom by an article prohibiting slavery and involuntary servitude forever.

In October, 1787 St. Clair was appointed governor. Reaching Marietta in July the following year, at once he created Washington County for which he appointed a number of magistrates and a Court of Quarter Sessions. The territory having been surveyed, and the public land nominally distinguished by the treaties of 1785 with the Indians from their remaining domains, Congress now proceeded to dispose of the public land first by the private sale of large tracts, and later by the sale or gift of the land in small parcels.

As might have been foreseen, the creation of a public domain by the partial extinguishment of the claims of the Ohio tribes, brought about a peculiarly difficult situation inasmuch as there were large tribal domains contiguous to tracts of public land between which there was a clear distinction in the law but not one that was visible to the preemptors. Nevertheless, the public lands were thrown open in 1787 to the whites, whereupon settlers began to swarm across the Ohio with little regard for the legal distinction between public and Indian lands, thus bringing about a condition of which Congress remained utterly heedless. To please the missionaries, however, on September 3, 1788, there were set apart out of the public domain three tracts of 4,000 acres each at Shoenbrun, Guadenhulten, and Salem, respectively, on the Musingum River in the Ohio country for the use of congregations of the Catholic Iroquois, Moravian, and Quaker sects, thus making provision for the Christian Indians.

The administration of the Indian Districts was a farce, for after the treaty of Fort Stanwix Congress had made what amounted to nothing less than a surrender of the position it had taken in 1784 at the instance of Washington. Thus, in an ordinance for the year 1786 it had declared that thereafter when transactions with the Indians who were within a state became necessary, the superintendent in whose district they were should act in conjunction with the state authorities in order that the rights of the states should not be infringed. Since it was impossible, of course, for the various agents to agree as to what Indians were in a state within the meaning of the existing law, Indian affairs were necessarily conducted in the most un-uniform, irregular and illegal manner. For instance, in 1786 New York ceded to Massachusetts the preemption of the Genesee country within the bounds of the state, thus settling an old dispute with Massachusetts. The following year the Six Nations leased the country to one Livingstone and his associates for ninety-nine years at $2,000 per annum. Afterwards Massachusetts sold its preemption right to Phelps and Gorham for $1,000,000. In 1788 Phelps and Gorham called a council of the Six Nations at Buffalo Creek, where the assembled Indians were induced to sell their remaining claim to the Genesee country for $5,000 and an annuity of $500. By reason of the sale to Phelps and Gorham, the Indians were next induced to reduce Livingstone's annuity by half. The state soon declared the Livingstone lease invalid, and in 1788 passed additional legislation over the Six Nations. Meantime, Georgia was not slow to take the cue from New York. In 1788 Congress was compelled to publish a proclamation enjoining further encroachments upon the lands of the Cherokees.

14

The Constitution and the New Status of the Indians

IT IS THE VERIEST CONCEIT to imagine that American independence more than nominally was won at Yorktown. Haldimand's successor, St. Leger, had been instructed to reenforce all the British garrisons and to exclude the Americans from the profitable fur trade of the North by restricting the use of the St. Lawrence River and the Great Lakes to British vessels, while on February 28, 1786, John Adams was notified by the British Secretary of State for Foreign Affairs that his government had no idea of relinquishing the posts on the Great Lakes until justice had been done British creditors and the guarantees of the treaty of peace were fulfilled. A state with foreign garrisons occupying it can not with truth be said to be independent.

The government created by the Articles of Confederation had proved itself inadequate to the needs of the American people, among whom a more perfect union had become necessary. A monarchy was being proposed in support of which a conspiracy had long existed to place a French nobleman upon the throne. The state troops were mutinous and threatening Congress. Disgusted with the government of the United States, the more powerful Indian tribes were on the verge of returning to British allegiance. In this situation a constitutional convention was at last called in 1786 and three years later the present Constitution was substituted for the old.

That the peculiar characteristics of the federal republic now created trace directly through Franklin, Adams and the other framers of the Constitution to the Iroquois League, has been pointed out by more than one student of political science.* The new republic, like the League of the Long House, was a superstate, or a superimposed sovereignty constituted of derivative powers, or powers delegated to it by constituent states which retained some of their sovereign powers, being the first state of its kind other than that of the Iroquois. And it is an established historical fact, though not finally admitted by the ethnologists until more than a century after Franklin and Adams had both testified to it, that the United States of America owes its governmental system and political organization to the natives of America and Sir William Johnson.

Often cited as evidence of the idealism of the American people, their new constitution in fact not only reflected little of the equalitarian philosophy expressed in the Declaration of Independence but less of the more humanistic tendencies of the time. Nor was it, as commonly declared, the result of three great compromises, namely those between centralists and decentralists, the large and the small states, and the advocates and the opponents of slavery. There was a fourth compromise between those who recognized the moral obligation of the nation to the Indians, and the Buffalo Party. Thus, notwithstanding all that had gone before, the unclarities of the Articles of Confederation with respect to the Indians were not done away with, while in order to insure its ratification by the states, the sole reference made in the Constitution to Indians was that contained in Section 1 of Article 8 in which it was merely provided that Congress should have the exclusive power "to regulate commerce with foreign nations and with the Indian tribes."

Such a provision not only did nothing to clarify the status of the Indians, but on the contrary, indicated clearly that they were not to be dealt with as part of the body politic to whom the guarantees of the Constitution applied. Plainly it was designed to permit the states to ratify the Constitution with such reservations as to their jurisdiction over the tribes within their boundaries as they might see fit.

What Kant said was only too true. In an age of broadened charity toward men, reacting alike against the narrow bigotry of Calvin and Loyola, the self-appointed champions of human freedom showed themselves powerless to liberate either the Indians or the Negroes from the tyranny of the modern democracy of man. The one was deliberately left subject to a political servitude that they might be exploited by the states, the other for purely economic reasons, in bondage, so that all the freedom the Constitution in fact guaranteed was reserved to the white man. With what prevision had spoken Voltaire, the old soothsayer whom intolerance branded the arch atheist of his age, when he said that in its application to

---

* The Bureau of Ethnology has adopted this view. See article by Hewitt, 18th Report, Smithsonian Institution, entitled: "A Constitutional Confederacy in the Stone Age" (1920).

the red race Jefferson's equalitarianism would amount to no more than the dogma of glittering rhetoric and borrowed phrases!

But at this there should be no surprise if we look at those who united to form this modern democracy instead of at their mere professions. They were still largely pioneers whose character had been formed in the melting pot of the American wilderness, a training school which at every turn had emphasized expediency at the expense of acute moral perceptions.

Entering the sea-worn edges of an illimitable forest, Englishmen, Scots, Irishmen, Dutchmen, Frenchmen, Germans, Scandinavians, all alike had come at once under an influence which in Europe had expended itself centuries before—the spell of that untamed nature which created primitive man. All the dim memories that lay deep in subconsciousness, all the vague shadows hovering at the back of the European mind, all the sense of encompassing natural power, and the need to struggle single-handed against it, the dangers lurking in the darkness of the forest, the brilliant treachery of the sunshine glinted through leafy secrecies, the strange voices in the illimitable murmur of the forest, the ghostly shimmer of its glades at night, the lovely beauty of the great gold moon; all the thousand wondering dreams that evolved of the elder gods—Pan, Cybele, Thor—all this waked again in the souls of the people who had come to make for themselves a home in the New World. And the influence of the forest itself was only intensified by the way they came into it—singly, or with but a wife and a child or two, or at best in very small company.

Nor were the surrounding presences limited to the spiritual world. A strange people, the children of the forest, soon as well armed as the pioneers themselves, quicker of foot and eye, more perfectly noiseless in their tread even than the wild beasts of the shadowy coverts, a people threatened from the first with expulsion from their God-given homeland— such invisible presences were watching them in a fierce silence ready at any moment to give over to the blood-curdling whoop that spoke of sudden death or worse. Thus, on the edge of a forest that was ever the same though always receding, there was a deadly need which made its invaders both more and less individual than they had been, releasing them in their daily life from the dictation of their fellows and at the same time enforcing upon them relentlessly a uniform mode of existence. As the unseen world became more and more familiar in its realities, the under-standing of it faded. Spirituality became chiefly a matter of emotional perception; scarcely at all a matter of philosophy. The morals of the pioneers became those of audacious, visionary beings loosely bound to-gether by the comradeship arising out of a common peril. Courage, cau-tion, swiftness toward an end, endurance, persistence, secrecy—these be-came the human virtues to be encouraged and which were accentuated more and more in each new generation. Dreaming of the welfare of humanity as a whole, like intellectual companionship and the softer, gentler, sweeter things of life, was a luxury to be enjoyed after, not before the New World environment had been made to yield a home that was safe.

Moreover, all sorts and conditions of men had been ensnared by the American wilderness, some of the worst as well as the best Europe had to offer, going into it to lose themselves, but all issuing in that posterity whose common purpose was expressed by the Constitution.

Manifestly, such a people could not have dealt for nearly two centuries with a subordinate race figuratively with a foot in the scale pan of justice, without an enduring effect upon their moral point of view. Once the idea had become fixed in their minds by the New World environment in the formative period of their character that necessity knows no law save that of the stronger, it was inevitable that the rights of a weaker race with much to yield should have been ignored by them in their attempt to establish guarantees of freedom for themselves.

Although the popular attitude in 1789 precluded justice to the Indians, there were those who knew only too well from past experience that the administration of Indian affairs, to be effective, must be centralized. This was absolutely necessary both to prevent the threatened alliance of the Six Nations with the English, and to enable the Iroquois and Cherokee domains to be maintained as buffer states. Nor was there any other way to insure the fulfillment of the guarantees of Indian rights contained in the Treaty of Paris, the disregard of which, as shown, was but affording the British imperialists an added pretext for a war looking to the recovery of the lost colonies.

At any rate, whether the Constitution was purposely so framed or not, which cannot be determined from the debates of the Convention, it contained provisions enabling the friends of federal control over the Indians to arrogate to the federal government powers which the framers had not dared expressly to confer upon it.

Section 10 of Article I expressly inhibited the states from entering into treaties, while Section 2 of Article II conferred upon the President the exclusive power, by and with the advice and consent of the Senate, to make treaties. Therefore, even though Section I of Article VIII did not confer upon Congress as the Articles of Confederation had done, the exclusive right and power to manage Indian affairs, inasmuch as the general government had had that power, and the tribes all along had been dealt with by treaties, the constitutional provisions mentioned taken together were construed by the federalists as conferring by necessary implication upon the federal government exclusive authority over the Indian tribes including those within the boundaries of any state. In other words commerce with the Indian tribes was interpreted by construction to include all relations with them.

The American people were dealing with a problem the legal aspects of which by reason of their novelty were not understood by them, and which was enormously complicated by the moral opaqueness of the times. Washington did not hesitate, immediately upon assuming office, to publish a proclamation declaring it was the constitutional function of the federal government as well as the duty of the nation to protect the tribes, and after ordering intruders off the Cherokee domains, characterized those who were seeking by violence to deprive the Indians of their rights as murderers and robbers.

At once loud protests arose against what was properly styled "the arbitrary and tyrannical assumption of power" by the federal government. All the warnings of the Anti-Federalists were recalled. "This is but the beginning of the end of popular government," it was said, "since Washington is doing exactly what the King attempted to do in 1763—take the Indian tribes and their lands out of the jurisdiction of the States." Nevertheless, the President was strong enough to compel a federalist Congress to adopt his interpretation, so that legislation was enacted in 1789 during its first session committing the conduct of Indian Affairs to the War Department. Since Indian treaties of allegiance to the new sovereignty were necessary, in the first appropriation act was included an item of $20,000 to defray the expense of negotiating treaties with the Indians, this being but a repetition of the appropriation made in 1787.

With such a one as Washington at its head the Indians generally felt that at last there was a government capable of protecting them and were disposed to cooperate in every way possible with the President. Accordingly on June 2, 1789, the Six Nations addressed to him a petition that is an epic. Pointing out that council fires had been kindled at too many different places to do business with them, with more intelligence than had been shown by the whites they asked that "one great council fire be kept burning." They related how by virtue of a so-called treaty negotiated by the state of New York in 1789 with a handful of Cayuga boys and girls headed by an Indian called Steel Trap, who was not a Cayuga, nor even a chief, but a mere warrior and the grandson of an Onondaga, the state was claiming lands set apart to the Six Nations by the United States in the treaty of Fort Stanwix.

Without regard to whom the fee in the Iroquois domain had passed, whether to the United States or New York, and whose was the right or preemption in the domain of the Six Nations, Washington knew that the organic law forbade the state of New York from entering into a treaty with the Indian tribes; that in 1784 the United States had in fact entered into the treaty of Fort Stanwix and guaranteed to the Six Nations the quiet enjoyment of their lands forever. Under these circumstances, even if New York owned the ultimate title, the dealings of which the Iroquois complained were flagrantly violative of the guarantees contained in the Treaty of Paris of 1783 and the treaty of Fort Stanwix of 1784, and were not to be tolerated.

Meantime, in September, 1789, St. Clair had transmitted to the Secretary of War the three treaties which at last he had negotiated with the Six Nations, the Wyandots, and a group composed of the Delawares, Ottawas, Chippewas, Potawatomis and Sacs, to take the place of those previously negotiated under the Articles of Confederation. These Washington referred to the Senate with the definite recommendation that the policy be adopted of requiring the ratification by the Senate of all Indian treaties just as in the case of other treaties, this in order to insure a full discussion of them on their merits. Of these treaties the tribes were already complaining, and publicity for them being the last thing the Senate wanted, despite the repetition by the President in a special message of May 25, 1789, of his recommendation, the Senate Committee on Indian Affairs resolved that since the treaties of Fort Harmor of January 9, 1789 were but confirmatory of those of Fort Stanwix and Fort McIntosh of 1784 and 1785, respectively, they should be executed and put into effect by the President without further discussion. Unwilling to be dictated to in this way and to do the bidding of the Senate, Washington simply refused to sign the treaties.

Encouraged by the attitude of New York, Georgia was determined to oust the Cherokees and Creeks. Therefore, despite Oglethorpe's statute forbidding the alienation of Indian lands, in August, 1790 there were no less than five hundred white families illegally settled upon the lands of the Cherokees. Cognizant of the purpose of both New York and Georgia, Washington had caused to be introduced in the first Congress two bills designed to protect the Indians, which were approved July 22, 1790. The first provided for the negotiation of treaties and the appointment of commissioners to manage Indian affairs. The second, known as the Indian Intercourse Act, was but a statutory adoption of Penn's policy, in all respects similar to the law which Penn and Oglethorpe had put into effect, and which had been embodied in the New York Constitution of 1777, providing in its terms that no sale of lands made by any Indians, or any nation or tribe of Indians, to any person or persons, or to any state, whether having the right of preemption to such lands or not, should be valid unless the same were made and duly executed at some public treaty, held under the authority of the United States. Here again Tamenend and Temo-chi-chi spoke!

Impeded by this statute, Georgia was now to resort to what was perhaps the most outrageous abuse of an Indian treaty that has ever occurred in the history of the United States, which appears from a fragmentary record long hidden but that may now be pieced together.

Long the state had wanted the Creek lands. Moreover, during the past century the English had built up with the aid of the Seminoles of southern Georgia and Florida a flourishing trade with the back country based on St. Marks, Pensacola and St. Augustine which it was now desired to divert to Savannah. Obviously, it would be highly beneficial if the Creeks could be induced to prevent the Seminoles from trafficking for the British merchants. The way to accomplish this was plain.

Gradually, as the avowed champion of democracy, the Saint Tammany Society of New York had become involved in politics, being the opponent of the alleged aristocratic tendencies of the Society of the Cincinnati. Organized after the manner of Tamenend's Delaware Confederacy in thirteen tribes, one for each state, each with a sachem subordinate to the Grand Sachem whose wigwam was Tammany Hall at the seat of Congress in New York, it was admirably adapted to use as an agency through which to control Congress. This the Buffalo Party in Georgia did not fail to see. Therefore, as soon as the Georgia politicians had insured the support of the state delegation in Congress, Tammany Hall was committed to the plan of bringing the Creek chieftains to New York under the auspices of Tammany for the purpose of cajoling them by shrewd entertainment into a treaty. Straightway, Colonel Marius Willett, a war hero and leader in Tammany, was despatched to the Creek country to deliver the invitation.

The head chief of the Creeks, Colonel Alexander McGillvray, a half-breed whose mother was Scottish, was a man of very great ability. With him a "friendly understanding" seems to have been quickly established by Willett. At any rate, in July, 1790, at the head of twenty-eight Creeks, McGillvray was on his way to New York accompanied by Willett.

Meantime, vast preparations had been made by the Tammany braves to entertain their visitors. "They brushed their Indian costumes, painted their faces, and pitched tents on the banks of the Hudson." The Creeks, it was said, "were wonderfully surprised and overjoyed, thinking they had found a new tribe of red men, giving vent to their excitement in loud whoops, which greatly startled, if they did not frighten, the Tammany braves. The 'Et-hah' song was sung, and then the Grand Sachem of Tammany, William Pitt Smith, assured the Indians in a speech that, though dead, the spirits of the two great chiefs, Tammany and Columbus, were walking back and forward in the wigwam."*

All this, of course, was designed to overawe the Creeks with the power of the federal government. They were taken to the theater, on excursions along the Hudson, and finally given a feast at which Washington himself appeared as host. All the while their minds were being prepared by the

* "History of the Tammany Society," Euphemia Vale Blake, quoted in *Tammany Hall*, Werner, Greenwood Publishers, 1932, p. 14.

subtle persuasions of the politicians to accede to the proposals of the government.

The trade between them and the British, they were told, was apt to be barred by the Spaniards at any time, leaving them with no market for their furs and no source of supply. The great and friendly government of the United States would not only guarantee them a supply of furs in the event of a war with Spain, but would give them money now, presently. All that was desired of them was an express promise not to deal with the British traders and the cession of a certain tract of land to Georgia! But despite all the pressure that could be put upon their "untutored minds" the delegation absolutely refused to relinquish all the land desired by the Georgians. The best they would do was to cede a part of it and this having been agreed upon, on August 10, 1790, they signed the treaty placed before them in which was inserted a secret clause obviously designed to encourage the Creeks to break up the trade conducted by the British through the Seminoles by interfering with the latter.

Well schooled in what was expected of them, the Creek chieftains returned to Florida and soon, as contemplated, the Seminoles were being provoked into those acts which were soon to furnish the politicians an excuse for the seizure of Florida.

"Tammany's Indian ritual served one very useful purpose, and was of assistance to President Washington and his government," in connection with the Creek treaty of 1790, says Werner, the historian.

There is much that is pitiful about the episode described. Especially sad is the reflection that a patriotic society designed to emulate the noble qualities of "Saint Tammany," a red man, should thus have been employed in its first venture into national politics to debauch his race. Exactly what, in fact, was Washington's part in this unconscionable transaction, to what extent he was informed of the purposes of the politicians, is obscure. Knowing his integrity with respect to the Indians generally, and his bold opposition to Georgia on other occasions, it is difficult to believe he was consciously a party to the secret diplomacy of which not alone Spain and Great Britain but the Creeks were the victims.

15

The Northwest Territory and the Indian Charter

Upon the creation of the public domain the public lands of the United States were disposed of by the sale to anyone who desired to purchase them, and in addition to such sales about 10,000,000 acres were granted in the Northwest and Southwest Territories during the next several decades to veterans of the Revolutionary War and the War of 1812, and to eminent persons like Lafayette who had rendered conspicuous public service to the American people. Land script was also employed to some extent to meet old military payrolls. While the land laws clearly distinguished between Indian country, or country to which the Indian claim attached, and the public domain, providing that Indian country formed no part of the latter and could not be made a part of it until the Indian right of occupancy was extinguished, it was useless for the Indians to protest against the irresistible demands that were being made upon them. The old trading posts in a twinkling became flourishing towns. More and more land was needed for them and for connecting rights of way through the tribal domains. It took time for the government to extinguish Indian titles by treaties. The settlers could not wait. Squatters soon claimed rights by adverse possession; dishonest Indian agents connived; illegal patents were issued; corruption developed in the disposition of the public lands; an increasing number of border ruffians made their appearance at the settlements; gamblers, bootleggers, sharpers of all kinds, cutthroats, all degrees and kinds of men continued to arrive. With the white man and whiskey came diseases of the vilest kind hitherto unknown to the Indians, followed by epidemics of fever which decimated the race. In

1781-82 smallpox had swept across the entire north country from Lake Superior to the Pacific, and a few years later another epidemic occurred, spreading with deadly effect from the Rio Grande to Dakota, adding greatly to the spirit of unrest among the Indians.

War with Great Britain now seemed almost sure to occur, so that Brant was urging the western tribes not to enter into any more separate treaties. He had spared no effort during the last two years to arouse both them and the Six Nations to resistance, with the result that in 1790 the Secretary of War reported upwards of fifteen hundred persons had been killed and much property destroyed during the past two years. In this situation, the government's persistent refusal to deal with the United Indian Nations, coupled with St. Clair's continuing efforts to force the western tribes into separate treaties extinguishing their remaining claims, could have but the one result which Washington had sought in vain to avoid by the rejection of the treaties of 1789. Finally, the Shawnees, Ottawas, Potawatomis, Delawares, Wyandots, Miamis, Mingos, and Chippewas, acting as the United Indian Nations, took to the warpath on September 19, 1790.

Reluctant as he may have been, there was nothing left Washington now but to direct St. Clair to dispatch a force against the tribes. Immediately, Colonel Josiah Harmar, a veteran of repute, with a force of fifteen hundred men, marched on the principal Miami village near the present Fort Wayne where the Miamis had assembled, under chief Mishickinikwa, or Little Turtle.

Setting fire to all his villages, Little Turtle, displaying great skill as a battle leader, met and defeated a large detachment under Major Fontaine on October 17th. Thereupon Harmar, seeing that his raw levies would not stand, ordered a general retreat during which many men were lost. Over this unexpected result the Indians were exultant. The effect of their victory upon the Six Nations at once became apparent.

It is not too much to say the fate of the infant republic now hung in the balance. The powerful Iroquois League was on the very verge of joining the western Confederacy. War with England was still threatening. Yet Washington did not dare encourage the Indians by showing a lack of decision. Instantly, therefore, he directed St. Clair to assemble and assume command of a new force of two thousand men and march upon the victorious Indians.

After assembling his troops and erecting Fort Hamilton and Fort Jefferson on the Miami River, St. Clair set out from his base at Cincinnati in October, 1790, pitching camp November 3rd following, on the Wabash a few miles west of Greenville. Meantime the Shawnees, Delawares, Wyandots and others had joined the Miamis, the now famous Little Turtle having been selected as general of the entire force which numbered about a thousand warriors. As St. Clair failed to advance to the attack of their well chosen position, on the morning of November 4th Little Turtle's forces made a furious assault upon his camp. Again the militia broke while even the better troops were hard put to it to defend the camp.

Blue Jacket, the Shawnee, and Buckangahelas, the Delaware chief, had conducted their parts of the attack with great skill, but now Blackfish, in direct command of the Miamis, came to the fore. "For me it is victory or death," he cried, and again his warriors rushed to the assault breaking down all resistance and completely routing St. Clair's entire command, he himself barely escaping. Fort Jefferson fell immediately and was plundered while thirty-eight officers and over six hundred men were killed by the Indians whose own losses were insignificant.

Something had to be done at once. Therefore, Washington decided to deal direct with the Senecas whose connections with the western tribes were still, as in the past, the most intimate, and endeavor to make them exert their influence upon the Confederacy to abandon the warpath while a force competent to deal with the Indians was being organized against their refusal to do so, since plainly it was useless to despatch any more militia against Little Turtle. Accordingly, he sent an invitation to the great Seneca chief, Cornplanter, the friend of peace, to attend a council in Philadelphia accompanied by a delegation from his tribe. This was promptly accepted.

So far, in the main, others have been allowed to speak for the Indians except in those cases where they have been quoted in a fragmentary way. Of their dignity, their eloquence, their aims and desires, the reasonableness of the native mind, nothing could be more conclusive than the addresses delivered by the Senecas during the negotiations which now ensued between them and Washington, and which evoked from the "Great White Father" a declaration that may well be taken as the Charter of Indian rights.*

On December 1, 1790, Cornplanter presented the case of his people before the Council as follows:

Father: The voice of the Seneca nation speaks to you, the great councillor, in whose heart the wise men of all the Thirteen Fires have placed their wisdom. It may be very small in your ears, and we therefore entreat you to hearken with attention: for we are about to speak of things which are to us very great. When your army entered the country of the Six Nations, we called you the town destroyer; and to this day, when that name is heard, our women look behind them and turn pale, and our children cling close to the necks of their mothers. Our councillors and warriors are men, and cannot be afraid; but

* The pleas are here given in full as a sample of Indian oration and legal argument. Cornplanter was, of course, merely the mouthpiece of Brant and the Hodenosaunnee whom New York and Congress would not recognize.

their hearts are grieved with the fears of our women and children, and desire it may be buried so deep as to be heard no more.

When you gave us peace, we called you father, because you promised to secure us in the possession of our lands. Do this, and so long as the lands shall remain, that beloved name will live in the heart of every Seneca.

Father: We mean to open our hearts before you, and we earnestly desire that you will let us clearly understand what you resolve to do. When our chiefs returned from the treaty at fort Stanwix, and laid before our council what had been done there, our nation was surprised to hear how great a country you had compelled them to give up to you, without your paying to us any thing for it. Every one said that your hearts were yet swelled with resentment against us for what had happened during the war, but that one day you would reconsider it with more kindness. We asked each other, What have we done to deserve such severe chastisement?

Father: When you kindled your thirteen fires separately, the wise men that assembled at them told us that you were all brothers, the children of one great father, who regarded, also, the red people as his children. They called us brothers, and invited us to his protection; they told us that he resided beyond the great water, where the sun first rises; that he was a king whose power no people could resist, and that his goodness was bright as that sun. What they said went to our hearts; we accepted the invitation, and promised to obey him. What the Seneca nation promise, they faithfully perform; and when you refused obedience to that king, he commanded us to assist his beloved men in making you sober. In obeying him we did no more than yourselves had led us to promise. The men who claimed this promise told us that you were children, and had no guns; that when they had shaken you, you would submit. We hearkened to them, and were deceived, until your army approached our towns. We were deceived; but your people, in teaching us to confide in that king, had helped to deceive, and we now appeal to your heart—Is the blame all ours?

Father: When we saw that we were deceived, and heard the invitation which you gave us to draw near to the fire which you kindled, and talk with you concerning peace, we made haste towards it. You then told us that we were in your hand, and that, by closing it, you could crush us to nothing, and you demanded from us a great country as the price of that peace which you had offered us; as if our want of strength had destroyed our rights; our chiefs had felt your power, and were unable to contend against you, and they therefore gave up that country. What they agreed to, has bound our nation; but your anger against us must by this time, be cooled; and, although our strength has not increased, nor your power become less, we ask you to consider calmly, Were the terms dictated to us by your commissioners reasonable and just?

Father: Your commissioners, when they drew the line which separated the land then given up to you from that which you agreed should remain to be ours, did most solemnly promise, that we should be secured in the peaceable possession of the lands which we inhabited east and north of that line. Does this promise bind you?

Hear now, we beseech you, what has since happened concering that land. On the day in which we finished the treaty at fort Stanwix, commissioners from Pennsylvania told our chiefs that they had come there to purchase from us all the lands belonging to us, within the lines of their State, and they told us that their line would strike the river Susquehanna below Tioga branch. They then left us to consider of the bargain till the next day; on the next day we let them know that we were unwilling to sell all the lands within their State, and proposed to let them have a part of it, which we pointed out to them in their map. They told us that they must have a whole; that it was already ceded to them by the great king, at the time of making peace with you, and was *their own*; but they said that they would not take advantage of that, and were willing to pay us for it, after the manner of their ancestors. Our chiefs were unable to contend, at that time, and therefore they sold the lands up to the line, which was then shown to them as the line of that State. What the commissioners had said about the land having been ceded to them at the price, and they passed it by with very little notice; but, since that time, we have heard so much from others about the right to our lands, which the king gave when you made peace with them, that it is our earnest desire that you will tell us what it means.

Father: Our nation empowered John Livingston to let out part of our lands on rent, to be paid to us. He told us that he was sent by Congress to do this for us, and we fear he has deceived us in the writing he obtained from us.

For, since the time of our giving that power, a man of the name of Phelps has come among us, and claimed our whole country northward of the line of Pennyslvania, under purchase from that Livingston to whom, he said, he had paid twenty thousand dollars for it. He said, also, that he had bought, likewise, from the council of the Thirteen Fires, and paid them twenty thousand dollars more for the same.

And he said also, that it did not belong to us, for that the great king had ceded the whole of it when you made peace with him. Thus he claimed the the whole country north of Pennsylvania, and west of the lands belonging to the Cayugas. He demanded it; he insisted on his demand, and declared that he would have it all. It was impossible for us to grant him this, and we immediately refused it. After some days he proposed to run a line at a small distance eastward of our western boundary, which we also refused to agree to. He then threatened us with immediate war if we did not comply.

Upon this threat, our chiefs held a council, and they agreed that no event of war could be worse than to be driven, with their wives and children, from the only country which we had any right to, and, therefore, weak as our nation was, they determined to take the chance of war, rather than to submit to such unjust demands, which seemed to have no bounds. Street, the great trader to Niagara, was then with us, having come at the request of Phelps, and as he always professed to be our great friend, we consulted him upon this subject. He also told us that our lands had been ceded by the King, and that we *must* give them up.

Astonished at what we heard from every quarter, with hearts aching with compassion for our women and children, we were thus compelled to give up all our country north of the line of Pennsylvania and east of the Genesee River, up to the fork, and east of a south line drawn from that fork to the Pennsylvania line.

For this land, Phelps agreed to pay us ten thousand dollars in hand, and one thousand a year for ever.

He paid us two thousand and five hundred dollars in hand, part of the ten thousand, and he sent for us to come last spring, to receive our money; but instead of paying us the remainder of the ten thousand dollars, and the one thousand dollars due for the first year, he offered us no more than five hundred dollars, and insisted that he agreed with us for that sum, to be paid yearly. We debated with him for six days, during all which time he persisted in refusing to pay us our just demands, and he insisted that we should receive the five hundred dollars; and Street, from Niagara, also insisted on our receiving the money, as it was offered to us. The last reason he assigned for continuing to refuse paying us was, *that the King had ceded the lands to the Thirteen Fires,* and that he had bought them from you, and *paid you for them.*

We could bear this confusion no longer, and determined to press through every difficulty, and lift up our voice that you might hear us, and to claim that security in the possession of our lands which your commissioners so solemnly promised us. And we now entreat you to inquire into our complaints and redress our wrongs.

Father: Our writings were lodged in the hands of Street, of Niagara, as we supposed him to be our friend; but when we saw Phelps consulting with Street on every occasion, we doubted of his honesty towards us, and we have since heard, that he was to receive for his endeavors to deceive us, a piece of land ten miles in width, west of the Genesee River, and near forty miles in length, extending to Lake Ontario; and the lines of this tract have been run accordingly, although no part of it is within the bounds which limit his purchase. No doubt he meant to deceive us.

Father: You have said that we are in your hand, and that, by closing it, you could crush us to nothing. Are you determined to crush us? If you are, tell us so; that those of our nation who have become your children, and have determined to die so, may know what to do.

In this case, one chief has said he would ask you to put him out of pain. Another, who will not think of dying by the hand of his father or of his brother, has said he will retire to the Chateaugay, eat of the fatal root, and sleep with his fathers in peace.

Before you determine on a measure so unjust, look up to God, who make *us* as well as *you.* We hope he will not permit you to destroy the whole of our nation.

Father: Hear our case. Many nations inhabited this country, but they had no wisdom, and, therefore, they warred together. The Six Nations were powerful, and compelled them to peace; the lands, for a great extent, were given up to them; but the nations which were not destroyed, all continued on those lands, and claimed the protection of the Six Nations, as the brothers of their fathers. They were men, and when at peace, they had a right to live upon the earth. The French came among us, and built Niagara; they became our fathers, and took care of us. Sir William Johnston came and took that fort from the French; he became our father, and promised to take care of us, and did so, until you were too strong for his king. To him we gave four miles round Niagara as a place of trade. We have already said, how we came to join against you; we saw that we were wrong; we wished for peace; you demanded a great country to be given up to you; it was surrendered to you, as the price of peace, and we ought to have peace and possession of the little land which you then left us.

Father: When that great country was given up, there were few chiefs present, and they were compelled to give it up, and it is not the Six Nations only that reproach those chiefs with having given up that country. The Chippewas, and all the nations who lived on those lands westward, call to us, and ask us, Brothers of our fathers, where is the place you have reserved for us to lie down upon?

Father: You have compelled us to do that which has made us ashamed. We have nothing to answer to the children of the brothers of our fathers. When, last spring, they called upon us to go to war to secure them a bed to lie upon, the Senecas entreated them to be quiet, till we had spoken to you. But, on our way down, we heard that your army had gone toward the country which those nations inhabit, and if they meet together, the best blood on both sides will stain the ground.

Father: We will not conceal from you that the great God, and not men, has preserved the Cornplanter from the hands of his own nation. For they ask, continually, Where is the land which our children, and their children after them, are to lie down upon? You told us, say they, that the line drawn from Pennsylvania to lake Ontario, would mark it forever on the east, and the line running from Beaver creek to Pennsylvania, would mark it on the west, and we see that it is not so. For, first one, and then another, come and take it away, by order of that people which you tell us promised to secure it to us. He is silent, for he has nothing to answer.

When the sun goes down he opens his heart before God, and earlier than that sun appears again upon the hills, he gives thanks for his protection during the night; for he feels that among men, become desperate by their danger, it is God only that can preserve him. He loves peace, and all he had in store, he has given to those who have been robbed by your people, lest they should plunder the innocent to repay themselves. The whole season he has spent in his endeavors to preserve peace; and, at this moment, his wife and children are lying on the ground, and in want of food; his heart is in pain for them, but he perceives that the great God will try his firmness, in doing what is right.

Father: The game which the Great Spirit sent into our country for us to eat, is going from among us. We thought he intended that we should till the ground with the plough, as the white people do, and we talked to one another about it. But before we speak to you concerning this, we must know from you whether you mean to leave us and our children any land to till. Speak plainly to us concerning this great business.

All the lands we have been speaking of belonged to the Six Nations; no part of it ever belonged to the King of England, and he could not give it to you.

The land we live on, our fathers received from God, and they transmitted it to us, for our children, and we cannot part with it.

Father: We told you that we would open our hearts to you. Hear us once more.

At fort Stanwix we agreed to deliver up those of our people who should do you any wrong, that you might try them, and punish them according to your law. We delivered up two men accordingly, but instead of trying them according to law, the lowest of your people took them from your magistrate, and put them immediately to death. It is just to punish murder with death, but the Senecas will not deliver up their people to men who disregard the treaties of their own nation.

Father: Innocent men of our nation are killed one after another, and of our best families; but none of your people who have committed the murder have been punished.

We recollect that you did not promise to punish those who killed our people, and we now ask, was it intended that your people should kill the Senecas, and not only remain unpunished by you, but be protected by you against the revenge of the next of kin,

Father: These are to us very great things. We know that you are very strong, and we have heard that you are wise, and we wait to hear your answer to what we have said, that we may know that you are just.*

This address, signed by chiefs Cornplanter, Half-Town, and Great-Tree, was a powerful indictment of the American people and an appeal that could not be ignored. Furthermore, it disposes forever of the common historical error that the Indians at this time were actuated primarily by sentiments of disloyalty inspired by British agents. It should also convince the most prejudiced mind that the Indians in 1790, as in 1786, were capable of grasping the moral, equitable, legal, and political principles of that civilization with which they were in contact; that the demands of civilization did not transcend their understanding.

In his reply to the Indians Washington was to resort to no subterfuges, no chicanery whatever. Taking ample time to consider their complaint in all its aspects, on December 29, 1790, he answered them in a formal written address, signed by himself, Jefferson, the Secretary of State, and Knox, the Secretary of War.†

I, the President of the United States, by my own mouth, and by a written speech, signed with my own hand, and sealed with the seal of the United States, speak to the Seneca nation and desire their attention, and that they would keep this speech in remembrance of the friendship of the United States.

I have received your speech with satisfaction, as a proof of your confidence in the justice of the United States, and I have attentively examined the several objects which you have laid before me, whether delivered by your own chiefs

* Am. State Papers, *Indian Affairs*, Vol. I, p. 140.
† Am. State Papers, *Indian Affairs*, Vol. I, p. 142.

at Tioga Point, in the last month, to Colonel Pickering, or laid before me in the present month, by the Cornplanter, and the other Seneca chiefs now in this city.

In the first place, I observe to you and request it may sink deeply into your minds, that it is my desire, and the desire of the United States that all the miseries of the late war should be forgotten and buried forever. That, in future, the United States and the Six Nations should be truly brothers, promoting each other's prosperity by acts of mutual friendship and justice.

I am not uninformed, that the Six Nations have been led into difficulties, with respect to the sale of their lands since the peace. But I must inform you that these evils arose before the present Government of the United States was established, when the separate States, and individuals under their authority, undertook to treat with the Indian tribes respecting the sale of their lands. But the case is now entirely altered; the General Government only, has the power to treat with the Indian nations, and any treaty formed, and held without its authority, will not be binding.

Here, then, is the security for the remainder of your lands. No State, nor person, can purchase your lands, unless at some public treaty, held under the authority of the United States. The General Government will never consent to your being defrauded, but it will protect you in all your just rights.

Hear well, and let it be heard by every person in your nation, that the President of the United States declares, that the General Government considers itself bound to protect you in all the lands secured to you by the treaty of fort Stanwix, the 22nd of October, 1784, excepting such parts as you may since have fairly sold to persons properly authorized to purchase of you. You complain that John Livingston and Oliver Phelps, assisted by Mr. Street, of Niagara, have obtained your lands, and that they have not complied with their agreement. It appears, upon inquiry of the Governor of New York, that John Livingston was not legally authorized to treat with you, and that everything that he did with you has been declared null and void, so that you may rest easy on that account. But it does not appear, from any proofs yet in possession of Government, that Oliver Phelps has defrauded you.

If, however, you have any just cause of complaint against him, and can make satisfactory proof thereof, the federal courts will be open to you for redress, as to all other persons. But your great object seems to be, the security of your remaining lands; and I have, therefore, upon this point, meant to be sufficiently strong and clear that in future you cannot be defrauded of your lands; that you possess the right to sell, and the right of refusing to sell, your lands; that, therefore, the sale of your lands, in future will depend entirely upon yourselves. But that, when you may find it for your interest to sell any part of your lands, the United States must be present, by their agent, and will be your security that you shall not be defrauded in the bargain you make.

It will, however, be important, that, before you make any further sales of your lands, you should determine among yourselves who are the persons among you, that shall give such conveyances thereof as shall be binding upon your nation, and forever prevent all disputes relative to the validity of the sale.

That, besides the before mentioned security for your land, and you will perceive, by the law of Congress for regulating trade and intercourse with the Indian tribes, the fatherly care the United States intend to take of the Indians. For the particular meaning of this law, I refer you to the explanations given thereof by Colonel Timothy Pickering, at Tioga, which, with the law, are herewith delivered to you.

You have said in your speech that the game is going away from among you, and that you thought it the design of the Great Spirit that you should till the ground; but before you speak upon this subject, you want to know whether the Union mean to leave you any land to till. You know that all the lands secured to you by the treaty of fort Stanwix, excepting such parts as you may since have fairly sold, are yours, and that only your own acts can convey them away. Speak, therefore, your wishes on the subject of tilling the ground. The United States will be happy in affording you every assistance, in the only business which will add to your numbers and happiness. The murders that have been committed upon some of your people, by the bad white man, I sincerely lament and reprobate; and I earnestly hope, that the real murderers will be secured and punished as they deserve. This business has been sufficiently explained to you here, by the Governor of Pennsylvania, and by Colonel Pickering, on behalf of the United States, at Tioga. The Senecas may be assured that the rewards offered for apprehending the murderers, will be continued until they are secured for trial; and that, when they shall be apprehended, they will be tried and punished as if they had killed white men.

Having answered the most material parts of your speech, I shall inform you that some bad Indians, and the outcasts of several tribes, who reside at the Miami village, have long continued their murders and depredations upon the frontiers lying along the Ohio. That they have not only refused to listen to my voice, inviting them to peace, but that, upon receiving it, they renewed their incursions and murders, with greater violence than ever. I have, therefore, been obliged to strike those bad people, in order to make them sensible of their madness. I sincerely hope they will hearken to reason, and not require to be further chastised. The United States desire to be the friends of the Indians upon terms of justice and humanity; but they will not suffer the depredations of the bad Indians to go unpunished. My desire is that you would caution all the Senecas, and Six Nations, to prevent their rash young men from joining the Maumee Indians; for the United States cannot distinguish the tribes to which bad Indians belong, and every tribe must take care of their own people. The merits of the Cornplanter, and his friendship for the United States, are well known to me and shall not be forgotten; and, as a mark of the esteem of the United States, I have directed Secretary of War to make him a present of ...... dollars, either in money or goods, as the Cornplanter shall like best; and he may depend upon the future care and kindness of the United States; and I have also directed the Secretary of War to make suitable presents to the other chiefs in Philadelphia; and also, that some further tokens of friendship be forwarded to the other chiefs, now in their nation.

Remember my words, Senecas! Continue to be strong in your friendship for the United States, as the only rational ground of your future happiness, and you may rely upon their kindness and protection. An agent shall soon be appointed to reside in some place convenient to the Senecas and Six Nations. He will represent the United States. Apply to him on all occasions. If any man bring you evil reports of the intentions of the United States, mark that man as your enemy; for he will mean to deceive you, and lead you into trouble. The United States will be true and faithful to their engagements.

Given under my hand, and the seal of the United States, at Philadelphia, this twenty-ninth day of December, in the year of our Lord one thousand seven hundred and ninety, and in the fifteenth year of the sovereignty and independence of the United States.

In view of these two fundamental documents, it would seem never to have been necessary for the historian, much less the courts, to look for any subsequent evidence to determine what were the rights guaranteed to the Indians by the United States, and what is the legal as well as the moral obligation of the United States to the aborigines over whom it asserted its guardianship. No legal quibble can alter the meaning of the words *fraud* and *protection* as employed by Washington, or relieve the United States from the moral obligation which it now assumed through its spokesman to protect these wards of the nation in all the "just rights" which they had possessed under the dominion of the British Crown. Those rights included a right of occupancy in the soil as sacred as the fee itself, which carried with it the right under the Constitution to be fairly compensated for any land taken from them and access to the courts of the United States to enforce those rights.

Washington's statesmanship and honesty not only accomplished its purpose in dealing with the Indians, as such a course always has, but in spite of the dreadful wrongs they had sustained, at once appealed to their characteristic sense of loyalty as shown by their reply of January 10, 1791.*

Father: Your speech, written on the great paper, is to us like the first light of the morning to a sick man, whose pulse beats too strongly in his temples, and prevents him from sleep. He sees it, and rejoices, but he is not cured.

You say that you have spoken plainly on the great point. That you will protect us in the lands secured to us at fort Stanwix, and that we have the right to sell or to refuse to sell it. This is very good. But our nation complain that you compelled us at that treaty to give up too much of our lands. We confess that our nation is bound by what was there done: and, acknowledging your power, we have now appealed to yourselves against that treaty, as made while you were too angry at us, and, therefore, unreasonable and unjust. To this you have given us no answer.

Father: That treaty was not made with a single State, it was the Thirteen States. We never would have given all that land to one State. We know it was

* Am. State Papers, *Indian Affairs*, Vol. I, p. 143.

before you had the great authority, and as you have more wisdom than the commissioners, who forced us into that treaty, we expect that you have also more regard to justice, and will now, at our request, reconsider that treaty, and restore to us a part of that land.

Father: The land which lies between the line running south from lake Erie to the boundary of Pennsylvania, as mentioned at the treaty at fort Stanwix, and the eastern boundary of the land which you sold, and the Senecas confirmed to Pennsylvania, is the land on which Half-Town and all his people live with other chiefs, who always have been, and still are, dissatisfied with the treaty at fort Stanwix. They grew out of this land, and their fathers' father grew out of it, and they cannot be persuaded to part with it. We therefore entreat you to restore to us this little piece.

Father: Look at the land which we gave to you at that treaty, and then turn your eyes upon what we now ask you to restore to us, and you will see that what we ask you to return is a *very little piece*. By giving it back again, you will satisfy the whole of our nation. The chiefs who signed that treaty will be in safety, and peace between your children and our children will continue so long as your land shall join to ours. Every man of our nation will then turn his eyes away from all other lands which we then gave up to you, and forget that our fathers ever said that they belong to them.

Father: We see that you ought to have the path at the carrying place from lake Erie to Niagara, as it was marked down at fort Stanwix, and we are all willing it should remain to be yours. And if you desire to reserve a passage through the Conewango, and through the Chataugue lake and land, for a path from that lake to lake Erie, take it where you best like. Our nation will rejoice to see it an open path for you and your children while the land and water remain. But let us also pass along the same way, and continue to take the fish of those waters in common with you.

Father: You say that you will appoint an agent to take care of us. Let him come and take care of our trade; for the agents which have come amongst us, and pretended to take care of us, have always deceived us whenever we sold lands; both when the King of England and when the States have bargained with us. They have by this means occasioned many wars, and we are therefore unwilling to trust them again.

Father: When we return home, we will call a great council, and consider well how lands may be hereafter sold by our nation. And when we have agreed upon it, we will send you notice of it. But we desire that you will not depend on your agent for information concerning land; for, after the abuses which we have suffered by such men, we will not trust them with anything which relates to land.

Father: We will not hear lies concerning you, and we desire that you will not hear lies concerning us, and then we shall certainly live at peace with you.

Father: There are men who go from town to town and beget children, and leave them to perish, or, except better men take care of them, to grow up without instruction. Our nation has looked around for a father, but they found none that would own them for children, until you now tell us that your courts are open to us as to your own people. The joy which we feel at this great news, so mixes with the sorrows that are passed, that we cannot express our gladness, nor conceal the remembrance of our afflictions. We will speak of them at another time.

Father: We are ashamed that we have listened to the lies of Livingston, or been influenced by threats of war by Phelps, and would hide that whole transaction from the world, and from ourselves by quickly receiving what Phelps promised to give us for the lands they cheated us of. But as Phelps will not pay us even according to that fraudulent bargain, we will lay the whole proceedings before your court. When the evidence which we can produce is heard, we think it will appear that the whole bargain was founded on lies, which he placed one upon another; that the goods which he charges to us as part payment were plundered from us; that, if Phelps was not directly concerned in the theft, he knew of it at the time, and concealed it from us; and that the persons we confided in were bribed by him to deceive us in the bargain. And if these facts appear, that your court will not say that such bargains are just, but will set the whole aside.

Father: We apprehend that our evidence might be called for, as Phelps was here, and knew what we have said concerning him; and as Ebenezer Allen knew something of the matter we desired him to continue here. Nicholson, the interpreter, is very sick, and we request that Allen may remain a few days longer, as he speaks our language.

Father: The blood which was spilled near Pine Creek is covered, and we shall never look where it lies. We know that Pennsylvania will satisfy us for that which we spoke of to them before we spoke to you. The chain of friendship will now, we hope, be made strong as you desire it to be. We will hold it fast; and our end of it shall never rust in our hands.

Father: We told you what advice we gave the people you are now at war with, and we now tell you that they have promised to come again to our towns next spring. We shall not wait for their coming, but will set out very early and show to them what you have done *for us*, which must convince them that you will do for them everything which they ought to ask. We think they will hear and follow our advice.

Father: You give us leave to speak our minds concerning the tilling of the ground. We ask you to teach us to plough and to grind corn; to assist us in building saw mills, and to supply us with broad axes, saws, augers, and other tools, so as that we may make our houses more comfortable, and more durable; that you will send smiths among us, and, above all, that you will teach our children to read and write, and our women to spin and weave. The manner of your doing these things for us we leave to you who understand them; but we assure you that we will follow your advice as far as we are able.

Having obtained an assurance of their loyalty, and enlisted the Senecas in the cause of peace between the Ohio Indians and the United States, Washington had no intention of allowing the chieftains to depart from Philadelphia under the slightest misapprehension as to their obligations under the treaty of Fort Stanwix. On January 19, 1791, therefore, he addressed them again in the following very clear words:*

Brothers: I have maturely considered your second written speech.

You say your nation complain that, at the treaty of fort Stanwix, you were compelled to give up too much of your lands; that you confess your nation is bound by what was there done, and acknowledging the power of the United States; that you have now appealed to ourselves against that treaty, as made while we were angry against you, and that the said treaty was, therefore, unreasonable and unjust.

But, while you complain of the treaty of fort Stanwix, in 1784, you seem entirely to forget that you, yourselves, the Cornplanter, Half-Town, and Great-Tree, with others of your nation, confirmed, by the treaty of fort Hamar, upon the Muskingum, so late as the ninth of January, 1789, the boundary marked at the treaty of fort Stanwix, and that, in consideration thereof, you then received goods to a considerable amount.

Although it is my sincere desire, in looking forward, to endeavor to promote your happiness, by all just and humane arrangements, Yet I cannot disannul treaties formed by the United States, before my administration, especially, as the boundaries mentioned therein have been twice confirmed by yourselves. The lines fixed at fort Stanwix and fort Hamar, must, therefore, remain established. But Half-Town, and the others, who reside on the land you desire may be relinquished, have not been disturbed in their possession, and I should hope, while he and they continue to demean themselves peacefully, and to manifest their friendly dispositions to the people of the United States, that they will be suffered to remain where they are.

The agent who will be appointed by the United States, will be your friend and protector. He will not be suffered to defraud you, or to assist in defrauding you of your lands, or of any other thing, as all his proceedings must be reported in writing, so as to be submitted to the President of the United States.

You mention your design of going to the Miami Indians, to endeavor to persuade them to peace. By this humane measure you will render those mistaken people a great service, and, probably, prevent them from being swept from off the face of the earth. The United States require, only, that those people should demean themselves peacably; but they may be assured that the United States are able, and will, most certainly punish them severely for all their robberies and murders. You may, when you return from this city to your own country, mention to your nation my desire to promote their prosperity, by teaching them the use of domestic animals, and the manner that the white people plough, and raise so much corn. And if, upon consideration, it would be agreeable to the nation at large to learn these valuable arts, I will find some means of teaching them, at such places within your country as shall be agreed upon.

I have nothing more to add, but to refer you to my former speech, and to repeat my wishes for the happiness of the Seneca nation.

* Am. State Papers, *Indian Affairs,* Vol. I, p. 144.

Firmness on the President's part only appealed to the Senecas, whose underlying loyalty had been secured by him. Before departing to rejoin their people, again, on February 7, 1791, they addressed him in a message which bespoke the utmost nobility of mind.*

Father: No Seneca ever goes from the fire of his friend, until he has said to him, "I am going." We therefore tell you, that we are now setting out for our country.

Father: We thank you, from our hearts, that we now know there is a country we may call our own, and on which we may lie down in peace. We see that there will be peace between your children and our children; and our hearts are very glad. We will persuade the Wyandots, and other Western nations, to open their eyes, and look towards the bed which you have made for us, and to ask of you a bed for themselves, and their children, that will not slide from under them.

We thank you for your presents to us, and rely on your promise to instruct us in raising corn, as the white people do; the sooner you do this, the better for us. And we thank you for the care you have taken to prevent bad men from coming to trade among us; if any come without your license, we will turn them back; and we hope our nation will determine to spill all the rum which shall, hereafter, be brought to our towns.

Father: We are glad to hear that you determine to appoint an agent that will do us justice, in taking care that bad men do not come to trade among us; but we earnestly entreat you that you will let us have an interpreter in whom we can confide, to reside at Pittsburgh; to that place our people, and other nations, will long continue to resort; there we must send what news we hear, when we go among the Western nations, which we are determined, shall be early in the spring. We know Joseph Nicholson, and he speaks our language so that we clearly understand what you say to us, and we rely on what he says. If we were able to pay him for his services, we would do it; but, when we meant to pay him, by giving him land, it has not been confirmed to him; and he will not serve us any longer unless you will pay him. Let him stand between, to entreat you.

Father: You have not asked any security for peace on our part, but we have agreed to send nine Seneca boys, to be under your care for education. Tell us at what time you will receive them, and they shall be sent at the time you shall appoint. This will assure you that we are, indeed, at peace with you, and determined to continue so. If you can teach them to become wise and good men, we will take care that our nation shall be willing to receive instruction from them.

There is perhaps, nothing more touching in American history than the repeated appeals made by the Indians at this time for education, agricultural aid, and instruction in the ways of the white man, which Washington was determined to give them if possible.

So soon as the council was concluded he issued detailed instructions of

* Am. State Papers, *Indian Affairs*, Vol. I, p. 144.

the wisest character to Colonel Thomas Proctor and Captain Houdin whom he had selected to accompany Cornplanter and the other chieftains on their mission to the Ohio Indians. Impressing upon them its grave importance, he urged the utmost despatch. Leaving Washington March 12, 1791, the commissioners arrived at the principal Seneca village near Buffalo about the middle of April. In the opinion of Cornplanter it was necessary to assemble his people near Erie before he could go farther, and late in May it was finally agreed by them that a deputation should be sent with Proctor and Houdin to Sandusky to treat for peace with the Ohio tribes, and to try to induce them to meet General St. Clair at a council at Fort Washington

16

Fallen Timbers and the Jay-Grenville Treaty of 1794

WASHINGTON'S PROPOSALS FOR THE civilization of the Indians were honest, definite and clear. The defense of the western borders should be accomplished in the most humane way. Unnecessary coercion should cease. Firmness, impartiality, justice were the best means to win over and attach the tribes. Commerce with them should be promoted by suitable regulations. A system "corresponding with the mild principles of religion and philanthropy" governing dealings with them would do credit to the country. But above all "that mode of alienating their lands, the main source of discontent and war, should be so defined and regulated as to obviate imposition and controversy." Adequate penalties for defrauding them were required to prevent the continuance of treaty violations which were certain provocatives of war.*

"Trusty" persons should be assigned to live among and guide them.† Over and over he urged provisions for the prevention of fraud and violence,‡ insisting that the establishment of government trading houses among them was most desirable.§ But still his voice was to go unheeded. With his efforts to pacify the rebellious tribes the frontier was disgusted. It did not understand the danger of risking more defeats. The government was no longer dealing with ill-directed bands of marauding Indians. The confederated tribes, now well organized and led, possessed a morale born

---

* Third Annual Message, Oct. 25, 1791.
† Fourth Annual Message, Nov. 6, 1792.
‡ Fifth Annual Message, Dec. 31, 1793; Seventh Annual Message, Dec. 8, 1795.
§ Sixth Annual Message, Nov. 1, 1794.

of two victories over the whites which made them worthy adversaries for seasoned troops.

Nevertheless, in the spring of 1791 the Kentuckians decided to take things in their hands as in the old days and in March a large force of "hunting-shirt" men assembled under one James Fallon for the purpose of invading the Shawnee country just at the time Washington was trying to restore peace. In his opinion, even where a tribe permitted its members to commit the first act, the whites of the locality wherein the offense occurred could not assume to take the law into their hands. In the passion of the moment it was impossible for the coolest heads to pass sound judgment, much less decree the penalty. That was for the government and the government alone to do. Retaliatory campaigns such as he had conducted in 1779, could only be officially authorized and directed. The Law of Nations recognized no other course. Therefore, he published a proclamation ordering Fallon and his irregulars to disperse.

Yet, the harm had been done. Speaking of the Americans, Buckongahelas, the Delaware chieftain, said at this time:

"Since they cannot enslave us, they kill us. They are not like the Red Men, who are enemies in war but in peace are friends. They take the Indian by the hand and at the same time destroy him." Therefore, Little Turtle and his head chieftains received Cornplanter, heard what he had to say, but knowing the danger the rebellious tribes were in, naturally refused to lay down their arms. Moreover, the various councils held by Proctor showed plainly that although the Indians trusted Washington implicitly they reposed no faith whatever in the "Thirteen Fires."

"You say our lands are secured to us and that the grant given by the Great Chief, General Washington, will last as long as the sun goes over us . . . and you tell us that there is a great paper in the hands of O'Beel for us. Now we want you to show with your finger how large the lands are which are given us." So spoke Little Bear.

Disaffection was also spreading to the Cherokees, the treaty of Hopewell of 1785 having been ignored by the Georgians. It would never do for them to join the British again on the most vulnerable flank of the republic. Yet, to Washington it was obvious the Georgians meant to have the domain set apart to the tribe in 1785. Accordingly, he proposed that they exchange it for another lying in North Carolina, Tennessee, Alabama and Georgia, and to this they agreed in the treaty of Holsten of July 2, 1791 to which in 1792 a protocol was added delimiting their new boundaries and granting them an annuity. From this treaty it clearly appeared that the intent of the parties was that the new domain was to be divided up among them as individuals for the purpose of cultivation, in order that they might attain to a higher stage of civilization and not revert to the hunter state from which by their own efforts they had already emerged while resident

in Georgia, subsequent amendments recognizing that the Cherokees, unlike the Iroquois, were the owners of the fee in their domain.*

While Washington was endeavoring to do justice to the Cherokees,
Brant had been hard at work in conjunction with British agents and Red
Jacket preparing the Six Nations to cooperate with the western tribes.
Some of the Iroquois chieftains like Red Jacket had even been found in
British uniforms, while modern firearms and even cannon had been fur-
nished them. British officers, also in uniform, were actually enlisting them
for the impending war.

Learning this, Washington saw that he must now deal not alone with
the Senecas but with the Iroquois as a whole. Therefore, he issued instructions to Colonel Pickering to summon the Six Nations to a great council to
be held at Painted Post in June. Learning at this council that Brant and
Fishcarrier regarded the treatment of the Mohawks and Cayugas, still
fearful of returning from Canada, as nothing more than a reprisal by the
state of New York in violation of the treaty of Paris, Washington appealed
to the governor to take instant action to placate them; and seeing that
unless the United Indian Nations would abandon the warpath they must
be crushed, at the same time he took the western situation in his own
hands, appointing General Wayne, a veteran of the border as well as of
the Revolution, to command the force of United States troops he had been
raising.

Yet, knowing that the fault was not really that of the Indians, in Washington's heart there was not the slightest bitterness against them. Thus, on
April 23, 1792, the Secretary of War wrote General Israel Chapin who
had been appointed his deputy to the Six Nations, that it was the desire of
the President that the utmost fairness and kindness be shown the Indian
tribes of the United States, that peace be preserved with them whenever
possible, and that the government comport itself as their guardian and
protector against all injustice. Moreover, while "Blacksnake" was drilling
his troops near Pittsburgh, beyond the danger of their being attacked before
trained and disciplined, in May, 1792 he sent General Rufus Putnam to
negotiate with the Indians in the Northwest and to inform them that the
United States was highly desirous of imparting to them the blessings of
civilization, was willing to bear the expense of teaching them to read and
write, and the agricultural arts of the white man, soon securing an act
providing teachers for the Oneida and Stockbridge tribes which had remained loyal and fought against the British during the Revolution. And in
1792 he also sent Randolph and Lincoln as a commission to the United
Indian Nations to impress upon them the certain consequences of continued hostilities.

At this time, in the opinion of the Secretary of War, the foreign situa-

---

* Amended in 1795 and 1798 by the treaties of Philadelphia and Tellico.

tion was "very ticklish," so it was doubly important the Six Nations be placated. Therefore, pending negotiations with the western tribes, Washington caused to be enacted an even more drastic law than that of 1790 The Indian Intercourse Act of March 1, 1793, which authorized the issue of domestic animals to the Indians, provided:

That no purchase or grant of lands, or any title or claim thereto, from any Indians, or nation, or tribe of Indians, within the bounds of the United States shall be of any validity in law or equity, unless the same be made by a treaty or convention entered into pursuant to the Constitution; and it shall be a misdemeanor, in any person not employed under the authority of the United States, in negotiating such treaty or convention, punishable by a fine not exceeding one thousand dollars, and imprisonment not exceeding twelve months, directly or indirectly to treat with any such Indians, nation or tribe of Indians, for the title or purchase of any lands by them held, or claimed: Provided, nevertheless, that it shall be lawful for the agent or agents of any State, who may be present at any treaty held with Indians under the authority of the United States, in the presence, and with the approbation of the Commissioner or Commissioners of the United States appointed to hold same to propose to, and adjust with the Indians, the compensation to be made for their claims to lands within such State, which shall be extinguished by the treaty.

Thus it was made plain that the compensations of Indians even for the relinquishment of mere claims to land was contemplated by Congress when the same were extinguished by the United States which alone could deal with them by treaty, and upon the passage of this act, Pickering was sent to the Six Nations to explain its terms, while the following December, Moses DeWitt advised Governor Clinton to assent to it and write the Mohawks and Cayugas to hold a council looking to the sale of their lands claimed by them.

Having advanced from Pittsburgh to Cincinnati some months before, in the summer of 1793, Wayne moved forward to Greenville, reestablished Fort Jefferson, built Fort Recovery twenty miles beyond West Branch on the Miami, and there settled down again to continue the process of disciplining his troops and accustoming them to the presence of the enemy. Besides two thousand regulars—the first United States troops ever sent afield—his force included sixteen hundred Kentucky militia in which, like all militia, neither he nor Washington had any confidence.

Such was the situation when on February 10, 1794, Lord Dorchester, Governor General of Canada, in a speech to the Seven Nations of Canada over which Brant presided, stated that "from the manner in which the people of the United States push on and act and talk, I shall not be surprised if we were at war with them in the course of the present year," whereupon the Secretary of State of the United States wrote Lord Dorchester that he considered his words "hostility itself." Yet, Washington's efforts to bring the Indians to terms by negotiations through Randolph and Lincoln had come to naught. The border railed and demanded action. Heeding not the outcry, in the summer of 1794 he caused Wayne to advance leisurely to the Maumee and build Fort Defiance at its junction with the Auglaize. In that position under instructions from Washington he opened final negotiations with the Indians looking to peace, though the only terms he was permitted to put forward upon the authority of Congress was the payment of $10,000 and goods of an equal value each year forever for the lands desired by the United States. "This they might take, and peace," said Wayne, or he would wrest from them what he could, by force of arms.

Little Turtle saw that both the Great White Father and "Blacksnake" were in earnest. He understood the ways of such men and respected them. Therefore, he now harangued the Indians at a big council, saying:

Twice we have beaten the enemy under separate commands. We cannot expect the same good fortune always to attend us. The Americans are now led by a chief who never sleeps. The night and day are alike to him. During all the time that he has been marching upon our villages, we have never been able to surprise him. Think well of it. Something whispers to me that it would be prudent to listen to his offers of peace.

Emboldened by past success Bluejacket and others taunted Little Turtle with cowardice. Finally he was goaded into joining them, with the result that on August 20, 1794, over the three hundred remaining warriors Wayne won a complete victory at Fallen Timbers almost under the walls of the British fort at the falls of the Maumee.

The Indian Intercourse Act of 1793, the opening of negotiations with the Cayugas by Clinton, the victory of Wayne, each exerted great influence upon the Iroquois. When the Jay-Grenville Treaty was concluded in November, 1794, there was no longer need for the British agents to agitate them. A new treaty with the Six Nations was quickly concluded at Canandaigua by which New York paid for the Cayuga lands.*

The Iroquois thoroughly understood the new act, and were in no way unreasonable in their expectations. When Governor Clinton was discussing with the Cayugas and Onondagas from Buffalo Creek and the Grand River at Albany on March 17, 1794 the sale of their reservations within the state of New York, the Little Cayuga chief who was present as one of the Grand River delegation, addressing the governor on behalf of the representatives of the assembled nations, said:

We wish the superintendent appointed by the United States, General Chapin, to be present and see justice done us and in our negotiations, as we look on

* November 11, 1794.

him as our father. We do not expect that he will confine his care to us wholly, but that he should be a mediator between both parties.

A more accurate, a finer conception of the principle of guardianship applying to the Indians, of the rights of the federal and state governments in respect to them, of their own rights in relation to both, and of the sense of equity and fair dealing incumbent upon all parties to a treaty, could not be imagined than was here expressed in the words of the Indians themselves. This expression fully embodied their view of the Constitution. Again, in their political morality, if not in their political conceptions, these early Indians appear to have been on a parity at least with those with whom they were dealing.

A treaty with the Wyandots, Delawares, Shawnees, Ottawas, Chippewas, Potawatomies, Miamis, Eel Rivers, Weas, Kickapoos, Piankishaws and Kaskaskias was soon concluded in which was ceded to the United States a tract comprising the eastern and southern portions of Ohio, embracing nearly two-thirds of that state, in addition to a large tract in Indiana.*

Now it only remained to satisfy the claims of the so-called Seven Nations of Canada. Accordingly, on May 31, 1796 the United States entered into a treaty with them through its commissioner Abraham Ogden, which was duly attested by three agents of the state of New York, whereby the state obligated itself as provided in the Statute of 1795 to compensate the Seven Nations of Canada for the lands in New York claimed by them in consideration of a full release of their claims, it being provided, however, that the land on the de Grasse River near the present site of the town of Massena, which had long been occupied by the St. Regis community, an early offshoot from the Mohawk tribe, should be reserved to their use forever, the St. Regis lands thus coming under the exclusive jurisdiction of the United States as an Indian reservation just as the domain reserved by the Six Nations in 1784.

Undoubtedly, this had much to do with the sudden rise of the democracy which was to bear Jefferson into office in 1799. It also explains in large measure the hostility to Washington of the violent young Democrat, Andrew Jackson of Tennessee, who at the time he refused with a few others in the Senate to endorse Washington's administration was the leader of the Buffalo Party. Behind his professed disapproval of the Jay Treaty was both the desire for vengeance against the British whom he held responsible for the resistance of the Indians, and his resentment of Washington's humane Indian policy.

Washington himself was not in the least deceived as to the adverse political effect of the latter. Nor was he shaken in his resolve to enforce it.

---

* Treaty of Greenville, Aug. 3, 1795.

On one occasion when the borderers addressed an urgent appeal to him that United States troops be sent to revenge an alleged local Indian massacre, investigation having showed the Indians had been goaded into resistance in order that they themselves might not be slaughtered and their lands appropriated, Washington declared the retaliation of the Indians lamentable but only natural. Thus, with utter fearlessness and disregard of the political consequences he called these things what they were in fact—murder.

In his Fourth Annual Message he said:

I cannot dismiss the subject of Indian affairs without again recommending to your consideration the expediency of more adequate provision for giving energy to the laws throughout our interior frontier and for restraining the commission of outrages upon the Indians, without which all pacific plans must prove nugatory. To enable, by competent rewards, the employment of qualified and trusty persons to reside among them as agents would also contribute to the preservation of peace and good neighborhood. If in addition to these expedients an eligible plan could be devised for promoting civilization among the friendly tribes and for carrying on trade with them upon a scale equal to their wants and under regulations calculated to protect them from imposition and extortion, its influence in cementing their interest with ours could not but be considerable.

The great statesman underwent no change of feeling toward the Indians while in office, for in his Seventh Annual Message he said:

While we indulge the satisfaction which the actual condition of our Western borders, so well authorizes, it is necessary that we should not lose sight of an important truth which continually receives new confirmations, namely, that the provisions heretofore made with a view to the protection of the Indians from the violences of the lawless part of our frontier inhabitants are insufficient. It is demonstrated that these violences can now be perpetrated with impunity, and it can need no argument to prove that unless, the murdering of Indians can be restrained by bringing the murderers to condign punishment, all the exertions of the Government to prevent destructive retaliations by the Indians will prove fruitless and all our present agreeable prospects illusory. The frequent destruction of innocent women and children, who are chiefly the victims of retaliation, must continue to shock humanity. . . .

To enforce upon the Indians the observance of justice, it is indispensable that there shall be competent means of rendering justice to them. If these means can be devised by the wisdom of Congress, and especially if there can be added an adequate provision for supplying the necessities of the Indians on reasonable terms (a measure the mention of which I the more readily repeat, as in all the conferences with them they urge it with solicitude), I should not hesitate to render a strong hope of rendering our tranquility permanent. I add with pleasure that the probability even of their civilization is not diminished by the experiments which have been thus far made under the auspices of the government. The accomplishment of this work if practicable, will reflect un-decaying luster on our national character and administer the most grateful consolations that virtuous minds can know.

Washington had created a wonderful opportunity for Congress. By justice alone the Six Nations had been constrained to peace, while the Indians generally, with the utmost firmness, had been compelled to accept it. The first organized resistance of the Indians against the United States had been put down by him in a way that had only enhanced their respect for the infant republic. Not only that, but the undying friendship for the government of the greatest Indian of his day—Little Turtle—had been won. Never again was he to bear arms against the whites.

"I was the first to sign, I will be the last to break the treaty," he said. Satisfied that further resistance was useless, the old seer was laboring peacefully for the welfare of his people in the councils of the whites.

The more one studies Washington the more remarkable seems the moral courage and wisdom which he brought to bear upon the complex problems that confronted him. At times he seems to have been inspired. That the things he proposed were possible is plain from the fact that later they were all done. Before the people had been misled by years of unconscionable neglect and encouragement into believing they had a right to despoil the Indians of their lands, it would have been easier for Congress to protect them than it was later. What might have been accomplished through the agency of such a one as Little Turtle is not difficult to imagine. One only need compare Washington's lofty statesmanship and humanitarianism with that of his contemporaries to see how truly great a man he was. That which instinctively the Indians perceived is confirmed by the closest scrutiny of his character. No wonder they called him Karondowanem, "the great tree," or an oak among men, as well as Hornandanigius, "the Conqueror." Even the bitterest Indian enemy of the republic—Red Jacket—wore with pride the peace medal which Washington gave him as the nation's pledge of good faith. This, like the undying admiration and trust of Little Turtle, was but evidence of Indian amenability to justice.

17

Jefferson, Louisiana and Tecumseh

THE INDIAN POLICY OF Washington's successor—John Adams—was worse than negligible. Although without personal experience of the border, the shrewd New Englander was highly versed in politics. He did not fail to sense the tremendous tide that had set in against the Indians—a tide which a dying Federalist dared not breast. Cold of heart and hot of prejudice, he stood supinely silent as the inundatory flood of civilization swept westward. The assurances given the Indians by Washington were ignored. During his administration Kentucky was to exemplify its name— "The Bloody Ground." Tennessee, which in 1796 had succeeded to the state of Franklin that had been created in 1784 out of the territory ceded to the union by North Carolina, was but another Kentucky. In both, and throughout the Southwest Territory, the history of the Northwest was duplicated during the four years following the retirement of Washington.

When Jefferson became President in 1801 the wail of the Indians was rising to the high heavens; the soil of the young republic was being smeared with their blood. Yet no one possessed a better understanding of the nature of the Indians and the human problem they presented than he had long since disclosed in his shrewd *Notes on Virginia*, while all his life, like Albert Gallatin, he was constantly recording his observations on the red race. As Secretary of State he had engaged in a ceaseless contest of wits with Brant. Furthermore, during the last eighteen years the venerable Miami chieftain, Little Turtle, the intimate friend of the French philosopher Volney and the patriot Kosciusko, had been visiting the "fourteen fires," winning the admiration of the country by his unremitting efforts to better the lot of his race. Obedient to his pleas, both the federal government and the legislature of Kentucky had passed the most stringent legislation against the sale of spirits to the Indians. The refusal of Ohio to enact Indian liquor prohibition laws had been a public scandal. Moreover, the very year he entered the White House intense interest in the Indians was aroused throughout the world by the brilliantly romantic pen of Chateaubriand.

Exiled from his native land by the outcome of events in 1789, this fanciful political adventurer had arrived in America in 1791 ostensibly, like Columbus, in search of the Northwest Passage! Preparatory to that quest, during the next year he penetrated deeply the American wilderness, passing from the Great Lakes down the Mississippi to Florida. Becoming deeply interested in the Indians, at one time he contemplated seriously studying the languages of the Iroquois and the Sioux, in order, he declares, to obtain "some notions of the Equimault" and to facilitate his ultimate journey. At any rate, being no ordinary traveller he acquired during his visitations to the Indians—"ces messieurs sauvages"—a knowledge of them which placed him in position to present the aborigines of America to his countrymen, and to dogmatize upon the customs and policy of the red man with all the authority of a La Salle or a Montcalm. After returning finally to France, he published in 1801 his masterpiece—*Atala*—a romantic novel portraying with affecting art the moral side of the Indian question. Ready for a Christian and idealistic reaction, France was profoundly moved by a work which anticipated Lamartine and Hugo. It was to France what Goethe's *Werther* was to Germany, at once giving to Chateaubriand the leadership in French letters. *Atala* was to be followed by *Les Natchez* and other so-called "Leatherstocking" writings. Two decades later Cooper and Simms idealized in immortal phrase the life of the American Indians.

Yet, during Jefferson's first administration not one word fell from his lips regarding the Indians for the obvious reason that, having been elevated to power by the Buffalo Party, he did not dare respond to any mandate his conscience may have imposed upon him. The good of the country, he no doubt argued, as is the way of presidents, demanded that he recognize the will of the people. He was apparently as opaque to the morals of *Atala* as if that appeal to the conscience of the American people had not been made. His policy was simply and solely one of concentrating the Indian tribes in order to make available their surplus lands. Thus, after the creation in 1801 of the so-called Indian Territory, consisting of the region of Ohio, Indiana, Illinois and Michigan, with William Henry Harrison as governor, during the next four years fifteen separate treaties were negotiated to that end.

It was an odd coincidence that just as the native inhabitants of the New World were being introduced to the people of France in a way that stirred their imagination, Napoleon was preparing to part with the remaining dominions of France in America. Willing enough to transfer Louisiana to the United States since he himself could not hold it, he hoped that its acquisition by the young republic would involve the American people in a war with Great Britain that would release the strangling hold of British sea power on the throat of France. Nor did Jefferson fail to respond to his proposal. As he stood upon the national promontory gazing westward, he did not fail to see the hand of destiny was beckoning the American people into the land of vast mysteriousness which lay beneath the setting sun of Napoleon. So it was that, impelled by its irresistible allure, in 1803, the champion of strict construction gave to the Constitution the most liberal interpretation it had yet received in order that he might purchase Louisiana.

Hardly had this been done when on the floor of Congress it was proposed to create an Indian Territory in Louisiana to which all the tribes east of the Mississippi should be removed, forcibly if necessary. This proposal was opposed by Jefferson on two grounds. First, it would almost certainly lead to war with Great Britain because it would violate the treaty guarantees of Indian rights. Second, his experience with Brant had shown him that it was not wise to create a powerful Indian state in the unsettled portion of the country which might prove disloyal in a crisis.

While there existed little knowledge of the trans-Mississippi it was known to possess a large native population, for in 1673 Marquette and Joliet had sailed down the Mississippi River to the mouth of the Arkansas and in 1681-82 La Salle had descended from the Great Lakes to the Gulf, both finding the people previously encountered by Coronado and other Spaniards; while in 1683 Father Hennepin had visited the Dakotas, leaving an illuminating account of them and their country.

In 1795, the year Jefferson resigned from Washington's cabinet with the purpose of organizing all the discordant Anti-Federalist elements into the Democratic Republican Party which bore him into office, Washington had appointed Leonard S. Shaw, an ethnologist, as deputy agent to the Cherokees, under instructions to study their language and collect material for an Indian history. Therefore, when Jefferson undertook to make a survey of the Louisiana Territory for the avowed purpose of "extending the internal commerce of the United States," he gave specific instructions to Captain Meriwether Lewis and Captain William Clark whom he selected for the purpose, to observe all the tribes they might encounter. They were to report the names of the tribes and their numbers, the extent and limits of their possessions, their relations with other tribes or nations, their language, traditions, and monuments, their ordinary occupations in agriculture, fishing, hunting, war, arts, and the implements for these, their foods, clothing, domestic accommodations, diseases, moral and physical circumstances, laws, customs, dispositions and articles of commerce they "need or furnish," and to what extent, "and considering the interest which every nation has in extending and strengthening the authority of reason and justice among the people around them, it will be useful," he said, "to acquire what knowledge you can of the state of morality, religion and information among them, as it may better enable those who endeavor to civilize and instruct them to adapt their measures to the existing notions and practices of those on whom they are to operate." Nothing could have been more conclusive that he was thoroughly alive to the fact that there was a widely dispersed population in Louisiana. Concerning it at least his curiosity was very great.

French trappers had uniformly reported that nothing was to be feared from these people. Accordingly on May 14, 1804, clothed with commissions of exploration issued by Jefferson, Lewis and Clark set out from St. Louis with a party of forty-five persons including fourteen United States soldiers, nine Kentucky volunteers, two French portageurs, an interpreter, a trapper, sixteen temporary carriers, and a Negro servant. Unimpeded by native resistance of any kind, the expedition passed up the Missouri River reaching the mouth of the Platte in the Sioux country July 26, and arriving at the long established villages of the Modocs, Mandans and Minetarees, about sixteen hundred miles distant, in October.* With the hospitable Yankton tribe of the great Siouan league of friends—the Dakotas—they spent the winter, and here occurred one of those casual incidents which affect history. A future Yankton chief—Struck-by-the-Ree—was born while the white visitors were the guests of the tribe. Lewis and Clark wrapped him in the nation's flag—then bearing but a slender galaxy. Little did the friendly Sioux at this time understand the portent of that banner, that it was to be bestarred in a dreadful travail—a travail that was to exact of them the life-blood of their race despite their own unbroken loyalty.

Upon the coming of spring onward pressed the explorers. Sighting the majestic Rockies in May, which they crossed in September, they reached the Pacific along the Columbia in November, 1805. Wintering there, they returned to St. Louis in September, 1806 after many physical hardships, having travelled altogether a distance of 8,500 miles.

The explorers had seen many things of interest—huge mountain ranges, snow-clad, rearing lofty peaks athwart their path to the very top of the world, between which lay vast abysmal, cavernous valleys, great rivers coursing southward and eastward through prairies of immeasurable ex-

* Possibly the descendants of early Welsh visitors to America.

tent, others flowing to the blue Pacific through virgin forests more priceless than the minerals that lay within their somber shades. Everywhere
new peoples were found, hospitable and kind, with whom but a single
conflict, and that of a trivial nature, occurred. Plainly, it was a realm of
untold wealth, of vast possibilities.

In their reported observations the utmost interest prevailed in the East.
With the delight of the pseudo-scientist Jefferson described in a message
to Congress what had been discovered. Now the reaching out of the
American people to the silver rim of the Pacific was as inevitable as the
diurnal plunge of the sun into its azure waters. Such being the case it
behooved Jefferson to pause upon the threshold of this newly acquired
empire, and pausing, ponder the human problem which it presented.

To no one could it have been more apparent than to the President who
himself had launched the American people upon a career of imperialism
sure to carry them to the Pacific, that unless something were done and
done at once to safeguard the helpless inhabitants of the trans-Mississippi
region, the old cruel process of preemption would extend to their lands.
Indeed, for them the situation was a real emergency. But an important
event had transpired during the exploration. The acquisition of Louisiana
had attracted the attention of a great adventurer to the Southwest. After
being defeated for the presidency by Jefferson, later for governor of New
York by the influence of Alexander Hamilton, and killing the latter in a
duel in 1805, Aaron Burr had gone there, it is now believed, in connection
with a plot which had been conceived to set up an independent state upon
Mexican territory over which he might rule. In 1806, however, it was
commonly believed, and by Jefferson, that he was implicated in the
treasonable design of separating the region of the trans-Mississippi from
the Union with the aid of the Indian tribes united in such a confederacy,
perhaps, as that of Pontiac and Brant. Therefore, callous as his indifference toward the Indians in his first administration may have been, now
reasons other than "good politics" on the part of a party leader dictated
his failure to provide for them. Blinded by fear and hatred of Burr,
laboring under the delusion that they were a real danger to the nation, he
could not rise to the moral height from which Washington saw that a just
and humane policy toward them was the surest way to insure their loyalty
to their fatherland. Thus, Jefferson, protestant against slavery, advocate of
its abolition, propounder of the fanciful doctrine that all men are born free
and equal, whom posterity delights to rate as a humanitarian along with
St. Pierre and Kant, but in fact unimbued with the higher humanism of
Vitoria, Roger Williams, Penn, Franklin and Washington, was to shut his
eyes to the situation of the Indians as completely as if King Philip and
Pontiac had never lived, as if even presently the Indian seer Little Turtle,
and Chateaubriand were not appealing to the conscience of the nation.

Thus he did absolutely nothing to protect half a million helpless people whom his own ambition had doomed.

There are those who by the sacrifice of moral right to present political expediency have achieved very large and popular results of great advantage. A policy of this kind has often caused one to reflect momentarily something of the dazzling luster of the large events of his time. It is, however, but a passing splendor, one that must inevitably wane in the cold gray light of after judgment. There come occasions in the history of a nation when its chief executive cannot excuse his failure to perform his moral duty by the plea that "the people are incapable of understanding that duty and of supporting him." Even in such a pass he must lead, not follow. He must point out to them what is their duty without thought of the effect upon his political career, not wait for them to develop of their own volition a higher sense of their obligations. He must have the spirit of self-sacrifice of the missionary. He is not permitted the policy of converting the heathen or the sinner, the ignorant or the wrongdoer, by compromising awhile the true faith. Fatal to his present political fortunes as the consequences may be, he must at such times boldly ascend to the most exalted platform and from it admonish the people, rather than encourage them in their momentary blindness by his silence, to pursue a course seemingly more advantageous to them than a sacrifice on their part to an ideal. If kings and presidents fail to uphold ideals, it is not to be expected that those over whom they are set will make those sacrifices which national integrity demands. They may recognize no neutrality between right and wrong. They must be misled by no vain hope of peace in a conflict between such forces unless the victory of right is complete. Their pride of present power must yield to duty, and in that sacrifice which the yielding entails, they must find their reward.

It is a cruel demand that an enduring fame makes upon them, this test that is applied to their moral courage. Yet, when in these respects they fail, though they may retain presently the popular favor, it is written that anon they shall be judged with the dispassion of a retrospect that knows not of favor, not of prejudice, that seeks no advantage. Then, their true characters will be distinguished with unfailing clarity from the largeness of their times, and posterity will rate them as mere opportunists. Few there are who in the history of politics stand out with the moral grandeur of a Washington, for the reason that few have possessed his moral courage. Lacking that attribute, the wisdom, the skill, and all the other qualities of the human heart and mind combined will not suffice to deceive the penetrating gaze of the future, to merit for their possessors at the hand of fame the title of "great." To merit that reward one must be not only wise but must dare to be nobler in his wisdom than others of his time. Like many another, Jefferson could not sacrifice his own interests to the cause of humanity. His humanitarianism failed under the test. It consisted of no more than "the dogma of glittering rhetoric and borrowed phrases."

"By what men doeth ye shall know them," is a maxim which the historian who, in his enthusiasm for democracy and its principle exponent in America, would compare Jefferson with Washington, may well ponder. The former's Indian policy, as lacking in vision as it was inhumane, was to fulfill absolutely the prediction of Voltaire.

In the negative Indian policy of Jefferson the Buffalo Party naturally found great encouragement. But while he was pandering to the frontier democracy upon whose support he depended for his power, the old Indian seer, Little Turtle, a real humanitarian, labored on ceaselessly for the good of both races.

"We are much more likely to obtain justice ultimately," he declared, "by maintaining the peace and adopting the white man's ways, than by giving over to passion and the bad council of hotheads who would have us resort to violence. The way to redress our grievances is in the courts as advised by the Great White Grandfather."*

Nevertheless, as the trade war with Great Britain which Jefferson's policy was provoking became more and more imminent, the agitation of the Indians by British agents increased, adding to their discontent. Still rankling over the loss of their colonies and resenting bitterly the acquisition of Louisiana by Jefferson, the British imperialists were determined to throttle, if possible, the growing American trade, so that they naturally looked upon the Indians of the United States as prospective allies to be prepared in every way possible against the eventuality of another war. In consequence, again the tribes found themselves between the millstones of diplomacy just as in 1754, 1763, 1776 and 1794.

Among the Shawnees, there was a young chieftain who did not fail to see the opportunity for the enforcement of a recognition of Indian rights which this situation presented. This was Tecumseh—The Meteor—whose name is significant of his brilliant, irresistible personality. Possessing all the prestige which high birth, personal beauty, charm, exceptional ability, character, and the rarest eloquence even for an Indian, could give him among his race, so beloved was he by his people it was said by his elders that when the Great Spirit delivered him into his mother's arms a sigh of joy rose from his native forest, that even the sturdiest oaks trembled with emotion.

This favored one was the son of the celebrated Shawnee chieftain, Cornstalk, who was killed in 1774 in the battle of Point Pleasant, and a Creek mother. Born at the Shawnee village of Piqua in 1768, and orphaned at the age of six, he was turned over to the guardianship of an

* Washington.

elder brother who, eight years after the tribal village was destroyed by the Kentuckians in 1780, was killed fighting against the whites on the Tennessee border. At twenty-six years of age he himself was a veteran warrior, another brother having been killed at his side in the battle of Fallen Timbers, where Tecumseh won the commendation even of Little Turtle.

Yet, the tragedies of his youth had not embittered his soul, for above all else he was distinguished for his humanity, having persuaded his tribe to abandon the practice of torturing their prisoners. Small wonder that as he matured, seeing the smoldering world upon which his eyes had first opened still smoldering, the young warriors about him being debauched with "firewater," his womenfolk violated, the stamina of his race weakening, its condition growing worse in every way, its woes increasing, such a one should have become possessed with a burning longing to better the lot of his race.

With its history he was eminently familiar. The memory of Pontiac was green among his own people, who had taken a leading part in the revolt of 1763. Often as a child he had been told by Cornstalk and his other elders that the Indians owned all the land in common, while he himself knew Brant well. Submission had brought nothing but misery. Since the battle of Fallen Timbers the primeval forest of Ohio had vanished, and with it the game which once there abounded. As if by magic the cleared lands had filled not with Indians but with the "Swannak"—a foe now more truculent than ever before, whose greed for Indian lands was apparently insatiable. Soon there would be no place left for the red man.

But what was this? On the floor of Congress it was being proposed to uproot all the tribes east of the Mississippi and remove them, forcibly if necessary, to Louisiana in order to give the whites their lands. And old Little Turtle merely continued to prate of seeking protection in the courts of the "Swannak" despite the guarantees contained in the treaties in which the Indians had acknowledged the sovereignty of the United States! Plainly something more than that was necessary. So thought Tecumseh. It was, therefore, the most natural thing in the world that Brant, still residing in Canada with a price on his head in the States, should regard Tecumseh as the one best qualified to succeed to the toga of Pontiac, to reorganize the United Indian Nations.

How Tecumseh acquired the unusual education and knowledge of political and military science he was to display during the next decade is not known, but certain it is he received every encouragement Brant could secure for him from the British to found a great Indian state with the Ohio River as its boundary. Yet, the purely selfish aims of British diplomacy were as much aside from the motives by which Tecumseh was animated as were those of the French in 1763 from the ultimate purpose of Pontiac. If in the enforcement of the recognition of the Indian state, and of a fuller measure of justice for Indian rights in general, it should be necessary to fight, his Confederacy would accept the proffered aid of the British for reasons of pure military expediency.

At the very outset, however, he was confronted like Pontiac with the necessity of arousing the spirit of Indian nationality essential to a reunion of the tribes, and again a prophet was necessary. The prophet should be his twin brother, Lalawéthika, until now a drunken sot and seemingly of little intelligence.

The appearance of this new prophet was adroitly arranged by Tecumseh to have the maximum effect. Thus, one day in November, 1805 at Wapakoneti, the ancient capital of the Shawnees, Lalawéthika fell into what appeared to be a trance while lighting his pipe. Immediately his fellow tribesmen and all their allies were assembled, whereupon to their amazement he rose and announced himself as the bearer of a new revelation from the Master of Life. He had been taken up to the spirit world and permitted to lift the veil of the past and the future—had seen the misery of evil doers, and learned the happiness that awaited those who followed the precepts of the Indian god.

A long harangue followed in which the prophet exhorted his people to give up witchcraft and medicine juggleries, declaring that those who continued such practices, and to use the accursed "firewater" of the whites, would be tormented after death. The young must cherish and respect the aged and infirm. *All property must be in common*, according to the ancient law of their ancestors. Indian women must cease to intermarry and cohabit with white men. The two races were distinct and must for their mutual good remain apart. The white man's dress with his flint and steel must be discarded for the old time buckskin and firestick. More than this, every tool and every custom derived from the whites must be put away, and the Indians must return to the methods the Master of Life had taught them. When they should do all this he promised that again they would be taken into the divine favor, and find the happiness their fathers had known before the coming of the "Swannak." Finally, in proof of his divine mission he announced that he had received power to cure all diseases and to arrest the hand of death in sickness or on the battlefield.

Just as the far-seeing Tecumseh planned, the excitement following the revelations of Lalawéthika was intense, spreading to all the Indians with lightning rapidity by that strange wire of communication connecting aboriginal peoples. The quondam sinner now changed his name to Tenskwatawa, signifying the new mode of life which he had come to point out to his people, and established himself at Greenville, Ohio, whither came representatives of all the tribes of the Northwest to hear his doctrines. And they called him the Prophet, and the fame of him went abroad, even unto the Seminoles in Florida, while emissaries travelled back and forth from one tribe to another.

Recalling with terror the stories of the earlier days, the frontier country was soon thoroughly alarmed. Tecumseh and the Prophet were of course fiends incarnate. Urgent appeals for aid were made to the government. Taking note of the situation in his second inaugural address, Jefferson said:

The aboriginal inhabitants of this country I have regarded with the commiseration their history inspires. Endowed with the faculties and the rights of men, breathing an ardent love of liberty and independence, and occupying a country which left them no desire but to be undisturbed, the stream of overflowing population from other regions directed itself on these shores; without power to divert or habits to contend against it, they have been overwhelmed by the current or driven before it; now reduced within limits too narrow for the hunter's state, humanity enjoins us to teach them agriculture and the domestic arts; to encourage them to that industry which alone can enable them to maintain their place in existence and to prepare them in time for that state of society which to bodily comforts adds the improvement of the mind and morals. We have therefore liberally furnished them with implements of husbandry and household use; we have placed among them instructors in the arts of first necessity, and they are covered with aegis of the law against aggressors from among ourselves.

But the endeavors to enlighten them on the fate which awaits their present course of life, to induce them to exercise their reason, follow its dictates, and change their pursuits with the change of circumstances have powerful obstacles to encounter; they are combatted by the habits of their bodies, prejudices of their minds, ignorance, pride, and the influence of interested and crafty individuals among them who feel themselves in the present order of things and fear to become nothing in any other. These persons inculcate a sanctimonious reverence for the customs of their ancestors; that whatsoever they did must be done for all time; that reason is a false guide and to advance under its counsel in their physical moral or political condition is perilous innovation; that their duty is to remain as their Creator made them; ignorance being safety and knowledge full of danger; in short,—among them also is seen the action and counter action of good sense and of bigotry. . . .

With these high-sounding words which constituted almost the sole contribution of Jefferson to the solution of a problem that was becoming more and more complex, the President sought to overawe Tecumseh. They had no effect. It was not commiseration which the Indians wanted. No one knew better than Jefferson that they were not covered by the "aegis" of the law; that not one cent had been appropriated since the days of Washington for their education and welfare. Therefore, undeterred by such sophistry, Tecumseh went his way.

On November 24, 1807 there occurred that which in Indian history was an event of great importance—the death of the great Thayendanegea, or Joseph Brant. Now Tecumseh at once became the moral successor of Pontiac with all that meant to his race.*

During Jefferson's second administration fifteen more treaties of concentration were negotiated with the tribes of the East, but the Georgians were not satified with this. They were determined to drive out the Cherokees, Creeks, Chickasaws, Choctaws and Seminoles and have their lands. Among the Cherokees there was, however, one who for over thirty years was to prove a stumbling block for Georgia—John Ross, or the hereditary chieftain, Coweescorvie, or Guwisgu, the Egret or Swan—now a mere lad.

Born at Rossville, Georgia, October 3, 1790, the son of a Scotch immigrant and a Cherokee woman, he had obtained a very good education at a white school in Kingston, Tennessee. Upon returning to his tribe he found that the land set apart to it by the treaty of Holsten was not only inadequate to its needs but unadapted in part to cultivation, especially in Alabama, where the transplanted Cherokee farmers of Georgia were being compelled to revert to hunting in order to find sustenance. With sorrow in their hearts these poor people saw the corn fields, the peach orchards, the vineyards they had planted, the churches, the sawmills, the bridges, the highways they had built in Georgia in the hands of the whites who had usurped them. In vain these planters had trekked the somber, moss-hung wilderness, catching only a glimpse now and then of the rising sun as they turned their faces westward. Contrary to the common belief, many of them, long engaged in peaceful pursuits, were no longer inured to the hardships of a remorseless struggle with nature. They had not dwelt in the misty shades of the primeval forest, amid the miasmas of the swamps in preference to the sunny highlands of Georgia. In exchange for sunlight and pure air, the blue peaks of the Appalachians and the dancing, sparkling streams of pure water that bounded from them seaward, they had received a land of dismal aspect, beneath the heavy mists of which the streams coursed sluggishly through malarial regions to the Gulf. Hundreds of them had perished. Laughter was no longer in their hearts.

While the Cherokees had joined Brant's Confederacy in 1782, Ross was not sympathetic with Tecumseh, believing in the submissive policy of Little Turtle. He had heard the talk about the creation of an Indian Territory in Louisiana, seeing in that a possible solution of the problem of

* Brant married three times, leaving numerous descendants, including a son who succeeded him as chief of the Mohawks. Buried in Ontario, a tomb was erected over his grave with the inscription:

"This tomb is erected to the memory of Thayendanegea or Captain Joseph Brant, principal chief and warrior of the Six Nation Indians, by his fellow subjects, admirers of his fidelity and attachment to the British Crown."

In 1879 the grave was desecrated

his people. At any rate, in 1808 he drafted a petition to the President pointing out that there would be no danger to the government whatever if the Cherokees should be allowed to remove to Louisiana, and praying that it assist them to locate sufficient good land there for their use to be exchanged for their eastern domain.

Upon this proposal Jefferson reflected a long time. The Georgians were pressing him to oust the Cherokees. Convinced by an investigation that the tribe by reason of its old blood feud with the Shawnees whom it had long since expelled from Georgia, like the Choctaws and Chickasaws was not well disposed toward Tecumseh, he concluded that to permit them to remove to an interior position in Louisiana out of contact with the Ohio tribes would not jeopardize the republic. Therefore, after causing a treaty to be negotiated in 1808 with the Great and Little Osages extinguishing their rights in a tract between the Arkansas and the White rivers the following year, he authorized the Cherokees to make a selection of land in this region. On this mission Ross himself went at once, finding some good prairie land along the Arkansas River, and when he reported this to be suitable for his people the government's agents began pressing the Cherokees to remove. All manner of arguments were employed. Finally, the more southern Cherokees under Chief Jolly, upon Ross's advice, decided at a great council to move at once, whereupon Jefferson caused Fort Clark to be established on the Arkansas River as an agency for them.

While Ross was engaged with the government's agents removing Chief Jolly's band, there occurred that which served to convince Tecumseh the more that nothing of justice for his race was to be expected by pursuing the policy of Little Turtle. This was a proceeding in the Supreme Court in 1810 in the case of *Fletcher v. Peck*, the first to come before the nation's most august tribunal involving Indian rights. The question was whether Georgia had a right to dispose of lands to which title had been acquired by an individual under a grant from an Indian tribe. John Quincy Adams and Joseph Story, representing the views of the Buffalo Party dressed up in legal verbiage, contended that the Indian tribes had no idea of a title to the soil; that it was overrun, rather than inhabited by them; that their right to it was not a true and legal possession; that it was a right not to be transferred but extinguished. Vattel, Montesquieu and Adam Smith were cited in support of their brilliant dissertations on the law pertaining to nomadic aboriginal tribes in the hunter states.

Such arguments, not being pertinent to the facts, did not mislead the court, which took notice of the true social and economic status of the Indians. Marshall above all others knew that many tribes had for long been occupying and tilling large tracts of land; that both among the Iroquois in New York and the southern Indians there were not only extensive areas that had been cultivated in rotation for years, but permanent settlements of long standing surrounded by vineyards and orchards set out by the native tenants; that among the tribes of Georgia hundreds of runaway Negro slaves had long been in possession of the Indians who had regularly employed them upon their lands; that those tribes, in the East at least, that were in a roving state in 1810 had been uprooted from the soil and compelled to rove by the whites; that it was in no sense true that the Indians generally merely "overran" instead of inhabiting their tribal domains.

But the issue in the case before him was not one which Marshall, like Washington, had hoped would come before the Court, since the question of the respective titles of the United States, a state, and a tribe was only a collateral issue. Therefore, speaking in the most guarded language for the majority of the court, the Chief Justice declared that while the nature of the Indian tribal title was such that it was certainly to be respected by all courts until legitimately extinguished, it was not such as to be *absolutely repugnant* to a seizin in fee on the part of a state. On the other hand, the minority, speaking through Mr. Justice Johnson, held that Indian tribes occupying land within a state were seized of a fee subject to a right of preemption on the part of a state which could convey that right to the United States.

Of the arguments made in this case Tecumseh learned with dismay. It seemed plain to him from the fact that two of the ablest and most renowned legalists in the country were employed to represent the defendants in error, that this was but a test case designed to prepare the way for the involuntary removal of the eastern tribes beyond the Mississippi. Therefore, having well prepared the soil of Indian nationalism during the past five years, he redoubled his efforts to complete the organization of the new Indian Confederacy in order to resist what he believed was surely impending.

18

Tippecanoe and the War of 1812

KNOWING NOT TECUMSEH'S PRIMARY motives, and convinced that he meant only war in alliance with the British, more than one friend sought to dissuade him from his course. Among others the old Indian Generalissimo under whom he had fought sixteen years before was sent to him. "Let me tell you," said Little Turtle, "should you defeat the American army, you have not done. Another will come; and if you defeat that, still another—one like waves of the Greatwater, overwhelming and sweeping you from the face of the earth." So too, General William Henry Harrison, Governor of the Indian Territory, an old friend, who admired Tecumseh, warned him that if he attempted to oppose the whites, "swarms of hunting-shirt men" would pour forth "thick as mosquitoes on the shores of the Wabash." To them all he replied: "The Sun is my father, the Earth my mother. I and my people have retreated enough. We will yield no more of our land." Declaring that he did not wish for war, that he would not strike the first blow, yet, he said if need be he and the other warriors of his race would do "like men," which, of course, was interpreted as evidence of his treasonable intent.

Early in 1811 it was apparent that despite the efforts of Madison to maintain peace, war with Great Britain was inevitable, for Clay, Calhoun and even Webster were now appealing to the patriotism of the country which had been almost completely disarmed by the supreme pacifism of Jefferson. In the far south the Creeks and their kinsmen, the Seminoles of Georgia, were in a ferment over a treaty into which the former had been compelled to enter in 1806 ceding more land, while the propaganda of the Prophet had long since reached them. Tecumseh did not fail to see the wonderful opportunity this situation presented him to strengthen his Confederacy by extending it to the Cherokees, Creeks, Choctaws, Chickasaws and Seminoles on the vulnerable southern flank of the United States where, if compelled to fight, the Indians could also be supported by the British. Accordingly, after having committed the powerful Dakotas and other great tribes in the far west to his Confederacy, leaving his affairs in Ohio in the hands of the Prophet, he proceeded to Georgia, establishing his new base among the Creeks among whom, by reason of his mother's blood, his influence was compelling. Yet, among the Cherokees, and the Chickasaws, who were confederated with them, and the Choctaws he seems to have made little headway against Ross, who continued to urge them to hold aloof from him.

At any rate, as it turned out, his visit to the South was a fatal mistake. Unfortunately for the success of his plans, the Prophet, who was not a statesman, consumed with self-importance, soon after Tecumseh's departure from Ohio, allowed several hundred Shawnees to assemble around him which seemed to indicate no other purpose on their part than war. The cry for aid which arose at once from the frontier could not be ignored by Madison. Therefore, he despatched to Ohio a force under General William Henry Harrison who was instructed to bring on the issue at once and crush the western tribes so thoroughly that they would be unable to aid the British. The southern tribes could be cared for later.

The evidence is convincing that the Indians under the Prophet until now had not the slightest idea of taking to the warpath until ordered to do so by Tecumseh, but as Harrison's force approached the inevitable occurred. They saw that his "hunting-shirt" men had but one purpose, so that swept away by enthusiasm the Prophet allowed himself to be drawn into a conflict on November 7, 1811, resulting in the complete route of the Indians. So came about the famous battle of Tippecanoe which at once gave its name to General Harrison, making of him a national hero, and elevating him to fame at the expense of his cherished friend Tecumseh.

Yet, always magnanimous, speaking later on the floor of the Indiana Legislature, Harrison said: "the utmost efforts to induce the Indians to take up arms would be unavailing, if one only of the many persons who have committed murders on their people could be brought to punishment.

But, like the people of the frontier generally, unable to credit the confederated tribes with any but a treasonable motive from the first, Madison was unscathing in his rebuke of Tecumseh and the Indians. Thus, in a special message to Congress on June 1, 1812, he said:

In reviewing the conduct of Great Britain toward the United States, our attention is necessarily drawn to the warfare just renewed by the savages on one of our Western frontiers. It is difficult to account for the activity and combinations which have for some time been developing themselves among tribes in constant intercourse with British traders and garrisons without connecting their hostility with that influence and without recollecting the authenticated examples of such interpositions heretofore furnished by the officers and agents of that Government.

Poor old Little Turtle! His heart was broken. He died near Fort Wayne during the summer of 1812, lamenting the ills which he conceived the rashness of Tecumseh had brought upon his people.

Having hastened back to Ohio, in vain Tecumseh endeavored to save the situation, but with the armed forces of the United States among them, further cooperation between the western tribes had proved impossible. Bitter was his anguish when he saw that all his work of the past decade had gone for naught by reason of the bad judgment of his brother. Moreover, since his own tribe had engaged in open hostilities against the United States, he himself, its chief, was under proscription as a public enemy, while the frontier was crying loudly for his head. Yet, to give himself up with his people would have accomplished nothing but their certain punishment. When, therefore, war was finally declared by the United States upon Great Britain, of necessity he passed over into Canada with about two thousand warriors who would not abandon him, where he was received with open arms. At once assigned to military duty with his command, he was to prove himself in the subsequent operations a consummate master of both Indian and European tactics. Thus, in August 1812 he routed a force of Ohio militia, and although defeated and wounded in a subsequent skirmish, was commissioned a brigadier-general in the British army.

This infuriated the American people so that in Madison's annual message for 1812 the vials of his wrath were poured upon the Indians generally.

Whilst the benevolent policy of the United States invariably recommended peace and promoted civilization among that wretched portion of the human race, and there was made exertions to dissuade them from taking either side in the war, the enemy has not scrupled, armed with the horrors of those instruments of torture which are known to spare neither age nor sex.

This, of course, was pure sophistry. The fact is, at the siege of Detroit in January, 1813, Tecumseh, true to his principles, prevented the massacre of the prisoners taken in the sortie, and no more honorable and humane soldier ever engaged in war. Here too, it is to be noted that Rising Moose, chief of the Medewakanton Sioux of Minnesota, had just joined the Americans with his tribe at St. Louis, receiving a military commission at the hands of General Clark, and it is well known that General Porter was constantly urging the Six Nations to take up arms against the British, notwithstanding the opposition of Red Jacket, Cornplanter and Blue Sky. Plainly Madison's protests were nothing more than a recruiting appeal.

At any rate, the disaffectation of the western Indians encouraged a new demand for the forcible removal of the eastern tribes, which was again seriously discussed on the floor of Congress when in 1812 the more settled part of the territory acquired in 1804 was admitted to the Union as the state of Louisiana and the remainder was erected into the territories of Missouri and Kansas. The creation of an Indian Territory under the existing circumstances being declared by Madison as most unwise, again the proposal was dropped.

After the decisive victory won by Perry on Lake Erie, Tecumseh protested in vain against the retreat of the British from Malden, and with his Indians covered the retirement until, declining to retire further, he compelled a stand on the Thames River near the present town of Charlton. With a presentiment of death he now discarded the scarlet tunic of a British general, and donned his native war dress. In the bloody battle which here occurred on October 5, 1813, the British and Indians were decisively defeated by General Harrison, now called Tippecanoe, and while bravely resisting the assault of the mounted Americans under Colonel Richard W. Johnson, Tecumseh was killed.

So perished The Meteor at the age of forty-five.

A Canadian historian has declared that it is probable he prevented the loss by Great Britain of Canada, and from all that is known of him there is no reason to doubt the accuracy of Trumbull's verdict that he was the most extraordinary Indian character in the history of the United States. Made of the stuff of all the great heroes of defeat—Hannibal, Vercengetorix, Schamyl, Scandeburg, Kosciusco, Lee—he was a patriot of the noblest ideals, an able, humane and heroic soldier, the most illustrious man yet produced by his race, an ornament to any nation, withal but another martyr to Indian liberty.

Although the real danger of the Indians passed at the battle of Tippecanoe, the far South did not fail to see the capital that could be made of their alleged disloyalty. For years the Georgians had been hearing of Tecumseh's emissaries passing back and forth among the Indians and of the presence of British agents among them, so that although a small force of Seminoles who assembled under King Payne just after Tecumseh's departure had been dispersed with ease in October, 1812 by Georgia troops under General Newman, it was not difficult to make it appear that both the Seminoles and the Creeks still intended war. In other words, the War of 1812 was to be made the occasion of doing that which had long been in their minds—the taking of the tribal lands, which would compel them all to follow Chief Jolly's Cherokees to the West.

So far the Creeks had committed no overt act, but knowing their sympathy for Tecumseh, General Andrew Jackson, an inveterate enemy of the Indian and known to the frontier by reason of his harsh methods as "Old Hickory," thirsting for glory like that which had come to Tippecanoe, was boasting that they should fare no better than the western tribes. Accordingly, in 1812 he was sent to Georgia with a force of 2,500 Tennesseans ostensibly to defend the southern frontier against the British. Having witnessed the breakdown of Tecumseh's Confederacy, naturally the Creeks were too wise to afford "Old Hickory" the opportunity for which he longed. After he and his troops had swaggered about among them a while, they were called away for service elsewhere.

During this military occupation of the Creek country countless offenses were given its occupants. Rankling with resentment over the way they had been treated, in 1813 a band of southern Creeks assembled under Lamotachee, or "Weatherford," the Red Eagle, a bold but thoughtless warrior. Things went from bad to worse. Finally they attacked Fort Mimms in Alabama, killing about three hundred men, women and children who had sought refuge there. After most of the soldiers had fallen, the blockhouse was set afire. Of the entire garrison but seventeen persons managed to escape to Fort Stoddard.

The holocaust of Fort Mimms is without doubt the greatest cruelty chargeable to the red man in all his long struggle with his merciless oppressors. Whether equal to the atrocities perpetrated on the Indians at Mystic, Kingston Swamp, Bedford, Pavonia, Bad Axe and other places is aside from the question. And without regard to whether it was provoked or not, it was another dreadful mistake on the part of the southern Creeks, for the country was horrified and even the Cherokees, Choctaws, Chickasaws, and northern Creeks were unanimous in condemning the act. Having hurried home from Arkansas in 1812 to help hold the Cherokees loyal, Ross now assisted in raising a regiment among the southern tribes in which he served until the end of the war as adjutant.

Old Hickory's chance had come at last. Now he marched upon the rebellious Creeks at the head of a force of 3,500 Tennesseans. Willing enough to employ the troops of "the execrable" red race, the government attached the Indian regiment of 700 warriors under the Creek chief, Colonel William McIntosh, to Jackson's command. After brushing aside Red Eagle's forces in several skirmishes Jackson decisively defeated them in November, 1813 in the battle of Talladega, whereupon their leader sued for peace. But Jackson's idea of dealing with Indian adversaries was that of the frontier, long since expressed by George Rogers Clark—"their families must perish before they will yield." Therefore, as he declared later, he was "determined to exterminate" the rebel Creeks. Nor did the vindictive Madison show any mercy for them though he himself continued to blame the British for their disloyalty. Thus, in his Fifth Annual Message on December 7, 1813, he said:

The systematic perseverance of the enemy in courting the aid of the savages in all quarters had the natural effect of kindling their ordinary propensity to war into a passion which even among those best disposed to the United States, was ready, if not employed on our side, to be turned against us.

Compelled to fight on though again defeated at Emuckfau and Enoto-chopco in January, 1814, in a last desperate stand on March 27th following at Tohopeka near Horse Shoe Bend on the Tallapoosa River, over five hundred Creeks were virtually massacred with the aid of the Indian contingent under McIntosh against which they had not a chance. Scores were drowned trying to swim the river after being surrounded. The Creeks had now lost over two thousand warriors, and no longer possessed the power of resistance. One day while Jackson was plotting to seize their leader and make an example of him, there appeared before him a tall warrior of fine presence, who said:

I am the chief who commanded at Fort Mimms. I am in your power—do with me as you please. There was a time when I had a choice. I have none now.

I have done the white man all the harm I could. Once I could animate my warriors. They hear me no longer—their bones are at Talassehatchi, Talladega, Emuckfau, Tohopeka.

While there was a chance I never asked for peace. But my warriors are gone; I ask it now—not for myself, but for my people.

About this incident there is much indicative of Indian character—love of race, unflinching courage, withal a frankness claiming not ungenuine motives. One cannot say if this noble act were purely voluntary, or if it were inspired by the advice of wiser Indians than Red Eagle who saw in his self-sacrifice the possibility of propitiating the conquerors. Yet, in the history of Europe there is nothing grander than the act of this savage of the American forest.

But while it may have excited the admiration of Jackson, it did not appeal to his compassion. As the Roman was wont to turn his thumb down upon the defeated gladiator whose courage he admired, that of Jackson was turned against the Creeks. No peace should be granted them unless they agreed to give up more than half their lands!

This perforce, during the summer of 1814 they did. Thus, the Georgians acquired at last that which by the treaties of 1790 and 1806 they had been unable to obtain from the Creeks.

Though Jackson, more cruel and relentless than the Indians, had made good his boast, the greed of the Georgians was by no means appeased as yet. With the acquisition of the lands of the southern Creeks their purpose was only partly achieved. The Seminoles were now to become the victims of the great political conspiracy in which Georgia was implicated.

The constitutional inhibitions against the slave trade between citizens of the United States and foreign countries after 1808 had driven New England out of that profitable enterprise into another, for meantime cotton ginning had been brought to an effective state of development leading directly to the transfer of New England's energy to textile manufacture and a tremendous demand upon the South for the staple. Suddenly, the institution of slavery which had become economically more and more injurious to the South had acquired a greater value than it had ever before possessed. But the concurrent industrial revolution in New England and agricultural revolution in the South was attended with the gravest political consequences. With a new wealth the population of New England increased apace and far more rapidly than that of the South. At once a demand for more equal representation in the affairs of the nation arose in the North, so that the cry for the abolition of slavery was becoming increasingly insistent, since representation in the South was based in part on slave population. But the South, whose material welfare had so suddenly come to depend upon slave labor, could not yield to this demand. To do so would be utter self-destruction. Moreover, in order to retain the institution of slavery, it must needs maintain the superiority of its representation in Congress, and this required the creation of new slave states to counterbalance the free states that were being rapidly erected out of the Northwest Territory. Such, in brief, were the politico-economic considerations that were to bring about a violent disturbance of the old relations of the states and one fraught with serious consequences for the Indians.

It was possible to create five nonslaveholding states out of the Northwest Territory. Kentucky and Tennessee had been admitted to the Union as slave states in 1792 and 1796, respectively, and Ohio as a free state, in 1803. But although Louisiana was admitted in 1812, it was apparent Indiana and Illinois would soon overbalance that accession with only Mississippi, Alabama and Missouri remaining as possible recruits for King Cotton while the anti-slaves still had Michigan, Wisconsin and Minnesota in prospect. Thus, it is seen that in 1812 the South was under the necessity of developing new slave territory if it were to maintain its ascendancy in Congress, for which purpose plainly Florida, a natural cotton country, promised the best field of exploitation.

In 1783 Spain had recovered Florida from Great Britain and it will be recalled that Georgia, whose eyes had long rested longingly on the land of the Seminoles, with the aid of Tammany Hall had succeeded in 1790 in bringing about a federal treaty with the Creeks designed in part, at least, to encourage them to strike at the British trade with the hinterland conducted through the Seminoles. In 1795 Spain had sold West Florida to France. Therefore, the next step of Georgia had been to obtain that year a federal treaty with Spain in which it was mutually agreed to suppress Indian depredations along the border. This was a further foundation for what was to follow.

Whether provoked or not, shortly after the capture of Red Eagle, the Seminoles of Florida marauded across the border. Instantly these forays were charged by the Georgians to the notorious British traders, Arbuthnot and Ambrister, while the Spanish treaty of 1795 binding Spain to prevent depredations was loudly invoked. "Plainly," cried the Georgians, "we must in self-defense take things in our own hands if the federal government will not protect us!"

At this juncture, Old Hickory, now the hero of New Orleans, whose methods of dealing with the Indians were popular with the Georgians, was rushed back to Georgia. Dashing across the border without delay, he seized St. Marks and Pensacola, hanged Arbuthnot and Ambrister on the drum head charge that they had been guilty of inciting the Indians to hostilities against the United States, destroyed the principal Indian towns, and in October, 1818 compelled the Seminoles to lay down their arms.

What were in fact his orders will probably never be known since the evidence is conflicting, but it seems certain he was expected by the politicians in Washington to proceed into Florida without regard to international consequences. Yet, it does not appear that President Monroe was in any way responsible for his acts. Not only did he deny that they were authorized by him, but like the entire cabinet with the exception of John Quincy Adams, he was horrified. Only by Adams, the Secretary of State, was Monroe finally persuaded to ratify them.

19

Monroe and Johnson v. McIntosh

MONROE WAS DISGUSTED with the whole Florida incident. As he surveyed the situation of the Indians they appealed to his humanitarian instincts even more than the free Negroes for whom he had done so much in the way of African colonization. Yet, history has been so much taken up with him as an internationalist it has utterly ignored his greater services to mankind. Oddly enough, not only was he to contribute largely to the creation of Liberia, whose capital was named after him, as a refuge for the black race, but he was to do more for the red race than any man of his time since the death of Washington.

So soon as the war of 1812 was over the government had begun to press all the Cherokees to move to Arkansas, knowing that if they did so the other southern tribes would follow. But good reports had not come from Chief Jolly and the emigrant Cherokees, so that although still willing enough to remove, the Cherokees in Georgia had no idea of following them until their rights were definitely fixed by treaty. Without this there was no reason to suppose a second move would be more permanent than the first.

Ross had formed an intimate acquaintance with Sam Houston, now an important man in Tennessee who as a mere lad had lived among the Cherokees, having also served with him in the late war. Now, in 1817, when he went to Washington as the tribal delegate, he became acquainted with the Honorable John Seargent of Philadelphia, one of the foremost constitutional lawyers of the country, later Speaker of the House of Representatives and vice-presidential candidate on the Whig ticket with Clay.

Seargent was the great-grandson of a famous Indian missionary of his name, and the grandson of the Reverend Elihu Spencer, an Indian missionary who in 1774 had visited the Cherokees in the effort to win them to the cause of independence. Beside an inherited attachment to the Indians he was a staunch Washingtonian who despised the Indian policy of Jefferson and Madison. It was, therefore, easy for Ross to enlist his great legal talents in the cause of the Cherokees. At any rate, with sound legal advice Ross and an associate chief, William Hicks, now drafted an unanswerable reply to the commissioners who came to arrange for the migration of the tribe.

The upshot was that although by the treaties of 1817 and 1819 the whole Cherokee Nation ceded a large portion of the land hitherto granted it east of the Mississippi in exchange for an equal area along the Arkansas and White Rivers, it was stipulated that they should have the perpetual ownership of the latter and be reimbursed for all improvements which had been made by them to the ceded lands. So too, when upon taking of a census in June, 1818 it appeared that two-thirds of the Cherokees either intended to remove to their western domain or had already done so, the accrued annuities due the Nation as a whole were apportioned accordingly, one-third to the eastern and two-thirds to the western Cherokees. By the treaty of 1819 a tract of land was also ceded to the United States with the understanding that it was to be sold as public lands and the invested proceeds held in trust by the United States as a school fund for the Cherokees who remained within the original domain east of the Mississippi. Thus, it is seen that in 1819 the Cherokee Nation voluntarily divided itself into two groups, the eastern Cherokees and the western Cherokees, each having acquired definite rights in the lands upon which they were seated.

In 1820 the latter instituted a government patterned by Ross after that of the United States, and the following year both branches adopted a remarkable alphabet invented by the Cherokee Sequoyia, especially adapted to the education of his people. In 1824 parts of the Bible were printed in this alphabet while four years later Sequoyia, then resident in Arkansas, began the publication in Cherokee and English of the *Cherokee Phoenix*, a weekly newspaper.

Monroe knew it was futile to speak of such a people as unamenable to civilization, just as it was barbarous to treat them as beasts. Truly the tragedy in which the Indians were involved was a national scandal. Driving them off into a desert would not solve the Indian problem any more than it would discharge the obligation of the nation. Therefore, in 1818 Monroe addressed Congress as follows:

The case of the Indian tribes within our limits has long been an essential part of our system, but unfortunately, it has not been executed in a manner to accomplish all the objects intended by it. We have treated them as independent nations, without their having any substantial pretensions to that rank. The distinction has flattered their pride, retarded their improvement and in many instances paved the way to their destruction. The progress of our settlements westward, supported as they are by a dense population, has constantly driven them back, with almost the total sacrifice of the lands which they have been compelled to abandon. They have claims on the magnanimity and I may add, on the justice of this nation which we must all feel. We should become their real benefactors; we should perform the office of their Great Father, the endearing title which they emphatically give to the Chief Magistrate of our Union. Their sovereignty over vast territories should cease, in lieu of which the right of soil should be secured to each individual and his posterity in competent portions, and for the territory thus ceded by each tribe some reasonable equivalent should be granted, to be vested in permanent funds, for the support of civil government over them and for the education of their children, for their instruction in the arts of husbandry, and to provide sustenance for them until they could provide it for themselves. My earnest hope is that Congress will digest some plan founded upon these principles, with such improvements as their wisdom may suggest, and carry it into effect as soon as it may be practicable.*

The proposal to set apart to Indians individually an inalienable estate was deemed "too absurd" for serious consideration. Indeed, the sequel of Jackson's invasion of Florida was now to occur.

In 1819 Spain ceded Florida to the United States just as it had long been determined that she should do, treaty of cession being unique. Jackson and the Buffalo Party saw to that. It was not difficult with John Quincy Adams, lately counsel for the Buffalo Party in the Indian litigation of *Fletcher v. Peck*, as Secretary of State, to cause to be omitted from the treaty any reference whatsoever to the Indians, and all guarantees of preexisting property rights of the 3,000 Seminoles who at this time inhabited the land and owned some 800 negro slaves.

In the three treaties with Great Britain—the Treaty of Paris of 1783, the Jay-Grenville Treaty of 1794, and the Treaty of Ghent of 1814—specific guarantees of Indian rights had been included. In the treaty of 1803 by which the Louisiana Territory had been acquired from France, even Bonaparte had insisted upon a guarantee of Indian rights. Yet, in the Spanish treaty ceding Florida it was merely provided that the "inhabitants" were to be incorporated in the Union "as soon as might be consistent with the Constitution," and then "admitted to the enjoyment of all the rights, etc., of citizens." Three years later Florida was organized into a territory later to be admitted into the Union with Missouri, Texas, Arkansas and Kansas

* Monroe's Second Inaugural Address, Nov. 16, 1818.

as a slave-holding state to balance the free states of Michigan, Wisconsin, Nebraska and Minnesota. Thus, did King Cotton achieve his purpose.

Nevertheless, as a result of Monroe's appeal, forty-four years after Congress assumed responsibility for the Indians it made in 1819 the first specific appropriation for Indian education in the sum of $10,000 which was repeated annually for many years without increase. At least Monroe had punctured the hide of its conscience.

The Choctaws of Alabama and Mississippi, weary of struggling with states determined to have their lands, were now ready to follow the western Cherokees to Arkansas. Therefore, in 1820 Monroe negotiated a treaty with them granting them a tract contiguous to that of the Cherokees in exchange for their immemorial domain, and binding the government to defray the cost of their removal.

Down to this time the original system of disposing of the public lands had continued in effect, but in 1820 the practice was instituted by Congress of allowing purchasers to select "cash lots." Knowing what was in the air, and having failed to make Congress provide for the Indians individually, Monroe appealed to the church to help him arouse the conscience of the nation, after appointing Eleazer Williams, a St. Regis Iroquois and missionary, Superintendent of the Northern Indian Department, he despatched the Reverend Jedediah Morse, a noted philanthropist, to the Indian country to study conditions and make an exhaustive report thereon. In 1820 Morse rendered a report to the Secretary of War which the following year was transmitted to Congress.

The Government, according to the law of nations, having jurisdiction over the Indian territory, and the exclusive right to dispose of its soil, the whole Indian population is reduced, of necessary consequence, to a dependable situation. They are without the privileges of self-government, except in a limited degree, and without any transferable property. They are ignorant of nearly all the useful branches of human knowledge, of the Bible, and of the only Savior of men therein revealed. They are weak and ready to perish; we are strong, and with the help of God, able to support, to comfort and to save them. In these circumstances the Indians have claims on us of high importance to them and to our own character and reputation as an enlightened, just, and Christian Nation. In return for what they virtually yield, they are undoubtedly entitled to expect from our honor and justice protection in all the rights which they are permitted to retain. They are entitled, as 'children' of the Government, for so we call them, peculiarly related to it, to kind paternal treatment, to justice in all our dealings with them, to education in the useful arts and sciences, and in the principles and duties of our religion. In a word, they have a right to expect and to receive from our civil and religious communities combined that sort of education, in all its branches, which we are accustomed to give to the minority of our own population, and thus to be raised gradually and ultimately to the rank and to the enjoyment of all the rights and privileges of freemen and citizens of the United States. This I conceive to be the precise object of the Government. If we fulfill not these duties, which grow naturally out of our relation to Indians, we cannot avoid the imputation of injustice, unkindness, and unfaithfulness to them—our national character must suffer in the estimation of all good men. If we refuse to do the things we have mentioned for the Indians, let us be consistent and cease to call them 'children,' and let them cease to address our President as their 'Great Father.' Let us leave to them the unmolested enjoyment of the territories they now possess and give back to them those which we have taken away from them.

As the Government assumes the guardianship of the Indians, and in this relation provides for their proper education, provision also should be made for the exercise of a suitable government and control over them. This government unquestionably should be in its nature parental—absolute, kind, and mild, such as may be created by a wise union of a well-selected military establishment, and an education family. The one possessing the power, the other the softening and qualifying influence, both combined would constitute, to all the purposes requisite, the parental or guardian authority.*

This report was particularly important inasmuch as it was the first attempt to apply the principles of international law applicable to partially civilized tribes to the American Indians, and although in substance the views of Morse were identical with those previously expressed by Washington and Monroe himself, coming from the church the argument against involuntary removals was unanswerable. Just provision must be made for the Indians irrespective of what their legal status might be.

Monroe had deliberately pitted himself against the Buffalo Party. After half a century of unbroken and successful reign it was indisposed to change its platform notwithstanding his moralizing. At any rate, whether with or without the connivance of Monroe, the issue between them was framed in a great test case known as *Johnson v. McIntosh*, which reached the Supreme Court in 1821 and in which the Indian interests were to have the aid of Webster, the Attorney General. The question was whether an Indian title purporting to have been granted by the Peankeshaw tribe of Ohio to a body of private individuals in colonial days was good as against the United States.

Marshall had not been at his best in 1810 when he delivered the first Indian decision of the Supreme Court in *Fletcher v. Peck*. In that case too, his opportunity had been limited to the legal range of the question involved. Since then, in the calm that came to him with the passing of Jefferson from the White House, he had found opportunity to reflect much upon the status of the Indians. In 1823, with an enhanced maturity of judgment, speaking for the Supreme Court he handed down its unanimous opinion. After exhausting the historical aspects of the Indian land ques-

* Quoted in *The Question of Aborigines in the Law and Practice of Nations*, Alpheus Snow, Afro-American Press, 1969.

tion from the standpoint of English and American law and the law of nations, he said:

The United States, then, have unequivocably acceded to that great and broad rule by which its civilized inhabitants now hold this country. They hold and assert in themselves the title by which it was acquired. They maintain, as all others have maintained, that discovery gave an exclusive right to extinguish the Indian title of occupancy either by purchase or conquest; and gave also a right to such a degree of sovereignty as the circumstances of the people would allow them to exercise.

Was there in that region whither had fled the spirits of Washington and Tecumseh some gentle stirring at these words? Surely, for the departed souls of men there is a message of their ultimate earthly victories.

What had been merely an executive policy inaugurated by Washington and supported by statutes, was now the supreme law of the land. Jefferson and Adams were overruled. Moreover, Marshall's great and fearless opinion could only be construed by Congress to imply that where the United States did not undertake to extinguish the Indian title, or when the Indians would not cede it and it became necessary to exercise the right of eminent domain over their lands, the Indians were entitled under the Constitution to just compensation. Congress might regulate their affairs as it saw fit, but in doing so it must not assume to confiscate their property any more than the property of white people.

Although for forty years Monroe had been trying to establish the rights of the Indians as now at last judicially construed, it was with amazement that the Buffalo Party heard the tribal title defined as a right to the perpetual occupancy of the land with the privilege of using it in such mode as the Indians saw fit until their right of occupancy had been surrendered to the government; that it was a property right; that no one could acquire the tribal title without the consent of the United States, by adverse possession or otherwise; that it could not be extinguished even by the United States save with the consent of the Indians. Moreover, Marshall had adopted entirely Monroe's and Morse's theory of governmental guardianship over the tribes in lieu of the specious theory of Adams and Story.

Yet, Monroe was not a dreamer. He alone could not create an Indian Territory merely by continuing for a few years to make contiguous treaty grants. The experience of the Cherokees in Georgia had shown how useless was the moving of a tribe from one place to another. Even now the southern Cherokees who had led the way to Arkansas were being ruthlessly ousted while Congress was conducting a war upon the Arickarees to clear the upper Missouri. It was, therefore, manifest that if removals were to result in anything but the destruction of the tribes in the process of migration, they must be assigned to permanent domains from which they could not again be uprooted, and to be permanent they must be protected against the old treaty processes. Therefore, in 1824, he addressed Congress as follows:

My impression is equally strong that it would promote essentially the security and happiness of the tribes within our limits if they could be prevailed on to retire west and north of our states and Territories on lands to be procured for them by the United States in exchange for those on which they now reside. Surrounded as they are, and pressed as they will be, on every side, by the white population, it will be difficult, if not impossible for them, with their kind of government, to sustain order among them. Their interior will be exposed to frequent disturbances, to remedy which the interposition of the United States will be indispensable, and thus their government will gradually lose its authority until it is annihilated. In this process the moral character of the tribes will also be lost, since the change will be too rapid to admit their improvement in civilization to enable them to institute and sustain a government founded on our principles . . . or to become members of a state, should any state be willing to adopt them in such numbers, regarding the good order, peace and tranquility of such State. But all these evils may be avoided, if these tribes will consent to remove beyond the limits of our present States and Territories. Lands equally good and perhaps more fertile, may be procured for them in those quarters. The relations between the United States and such Indians would still be the same.

Considerations of humanity and benevolence, which would have now great weight, would operate in that event with an augmented force, since we should feel sensibly the obligation imposed on us by the accommodation which they thereby afforded us. Placed at ease, as the limited states would then be the improvement of those tribes in civilization and in all arts and usages of civilized life would become a part of a general system which might be adopted on great consideration, and in which every portion of our Union would then take an equal interest. . . .

I submit this subject to the consideration of Congress under a high sense of its importance and of the propriety of an early decision on it.*

For the handling of Indian affairs the old administrative system of two Indian Districts created in 1786 which had endured for half a century was now hopelessly inadequate to Monroe's extensive plans. Accordingly he had addressed himself energetically to the creation of a special department for the administration of the increasingly important Indian affairs of the federal government. To this, however, Congress was not to consent. But it compromised with Monroe by creating in 1824 a fiscal bureau in the War Department called the Bureau of Indian Affairs which was given no real administrative powers.

Finally, after addressing a special message to Congress in 1825 designating the general region in which the Cherokee and Choctaw grants had

---

* Message to the Senate and House of Representatives, dated March 30, 1824.

been made as the most suitable for an Indian Territory, in his last annual message Monroe said:

The removal of the tribes from the territory which they now inhabit to that which was designated in the message at the commencement of the session, which would accomplish the object for Georgia, under a well digested plan for their government and civilization, which should be agreeable to themselves, would not only shield them from impending ruin, but promote their welfare and happiness. Experience has dearly demonstrated that in their present state, it is impossible to incorporate them in such masses, in any form whatever, into our system. It has also demonstrated with equal certainty that without a timely anticipation of and provision against dangers to which they are exposed, under causes which it will be difficult, if not impossible to control, their degradation and extermination will be inevitable.*

The principal remaining objection to the formation of an Indian territory seems to have been that once set apart as such, land that might later be desired could not be taken from the Indians. At any rate again Congress rejected Monroe's wise proposals in favor of its old policy of merely narrowing the tribal domains in the West wherever they might be, to reservations, and of transferring the eastern tribes that could be compelled to migrate.

* Message to the Senate and House of Representatives, dated January 27, 1825.

20

Adams and Jackson:
the Rape of the Eastern Tribes

John Quincy Adams had none of the humanitarian instincts of Monroe. It has been shown that with Story he had already served as the representative of the Buffalo Party in its legal conflict with the Indians when he became President in 1825. This did not promise well for the Indians.

Meantime, however, a powerful pen had come to the aid of the hard-pressed tribes, for James Fennimore Cooper, following the lead of Chateaubriand, was idealizing the Indian and Indian character, laying the foundation for Longfellow's immortal song of Hiawatha. After all, was there not something in the nature of these savages nobler than what the frontiersman depicted? Many a person in the East, removed from contact with Indian dangers, was now asking that question. Were these people, after all, no more than wild animals? Mere appendages to the soil? Was it true that they were without human instincts, incorrigible to civilization?

On the other hand the frontier scoffed at Marshall's recent decision. He and Monroe and the preachers might prate of Indian legal rights just as Washington had done. Let them have their theories. As a plain common sense proposition the "savages" must be made to yield.

"Old Man Eloquent" was a true son of his father. In his heart there was no room for Indians. The *Last of the Mohicans* had no more appeal for him than *Atala* for Jefferson. New ideas did not come to him from novelists—mere romancers. Inconsistency was a weakness to be abjured. As President he must maintain the views he had entertained as a lawyer. Marshall's recent decision had not changed his opinion in the least. He could not believe the Indians had any real title to the soil. Like others he believed it quixotic to create an Indian Territory. Therefore, he was the willing tool of Congress which, to circumvent the recent decision of the Supreme Court, was demanding that treaties of removal on the part of the eastern tribes be obtained by one means or another, and that the western tribes be concentrated so that their surplus lands might accommodate the migrants.

Amply provided with funds, Adams' commissioners set to work knowing from past experience how to obtain the desired treaties. Thus, in 1825 and 1826 the peaceful Dakotas, Osages, Kansas, Chippewas, Sauk, Fox, Winnebagos, Miamis, Ottawas, and Potawatomis, were all led into treaties, agreeing to concentrate, while the Shawnees of Missouri were pressed westward to Kansas to be joined soon by those who remained in Ohio and another band expelled from Texas.

Sufficient land from the Osages and Kansas Indians having been obtained to accommodate the Georgia tribes, in 1825 a treaty of removal was also obtained from the Creeks.

The Indians now saw that treason, something even worse than the pillage and murder of the past, was threatening them from all sides. The traitorous habits of the white man born of the lust which private property hand engendered among them were invading their own ranks. Against such an enemy no boundary was secure. They knew of the doings of Reverend Eleazer Williams among the Oneidas. There were numerous Indians engaged in tricking the tribes into treaties, bribing their chieftains, betraying them like Williams who, Judas-like, with his smug hypocrisy, had preached Christianity while practising theft as Superintendent of the Northern Department.*

This hydra-headed thing which had made its appearance among them must be destroyed. The missionaries had told them of Iscariot. But lately they had witnessed the "two-tongues" of Arnold. While in the flesh, at least, both had escaped punishment, yet for their kind the white man's penalty was death. Was the Great Spirit of the Indian to be less exacting in retributive justice than the white man's Jehovah? Now more than one tribe prescribed death as the penalty for a chief who without authority presumed to release the territory of his people to the whites. When, therefore, the veteran Creek chieftain, William McIntosh, in whom his tribe had long reposed the fullest confidence, was convicted of having accepted a bribe to secure the signature of the requisite number of chieftains to the treaty of 1825, the Creeks repudiated the treaty and executed McIntosh. Thus they compelled the government to recognize in the new treaty of 1826 the rights conferred on them by that of 1804.

The execution of McIntosh was widely exploited as evidence of the

* In 1853 this remarkable man assumed the role of the Dauphin of France, or the lost Prince, Louis XVII, winning many adherents in Paris.

encouragement which the Supreme Court had afforded the "murderous savages." Nor was the state of Georgia to be thwarted by the treaty of 1826. Without delay, it denied the validity of that treaty and threatened to resist with arms the federal agents who were designated to resurvey the Creek domain. Neither Congress nor Adams made any effort to help the Creeks, leaving Georgia to settle her own difficulties with them.

Such was the first instance in which a state threatened to use force to resist the federal government. Nullification was an accomplished fact.

So far, litigation by the Indians had served but one purpose—to unmask hypocrisy. Forced into the open by the Supreme Court, the state of Georgia had merely been compelled to disclose its real purpose—the taking of the land of the Indians without regard to the guarantees either of the federal Constitution or its own statutes, and for all this John Quincy Adams was responsible. The stains upon his own character and the honor of the nation, by his failure to enforce the federal law and his consent to its overriding by Georgia, can never be eradicated. His conduct in this one instance alone brands him forever as a political weakling, irrespective of whether he was actuated by ill design or moral cowardice. It was a time when his duty as President demanded action that he called upon Congress and through it upon the nation, to manifest their political self-respect. This, however, was entirely beyond him. In 1827 he proceeded to establish Fort Leavenworth in Kansas as a base for the "pacification" of the far West, while in the last year of his administration a war was waged upon the Chippewas and Winnebagos of Wisconsin in order to "dispose" them to remove.

About 1826 a new element of value entered into the lands of the Cherokees through the discovery of gold in their domains, whereupon white adventurers swarmed to the mines. Seeking to protect themselves, the eastern Cherokees adopted the Constitution drawn up for them by Ross with the aid of Seargent, and declared themselves a sovereign, independent nation.

Immediately the Legislature of Georgia passed resolutions declaring the Cherokee lands belonged to the state "absolutely"; that the Indians were only "tenants at her will"; that Georgia had the right to, and would, extend her "conventional limits" and coerce obedience to them "from all descriptions of people, be they white, red, or black." Whereupon laws taking over the Cherokee lands were enacted, dividing them into counties, and purporting to annul "all laws, usages and customs" of the Indians, forbidding them to hold courts, or to make laws or regulations for the tribe. White persons found in the Cherokee country without a license from the government were, upon conviction, however, to be imprisoned at hard labor for four years, and a State Guard was established to "protect" the mines and arrest anyone detected in a violation, not of the rights of the

Indians who were deemed to have none, but of the recently enacted state laws.

Such was the situation when Jackson, the accepted leader of the Buffalo Party, was elected President in 1828.

Nevertheless, upon the advice of Governor Houston of Tennessee the Cherokees appealed at once to the President who, true to his character, rebuffed them, upholding Georgia and urging the eastern Cherokees to exchange their country for lands in Arkansas.

Unfavorable reports having come from the western Cherokees, Ross sent another commission to survey the region available for his people which reported that it was unfit for agriculture and that, once there, they, like those who had preceded them, would have to fight savage tribes. Again they appealed to the President; and again Jackson told them Georgia had absolute authority over them.

One of the strangest things in American history was now to occur. Among the Cherokees, before the migration to Arkansas in 1810 of the first contingent under Chief Jolly, there had lived as a lad one of the most remarkable men in American history in the person of Sam Houston to whom they gave the name "Blackbird." Born near Lexington, Virginia, in 1793, he was the son of a Scotch-Irish father and Revolutionary soldier upon whose death in 1806 Houston removed with his mother to Blount County, Tennessee. When about fifteen he ran away from home, taking up his residence with the Cherokees and becoming a member of Chief Jolly's household where he formed a lifelong friendship with John Ross. While with the tribe he learned to love them. On one occasion he severely chastised the agent of the Indian Department who while attending a council at which the Cherokees were urged to move again, attempted to ravish Jolly's little daughter whom Blackbird regarded as his sister.

Out of this incident arose such hostility on the part of the Indian officials in Washington that upon Chief Jolly's advice, at the age of eighteen Houston returned to his home in order to avoid arrest and persecution, but during the War of 1812 in which he served with great gallantry under Jackson, again he came in contact with his old friends, the Cherokees. Fighting beside them at the battle of Tohopeka, he was severely wounded. At this time he won the commission of Lieutenant and the lifelong friendship of "Old Hickory," with the result that in 1817 he was appointed subagent in charge of the removal of the Cherokees from East Tennessee to Arkansas.

While serving in this office he was so severely rebuked by Calhoun, the Secretary of War, who no doubt reflected the hostility of the department, for appearing before him in Indian dress, that in 1818 he resigned. Thereupon he entered a law office in Nashville. After being admitted to the bar he became District Attorney, and from 1823 to 1827 represented his district in Congress, being elected governor of Tennessee in 1827 as a Democrat and ardent supporter of Jackson. Thus he found himself in a position to intervene in Indian affairs with an authority that enabled him to serve the interests of the Cherokees in many ways, not least by direct appeals in their behalf to the new President.

In 1829 Governor Houston married, but three months later, left the young bride who had confessed she did not love him. At this time he was at the very pinnacle of success, being regarded as an almost sure successor to Jackson in the White House. But so crushed was he with grief that resigning as governor, he proceeded at once to join Chief Jolly, the Indian friend of his childhood, in the wilds of Arkansas, became a member of the western Cherokees, and resumed their mode of life.

There was much in the situation of these poor people that appealed to him, to make him forget his own troubles. In their migration they had passed through a dreadful ordeal. Beset now on every side by dishonest government officials, they were being bled white. A former Congressman himself, at these things his blood boiled. Now too, romance entered into Houston's life anew for having been divorced by his Tennessee bride, he married Talahina, the mission-educated daughter of a Cherokee mother and an English-born American who had served as an Officer with the Cherokees in the Creek War.

In 1830 there came to him from "Old Hickory" a letter expressing sympathy for his domestic crash. The appeals he had made to the government on behalf of his adopted people had gone in vain. So he went to Washington as the accredited delegate of the Cherokees accompanied by two chieftains and in person appealed to his old friend, the President. Jackson listened, weighed the evidence collected by Houston against the Indian agents who had been exploiting the western Cherokees, and with characteristic promptness dismissed four of the worst and the head of the Indian Bureau as well. But as to the Georgia Cherokees he was adamant.

The dismissals provoked a storm of protest in and out of Congress, but the former Governor of Tennessee was not dismayed. Loyal to his old chief, before returning to Arkansas he would appeal to the country to support Jackson, Thus, from Washington to Baltimore, Philadelphia, New York and Boston he went, a magnificent figure, standing six feet three inches, and clad in picturesque Cherokee dress, appealing to the conscience of the nation, ignoring the groups of enemies who called him a renegade squawman, vagabond, scoundrel, and self-advertiser, even accusing him of profiting at the expense of the Indians. Everywhere he was met with a tumult of enthusiasm. Women by the hundreds came to his meetings out of curiosity to see the man who had renounced a governorship because of a broken heart and put on "savage" trappings to fight single-handed against the rapacious oppressors of a weak and helpless race.

They proclaimed him a hero, a real knight-errant. In consequence of the public opinion aroused by him, and the impassioned poems of John Howard Payne, there was exhumed from the files of Congress the long forgotten report of the Reverend Jedediah Morse of 1820 which was spread on the record, while the Committee on Indian Affairs reported to Congress as follows:

The Indians are paid for their unimproved lands as much as the privilege of hunting and taking game upon them is supposed to be worth, and the Government sells them for what they are worth to the cultivator. . . . Improved lands or small reservations in the States are in general purchased at their full value to the cultivator. To pay an Indian tribe what their ancient hunting grounds are worth to them after the game is fled or destroyed as a mode of appropriating wild lands claimed by Indians has been found more convenient, and certainly it is more agreeable to the forms of justice, as well as more merciful, than to assert the possession of them by the sword. Thus, the practice of buying Indian titles is but the substitute which humanity and expediency have imposed in place of the sword in arriving at the actual enjoyment of property claimed by the right of discovery and sanctioned by the national superiority allowed to the claims of civilized communities over those of savage tribes. . . .*

How much of "humanity" and how much of "expediency" there was behind a system under which the guardian acquired the land of its wards at a "presumed" and "estimated" price without private competition can well be imagined.

Nevertheless, the unanswerable arguments contained in Morse's report compelled the Act of May 28, 1830, empowering the President to exchange territorial lands of the United States for those owned by the tribes within the state limits, thus doing away with the necessity of resorting to the fiction of Indian treaties for this purpose with all the frauds and delays incident to that process. In other words, when a tribe was now ready to move, the arrangement could be made with a maximum of expedition and a minimum of politics. This in itself was a great advance.

But those who heard Houston could not absolutely control a Congress hostile to Jackson. Moreover, Houston had urged the President to award him a contract for rationing the migrating Cherokees and Creeks in order that he might prevent their being exploited by the Indian agents. Jackson's enemies got wind of this and instantly it was cited as proof of the President's corruptibility and the real motives of Houston, so that dramatic and stirring as the latter's efforts had been, the dismissed officials were reinstated. Soon after this Houston returned to Arkansas and nothing more was done about creating the Indian Territory. In vain Houston raged

* H. R. Rep. No. 227, Feb. 24, 1830, 21st Cong., 1st sess.

against the government. His letters of protest against its methods remained unanswered.

Seeing that even Houston could accomplish nothing with Jackson on behalf of the eastern Cherokees, Ross employed John Seargent as counsel who, being an ardent Whig and political opponent of the President with no liking for his form of democracy, had no scruples against bearding him. Upon the advice of William Wirt, Seargent, and Houston, Ross decided to institute an injunction suit to restrain Georgia from executing its tyrannical statutes.

Such was the origin of the case of the *Cherokee Nation v. the State of Georgia.* At Wirt's request Judge Dabney Carr laid the whole matter before Marshall, Wirt having determined to proceed with it or drop it as the Chief Justice should advise. Marshall, of course, declined to express any opinion on the legal questions involved:

I have followed the debate in both houses of Congress, with profound attention and with deep interest, and have wished, most sincerely, that both the executive and legislative departments had thought differently on the subject. Humanity must bewail the course which is pursued, whatever may be the decision of policy.

Before the case could be heard by the Supreme Court, Georgia availed herself of an opportunity to show her contempt for the national judiciary and to assert her "sovereign rights." A Cherokee named George Tassels was convicted of murder in the Superior Court of Hall County, Georgia, and lay in jail until the sentence of death should be executed. A writ of error from the Supreme Court was obtained, and Georgia was ordered to appear before that tribunal and defend the judgment of the State Court.

The order was signed by Marshall. Georgia's reply was as insulting and belligerent as it was prompt and spirited. The Legislature resolved that "the interference by the Chief Justice of the Supreme Court of the United States, in the administration of the criminal laws of this state, is a flagrant violation of her rights"; that the governor "and every other officer of this state" be directed to "disregard any and every mandate and process . . . purporting to proceed from the Chief Justice or any Associate Justice of the Supreme Court of the United States"; that the governor be "authorized and required, with all the force and means . . . at his command . . . to resist and repel any and every invasion from whatever quarter, upon the administration of the criminal laws of this state," that Georgia refuses to become a party to "the case sought to be made before the Supreme Court"; and that the governor, "by express," direct the sheriff of Hall County to execute the law in the case of George Tassels.

Five days later, Tassels was hanged, and the Supreme Court of the United States, powerless to vindicate its authority, defied and insulted by a "sovereign" state, abandoned by the administration, was humiliated and helpless.

When he went home on the evening of January 4, 1831, John Quincy Adams, now a member of Congress, wrote in his diary:

The resolutions of the Legislature of Georgia setting at defiance the Supreme Court of the United States are published and approved in the Telegraph, the Administration newspaper at this place. . . . The Constitution, the laws and treaties of the United States are prostrate in the State of Georgia. Is there any remedy for this state of things? None. Because the Executive of the United States is in League with the State of Georgia. . . . This example . . . will be imitated by other States, and with regard to other national interests—perhaps the tariff. . . . The Union is in the most imminent danger of dissolution. . . . The ship is about to founder.

Is it conceivable that in noting such an observation upon the political import of the situation that "Old Man Eloquent" failed to understand his direct responsibility for it? Is it possible that vanity or any other influence could so blind his eyes to the facts? Did he not see that he himself by his own neglect had induced that which now occurred? Neutrality on his part in the conflict between state and nation had but emboldened Jackson and Georgia to violate the Constitution.

Ross, supported by Seargent, was not to be intimidated. At once they brought suit in the Supreme Court in the name of the Cherokee Nation to enjoin the state from executing its laws, and at the February term of 1831 the cause was argued for the Indians by Wirt and Seargent. Georgia disdained to appear—not for a moment would that proud state admit that the Supreme Court of the nation could exercise any authority whatever over her.

On March 18, 1831, Marshall delivered the opinion of the majority of the court, and in it he laid down the broad policy which the government has unwaveringly pursued ever since. At the outset the Chief Justice plainly stated that his sympathies were with the Indians, but that the court could not examine the merits, or go into the moralities of the controversy, because it had no jurisdiction. The Cherokees sued as a foreign nation, but, while they did indeed constitute a separate state, they were not a foreign nation. The relation of the Indians to the United States is "unlike that of any other two people in existence." Their territory comprises a "part of the United States." Summarized, his opinion was, that in foreign affairs and commercial regulations, the Indians are subject to the control of the national government. "They acknowledge themselves in their treaties to be under the protection of the United States." They are not, then, foreign nations, but rather "domestic dependent nations. . . . They are in a state of pupilage." Foreign governments consider them so completely under our "sovereignty and dominion" that it is universally conceded that the acquisition of their lands or the making of treaties with them would be "an invasion of our territory, and an act of hostility." By the Constitution power is given Congress to regulate commerce among the states, with foreign nations, and with Indian tribes, these terms being "entirely distinct."

The Cherokees not being a foreign nation, the Supreme Court has no jurisdiction in a suit brought by them in that capacity. Furthermore, the court was asked "to control the Legislature of Georgia, and to restrain the exertion of its physical force"—a very questionable "interposition," which "savors too much of the exercise of political power to be within the proper province of the juridical department." In "a proper case with proper parties," the court might, perhaps, decide "the mere question of right" to the Indian lands. But the suit of the Cherokee Nation against Georgia is not such a case.

Marshall closed with a reflection upon Jackson in terms much like those with which, many years earlier, he had so often rebuked Jefferson:

If it be true that the Cherokee Nation have rights, this is not the tribunal in which those rights are to be asserted. If it be true that wrongs have been inflicted, and that still greater are to be apprehended, this is not the tribunal which can redress the past or prevent the future.

As a "foreign state" the Indians had lost, but the constitutionality of Georgia's Cherokee statutes had not been affirmed. Wirt and Seargent had merely erred as to the method of attacking such legislation. Another proceeding by Georgia soon brough the validity of her expansion laws before the Supreme Court. Among the missionaries who for years had labored in the Cherokee Nation was one Samuel A. Worcester, a citizen of Vermont. This brave minister, licensed by the national government, employed by the American Board of Commissioners for Foreign Missions, appointed by President John Quincy Adams to be postmaster at New Echota, a Cherokee town, refused, in company with several other missionaries, to leave the Indian country.

Worcester and a Reverend Mr. Thompson were arrested by the Georgia guard. The Superior Court of Gwinnett County released them, however, on a writ of habeas corpus, because, both being licensed missionaries expending national funds appropriated for civilizing Indians, they must be considered as agents of the national government. Moreover, Worcester was postmaster at New Echota. Georgia demanded his removal and inquired of Jackson whether the missionaries were government agents. The President assured the state that they were not, and removed Worcester from office. Thereupon both Worcester and Thompson were promptly ordered to leave the state. But they and some other missionaries remained, and were arrested, dragged to prison—some of them with chains around their necks, tried and convicted. Nine were pardoned upon their promise to depart forthwith from Georgia. But Worcester and one Elizur Butler sternly rejected the offer of clemency on such a condition and were put to hard labor in the penitentiary.

From the judgment of the Georgia court, Worcester and Butler appealed to the Supreme Court of the United States. Once more Marshall and Georgia confronted each other; again the Chief Justice faced a hostile President far more direct and forcible than Jefferson, but totally lacking the subtlety and skill of that incomparable politician. Thrilling and highly colored accounts of the treatment of the missionaries had been published in every Northern newspaper; religious journals published conspicuous soul-stirring narratives of the whole subject; feeling in the North ran high; resentment in the South rose to an equal degree.

This time Georgia did more than ignore the Supreme Court. She formally refused to appear; formally denied the right of that tribunal to pass upon the decision of her courts. Never would Georgia so "compromise her dignity as a sovereign state," never so "yield her rights as a member of the Confederacy." The new Governor, Wilson Lumpkin, vowed that he would defend those rights by every means in his power. When the case of *Worcester v. Georgia* came on for hearing before the Supreme Court, no one answered for the state. Wirt, Seargent, and Elisha W. Chester appeared for the missionaries as they had for the Indians. Seargent made an extended and powerful argument, incomparably superior to those of Story and Adams in 1810, and was ably supported by Wirt—an argument that was to cost him the everlasting enmity of Jackson and the Buffalo Party.

Marshall's opinion, delivered March 3, 1832, and closely following Seargent's theory of the law, is one of the noblest he ever wrote. "The legislative power of a State, the controlling power of a Constitution and laws of the United States, the rights, if they have any, the political existence of a once numerous and powerful people, the personal liberty of a citizen are all involved," begins the aged Chief Justice. Does the act of the legislature of Georgia, under which Worcester was convicted, violate the Constitution, laws, and treaties of the United States? That act is "an assertion of jurisdiction over the Cherokee Nation."

He then goes into a long historical review of the relative titles of the natives and of the white discoverers of America; of the effect upon these titles of the numerous treaties with the Indians; of the acts of Congress relating to the red men and their lands; and of previous laws of Georgia on these subjects. This part of his opinion is the most extended and exhaustive historical analysis Marshall ever made in any judicial utterance, except that on the law of treason during the trial of Aaron Burr.

Then comes his condensed, unanswerable, brilliant conclusion:

A weaker power does not surrender its independence, its rights to self-government, by associating with a stronger, and taking its protection. A weak state, in order to provide for its safety, may place itself under the protection of one more powerful, without stripping itself of the right of self-government, and ceasing to be a state. . . . The Cherokee Nation . . . is a distinct community, occupying its own territory . . . in which the laws of Georgia can have no right to enter but with the assent of the Cherokees themselves, or in conformity with treaties, and with the acts of Congress. The whole intercourse between the United States and this nation is by our Constitution and laws vested in the Government of the United States.

The Cherokee Acts of the Georgia Legislature "are repugnant to the Constitution, laws and treaties of the United States. They interfere forcibly with the relations established between the United States and the Cherokee Nation." This controlling fact the laws of Georgia ignore. They violently disrupt the relations between the Indians and the United States; they are equally antagonistic to acts of Congress based upon these treaties. Moreover, "the forcible seizure and abduction" of Worcester, "who was residing in the nation with its permission and by authority of the President of the United States, is also a violation of the acts which authorize the Chief Magistrate to exercise this authority."

The great Chief Justice closed with a passage of eloquence almost equal to, and of higher moral grandeur than, the finest passages in *McCulloch v. Maryland* and in *Cohens v. Virginia*. So the decision of the court was that the judgment of the Georgia court be "reversed and annulled."

Two years before, the interest of the country in Indian affairs had been deeply stirred by Houston's dramatic appeals. Now it was swept by a tempest of popular excitement. South and North took opposite sides. The doctrine of state rights was at once invoked by the Buffalo Party on behalf of Georgia. Congress was thrown into veritable furor; Georgia was enraged; The President was agitated and belligerent. In a letter to Ticknor, written five days after the judgment of the court was announced, Story accurately portrays the situation:

The decision produced a very strong sensation in both houses; Georgia is full of anger and violence. . . . Probably she will resist the execution of our judgment, and if she does I do not believe the President will interfere. . . . The Court has done its duty. Let the Nation do theirs. If we have a government let its commands be obeyed; if we have not it is as well to know it at once, and to look to consequences."

Story's forecast was justified. Georgia scoffed at Marshall's opinion, flouted the mandate of the Supreme Court. "Usurpation!" cried Governor Lumpkin. He would meet it "with the spirit of determined resistance." Jackson defied the Chief Justice. "John Marshall has made his decision— *now let him enforce it!*" the President is reported to have said.

But old Chief Jolly, whom Houston had actually struck on one occasion while intoxicated, had never lost faith in him. Now, under Chief Jolly's influence he curbed his drinking and making a great effort to pull himself together, agreed to go to Washington again as a delegate of the tribe. Soon after his arrival in the Capital, a representative from Ohio bent on defeating his mission indirectly charged Jackson on the floor of Congress with having removed the head of the Indian Bureau in 1830 upon the trumped-up charges of Houston in order that the latter might be given a valuable contract. This reawakened all that was high and fine in Houston's nature, appealing to the deep loyalties that were in him. Then and there he challenged the veracity of his accuser who though refusing to respond to him allowed him to be subjected to a penalty under the House rules which afforded Houston an opportunity to be heard.

Seeing that not only his own political reputation but that of the President was at stake, quietly he set about preparing his defense with Francis Scott Key, the author of the Star Spangled Banner, as his counsel, holding conferences with Jackson's Congressional leaders—Senators Grundy, Polk and Peyton.

"But Sam," said the President, "unless you shed that savage rig you've been wearing, they'll call you a barbarian when you get up to speak!"

When unabashed Houston explained that he had no other and no money to buy new clothes, the President furnished him with the requisite funds. So it was that on the day in April 1832 appointed for the Cherokee Delegate to defend himself, Houston appeared, not in Indian costume, but in all the elegance of apparel which Jackson's pocketbook afforded. Once more in blue broadcloth, he was in appearance the superb Governor of Tennessee and former member of Congress!

"By God," said he to Senator Polk, "I've been in hell for nearly three years, and here's my chance to rise out of it and be a man once more."

When the session was called to order all of Washington that could be there was present to witness what was now seen to be a defense of the President. Speaking with majestic poise first he made tribute to Jackson, then passed to the theme nearest his heart. Such a story of wrongs to the Indians as he now poured forth had never been heard before.

It was a triumph of combined eloquence and reason. The fine of $500 imposed by the House was remitted by Jackson the next day, for Houston had saved his reputation while defending the Indians. During his subsequent stay in Washington Houston was to bring about two acts of great political importance affecting the Indians. First, he compelled Congress through Jackson to create by the Act of July 9, 1832 a Commissioner of Indian Affairs of the War Department, in lieu of the old bureau which was nothing more than a fiscal agency, to whom was entrusted "the direction and management of Indian affairs and of all matters arising out of Indian relations." This was a preliminary measure to the creation of the long-promised Indian Territory "of which there can be not the least doubt," he told the President. Second, he told Jackson all he had learned while living among the Arkansas Cherokees of the possibility of seizing Texas for the United States. Indeed, it was far from likely that Burr had contemplated no more than setting up a Republic in Texas and California with the aid of the Commanches, Pawnees, Kiowas, Shoshones, Navajos, Apaches and other powerful southwestern tribes. Even now the Mexican Indians were raiding across the Rio Grande from Cohiula affording ground for intervention. "Why should Texas not be taken?" enquired Houston.

Fascinated, Jackson gazed hard at Houston's map. Surely he must have asked himself if after all there were not foresight in Burr's plan. Plainly, Texas would furnish new territory for King Cotton, much needed since the Missouri compromise of 1820. There would be no difficulty enlisting the support of the party in power—the Democratic Party dominated by the South.

"Appoint me commissioner to negotiate treaties with the Texas tribes ostensibly for the purpose of protecting American trade interests," urged Houston, "and I'll do the rest!"

From that moment the fate not only of Texas but of California was sealed, for Polk, already a risen power in the Senate, was as quick as "Old Hickory" to envisage the future. In December, 1832 Houston left the Capital for Texas, stopping on the way only long enough to bid his Indian wife farewell until the second great mission of his life had been fulfilled. "Sam's good medicine for Indians," said old Chief Jolly as he grasped his hand. "I knew all time you would make big council help Cherokees."

21

Black Hawk, Osceola
and the Eventual Removal West

WHILE HOUSTON WAS ENGAGED stirring the country to its depths, a very able Indian lawyer in the person of John W. Quinney, of the Stockbridge tribe, was cooperating with Ross, Seargent and Houston in the effort to enforce a recognition of Indian rights.

Born in 1797, he was one of three members of his tribe to receive a common school education at the hands of the United States in New York where his tribe then resided, gradually, as hereditary chief acquiring a complete control of tribal affairs. When in 1822 conditions for the Indians in New York became intolerable, he visited Wisconsin with a deputation appointed at his suggestion to negotiate a treaty with the Menominees of the Green Bay country pursuant to which sufficient land was purchased there to accommodate his tribe, and in 1825 procured the passage of a statute by New York granting it full value for its lands. Meantime, however, the United States intervened and purchased the Menominee lands sold to Quinney, so that in 1828 he was compelled to present a petition to Congress that the prior purchase by his people be respected. Denying this, Congress reimbursed the Stockbridge tribe for their expenditures, whereupon, in 1832, Quinney obtained from Wisconsin the grant of two townships at Lake Winnebago whither the tribe removed, and in 1833 drafted for his people a constitution as the basis of the tribal government under which for many years they continued to live.

As evidence of how desperately the Indians were seeking at this time to accommodate themselves to their fate should also be cited the efforts of the New York Iroquois, who for years had seriously contemplated a removal to Wisconsin where they might set up a new state. In the end only the Oneidas and a few others removed, settling in the Green Bay country in 1832. Senecas, Tuscaroras and Onondagas were determined to hold on to their lands in New York at any cost.

Meantime Catlin had toured the West and made what was, perhaps, the most extensive study so far of the western tribes. The author of numerous subsequent works on the Indians of the greatest value, in order to introduce the American native to Europe he visited England, France, and Belgium with an Indian Museum which aroused the deepest interest. The dragon's teeth sowed by Adams and Jackson, however, were now to yield the inevitable whirlwinds. Georgia's successful nullification on two occasions had only encouraged Alabama to set at naught the federal statutes applicable to the processes of the criminal law, necessitating the Force Bill of 1833. Within a year a federal soldier and a deputy marshal, while enforcing the federal law with respect to the Creek Indians, killed a citizen of Alabama. Indicted for murder by the state, he removed his case into Federal Circuit Court. The most intense excitement was aroused by the United States assuming to assert its exclusive jurisdiction. The idea that the federal government should assume to usurp state authority over the Indians was itself declared by the Alabama press to be preposterous. That it should now deny the criminal jurisdiction of the state over "murder" was intolerable! Strongly supported by public sentiment, the press, not only of Alabama but of Georgia and Mississippi, declared that Alabama would secede from the Union rather than suffer such injustice. The episode even became the subject of heated speeches on the floor of Congress in 1834, in which it was declared that "to control the jurisdiction of a state over persons charged with the commission of crimes within its limits was an assumption of a higher power than had yet been exercised" by the federal government over the states. Senator Gilmer of Georgia expressed the hope that

the House as well as the people of every part of the country were now ready to unite with him in the fervent wish that this may be the last attempt to enforce that Act (of 1833), and that this excrescence upon the body politic which had been thrown out from the exuberant feeling excited by the late collision between the General and State Governments would soon be cut off.*

Whether or not Jackson perceived that Adams and he were responsible for all this, he did understand the inevitable consequence should the case reach the Supreme Court—another nullification of its mandate. Accordingly, he deputed Lewis Cass to negotiate a compromise with the Governor

* Washington Globe and Mobile Commercial Advertiser, Aug., 1833, to Jan., 1834, passim. See also 23rd Cong., 1st Sess., pp. 2302, 2311, 2709, Jan. 6, Feb. 15, 18, 1834, esp. speeches in the House of Dixon H. Lewis of Alabama and Seaborn Jones of Georgia, and of Geo. R. Gilmer in the Senate.

of Alabama and caused the case while in the Circuit Court of Appeals to be withdrawn.*

But this was only one of the whirlwinds which Jackson was to reap, for in the history of society there is a principal that is invariable—the excessive repression of one race by another however weak the former may be produces reaction whether by strengthening the oppressed or in the moral attitude of the world. Among the tribes of Illinois and Wisconsin there was still the tinder of the spirit of Pontiac to which a rash leader named Black Hawk was to apply the spark in April, 1832, or the year following the enforced removal to Kansas of the remaining Shawnees.

Born at the Sac or Sauk village near the mouth of Rock River in 1767, at the age of fifteen he had distinguished himself in Indian warfare and before he was seventeen had taken an Osage scalp, later serving under Tecumseh. After the War of 1812 he had represented a faction of his tribe which, under Keokuk, was friendly to the whites, and for long, like Little Turtle, had advocated implicit obedience to the government.

His fate was the usual one for a friendly Indian. By the treaty of St. Louis of 1804 under Keokuk's influence the Sacs and Foxes had agreed to surrender all their lands on the east side of the Mississippi but had been left undisturbed until the country should be thrown open to settlement. After the War of 1812, when the stream of settlers began to flow westward once more, bowing to the inevitable, Keokuk and the majority of his people migrated into the present state of Iowa.

Black Hawk and his following, however, insisted, and with reason, that when they signed the treaty of 1804 they had been deceived as to its terms. Regarding it as useless to trek further into the western wilderness, which past experience showed would also soon be wrested from them, they proposed as Tecumseh had done to establish a final boundary for the Indian country. Emboldened by Marshall's decisions, he could not as Ross had done interpret the signs of the times aright—that armed resistance was futile. Therefore, he set about negotiations with the Winnebagos, Potawatomie, and Kickapoos to enlist them in concerted opposition to the aggressions of the whites, and by the spring of 1831, so much friction had taken place between the settlers and the Indians in consequence of his agitations that Governor Reynolds of Illinois was induced to call out the militia. Desiring to avoid the expense of active operations, General Gaines summoned Black Hawk and his friends to a conference at Fort Armstrong at which a violent altercation ensued. Thereupon the negotiations for peace terminated, and on June 15th the militia left their camp at Rushville

* Harvard Law Review, March, 1925, Federal Criminal Laws and State Courts, Charles Warren, pp. 590-91. For this interesting incident the author is personally indebted to Mr. Warren. As declared by him in the above illuminating article, the Alabama case is one "not hitherto noted by legal historians."

bent on dispersing Black Hawk's assembled forces. Upon reaching Black Hawk's village and finding that he and his people had departed, they burned it to the ground. Immediately afterwards Gaines demanded that all the hostile warriors present themselves for a peace talk, and after a few minor conflicts, on June 30, 1831, Black Hawk and a mere handful of followers signed a treaty with Governor Reynolds by which they agreed to abstain from further hostilities and join their tribe in Iowa.

During the following winter Black Hawk, like his great Shawnee predecessor, Tecumseh, sent emissaries in all directions to win various tribes to his interest, and is said to have endeavored, though unsuccessfully, to destroy the authority of his own head chief, Keokuk, or commit him to war against the whites. On April 1, 1832, General Atkinson received orders to demand from the Sacs and Foxes the chief members of a band who had massacred some Menominees the year before. Arriving at the rapids of Des Moines river on the 10th, he found that Black Hawk had recrossed the Mississippi four days previously at the head of a band estimated at 2,000 of whom more than 500 were warriors. Again the militia were called out, while Atkinson sent word to warn the settlers, and collected all the regular troops available.

Black Hawk had proceeded up Rock River, expecting that he would be joined by the Winnebagos and Potawatomies, but only a few small bands had responded. Regiments of militia were by this time pushing on in pursuit of him, but were poorly disciplined and unused to Indian warfare, while jealousy existed among the commanders. Two brigades under Isaiah Stillman were met by three Indians bearing a flag of truce. Other Indians showed themselves nearby, treachery was feared, and in the confusion one of the bearers of the flag was shot down. A general and disorderly pursuit of the other Indians ensued. Suddenly on May 14, 1832, the pursuers were fallen upon by Black Hawk at the head of forty warriors and driven from the field in a disgraceful rout.

Black Hawk now let loose his followers against the frontier settlements, many of which were burned and their occupants slain, but although able to cut off small bands of Indians the militia and regulars were for some time able to do little in retaliation. On June 24, he made an attack on the Apple River fort but was repulsed, and on the day following defeated Major Dement's battalion, though with heavy loss to his own forces. On July 21, while trying to cross to the west side of Wisconsin River he was overtaken by volunteers under General James D. Henry and crushingly defeated with a loss of sixty-eight killed and many more wounded. With the remainder of his force he retreated to the Mississippi, which he reached at the mouth of Bad Axe River, and was about to cross when intercepted by the steamer *Warrior*, which shelled his camp. The following day, August 3rd, the pursuing troops under Atkinson came up with his band and after a desperate struggle killed or drove into the river more than 150, while forty were captured. Most of those who reached the other side were subsequently cut off by the Sioux. Black Hawk and his principal warrior, Neapope, escaped to the northward, whither they were followed and captured by some Winnebagos. He was then sent East and confined for more than a month at Fortress Monroe, Virginia, when he was taken on tour through the principal Eastern cities, everywhere proving an object of the greatest interest. In 1837 he accompanied Keokuk on a second trip to the East, after which he settled on the Des Moines River near Iowa-ville, dying there October 3, 1838. His remains, which had been placed upon the surface of the ground dressed in military uniform presented by General Jackson, accompanied by a sword also presented by Jackson, a cane given by Henry Clay, and medals from Jackson, John Quincy Adams, and the city of Boston, were stolen in July, 1839, and carried away to St. Louis, where the body was cleaned and the bones sent to Quincy, Illinois, for articulation. On protest being made by Governor Lucas of the territory of Iowa, the bones were restored, but the sons of Black Hawk were satisfied to let them stay in the Governor's office, where they remained until removed to the collections of the Burlington Geological and Historical Society. There they were destroyed in 1855 when the building containing them was burned.

Morally Black Hawk was the brother of Red Eagle. He was not the offspring of Tecumseh. His attempt to arouse the Indians to concerted action was a complete failure. Without the united support of his own people, he lacked both the following and ability to accomplish more than a spectacular result.

Be it said to his credit, Lewis Cass, Governor of the Northwest Territory, cooperated to the full extent possible under the Indian system with Quinney and the tribes generally, administering his office with humane regard for their rights. Yet, there was just so much of justice possible. The plight of the Indians in the Northwest at this time was well described by Hole-in-the-Day, a celebrated Chippewa chief who cried out in despair: "We are not near so ignorant as white men think us to be. When we view the country once owned by the Indians we feel like a wild beast driven into a hollow tree with no escape."

Similar conditions prevailed in the South, so that Congress, under the pressure of the Buffalo Party and desiring no more wars in the East, determined to create the long contemplated Indian Territory and bring about the removal thereto of the eastern tribes as quickly as possible. Already large tracts had been granted the Cherokees, Choctaws, Creeks and Chickasaws out of lands obtained from the Osages and Kansas Indians in 1825. Now by the Act of June 30, 1834, there was created out of the old territory of Missouri the territories of Arkansas and Oklahoma, and the Indian Territory which was made to embrace these grants and comprising approximately 30,341 square miles bounded on the north by Kansas and Oklahoma, the east by Missouri and Arkansas, the south by Texas, and the west by Oklahoma.

Comprehensive provisions were made in the same act for the administration of the Indian Territory and all Indian affairs, while a Bureau of Indian affairs within the War Department to be presided over by a Commissioner was created.

Washington's Indian Intercourse Act of 1795 had remained intact many years, but after the fundamental decisions in *Johnson v. McIntosh*, *Cherokee Nation v. Georgia*, and *Worcester v. Georgia* were handed down by Marshall, it had become necessary for those who wished to defraud the Indians out of their land to get around these decisions. Since grants from individual Indians had been held not to be binding upon them, nor good against the United States, resort to the scheme of long leases had been had. In order to guard the Indians against such devices in the Act of 1834, the prior statute requiring purchases and grants of land from an Indian to be evidenced by a treaty or convention entered into by the United States with his tribe was extended to include leases and other conveyances, while it was expressly provided, in Section 22 of the Act, that in all treaties about the right of property in which an Indian may be a party on one side, and a white person on the other, the burden of proof shall rest upon the white person, whenever the Indian made out a presumption of a title in himself from the fact of previous possession or ownership.

This last provision was construed to have conferred power upon the Indians to sue and be sued in order that they might be under no disability and consequent disadvantages in the unequal contest over their lands either against the United States or others.*

As to the abuses of the Indians and their property that had been committed in the past, the regulatory provisions of this new act are themselves conclusive evidences, since they were designed to restrain all manner of fraud and violence. Nevertheless Black Hawk's revolt was followed in 1834 by a campagin against the Indians of Ohio and Michigan known as the Toledo War, and a war with the Pawnees on the Texas border.

Even before the creation of the Indian Territory the general migration of the Georgia Indians had commenced. Thanks to chief Ross, in the treaty of 1828 the Cherokees were granted, in exchange for their home-

---

* *Felix v. Patrick*, 34 Fed. 651.

The Congressional legislation of importance affecting Indians during the taking of the East consisted of the following acts: August 20, 1789, July 22, 1790, March 1, 1793, May 19, 1796, January 17, 1800, April 22, 1801, March 30, 1802, March 26, 1804, April 29, 1816, March 3, 1817, May 6, 1822, June 30, 1834, which together disclose the evolution of the statutory policy of the United States down to the middle of the nineteenth century.

land, the fee in 7,000,000 acres of "good lands" between the White and the Arkansas Rivers for the lump sum of $30,000, an annuity of $2,000 and the right to organize and conduct their own government. To the Choctaws in 1820 had been granted a tract between the Arkansas and the Red Rivers; and to the Creeks and Chickasaws in 1826 and 1832, respectively, tracts between the Arkansas River and the Canadian Fork. It now developed that although the creation of the Indian Territory had been contemplated since 1804, and it had been assumed the entire region was open rolling prairie land such as that selected by Ross for the Cherokees, in fact it had not been explored in its entirety and no principle of equality whatever had been observed in the several grants. Thus, the Choctaws and Chickasaws whose migration commenced in 1832 found their domain mountainous and heavily wooded along the Canadian River while the Creeks who began to arrive in 1836 found theirs much like that of the Cherokees to the North.

Their new domains were, of course, in no sense reservations, though constantly spoken of as such, since they were acquired by specific grants out of the public domain in exchange for those relinquished in the East. These grants aggregated 19,475,614 acres, or about 30,431 square miles, an area equal to that of South Carolina and equivalent to 230 acres for each man, woman, and child of the entire population of the five tribes (84,507) as estimated in 1834.

About 6,000 Cherokees and a number of Chickasaws had already arrived before the general migration set in. Between 1832 and 1834 upwards of 25,000 Choctaws and 3,500 more Chickasaws, and between 1836 and 1840 from 15,000 to 20,000 Creeks arrived, the movement being in charge of the military. Meantime, among the eastern Cherokees occurred that which was to postpone their migration until 1838.

After the decisions of the Supreme Court of 1832 fixing a title in them beyond the peradventure of a doubt, they objected to the treaty of 1828 in which the grant was not made to the tribe in common in accordance with their system of ownership. Therefore, they had insisted upon a new treaty which was given them in 1833, granting the land to the Cherokee Nation as a whole. This treaty fully disclosed in the following article the circumstances of the migration and the general provisions therefor which were more or less common in the case of all the tribes.

The Cherokee Nations, West of the Mississippi having freed themselves from the harassing and ruinous effects consequent upon a location amidst a white population, and secured to themselves and their posterity, under the solemn section of the guarantee of the United States, as contained in this agreement, a large extent of unembarrassed country; and that their brothers yet remaining in the States may be induced to join them and enjoy the repose and blessings of such a state in the future, it is further agreed on the part of the United States, that to each head of a Cherokee family now residing within the chartered limits of Georgia, or of either of the states, east of the Mississippi, who may desire to remove west, shall be given, on enrolling himself for emigration, a good Rifle, a Blanket, and Kettle, and five pounds of tobacco; (and to each member of his family one Blanket,) also, a just compensation for the property he may abandon, to be assessed by persons to be appointed by the President of the United States. The cost of the emigration of all such shall be borne by the United States, and good and suitable ways opened, and provisions procured for their comfort, accommodation, and support, by the way, and provisions for twelve months after their arrival at the Agency; and to *each* person, or head of a family, if he take along with him four persons, shall be paid immediately on his arriving at the Agency and reporting himself and his family, or followers, as emigrants and permanent settlers, in addition to the above, provided he and they shall have emigrated from within the chartered limits of the state of Georgia, the sum of fifty dollars, and this sum in proportion to any greater or less number that may accompany him from within the aforesaid chartered limits of the State of Georgia.*

Still there were many Cherokees who were unwilling to remove from the lands upon which they had expended so much labor and go through the terrible ordeal of a move to the West through which the western Cherokees had passed. Nor were they convinced they would be welcomed by the latter. Finally, however, under the increasing pressure brought to bear upon them by the Georgians, they called a tribal convention to meet in 1835 at New Echota, Georgia, at which delegates from the western Cherokees appeared to deliver a formal invitation to their eastern brethren to join them, whereupon on December 29, 1835 they entered into the treaty of New Echota.

Whereas the Cherokees are anxious to make some arrangement with the Government of the United States, whereby the difficulties they have experienced by a residence within the settled parts of the United States under the jurisdiction and laws of the State Governments may be terminated and adjusted; and with a view to reuniting their people in one body and securing a permanent home for themselves and their posterity in the country selected by their forefathers without the territorial limits of the state sovereignties, and where they can establish and enjoy a government of their own device and perpetuate such a state of society as may be most consonant with their views, habits and condition; and as may tend to their individual comfort and their advancement in civilization. . . ."

So ran the preamble of this new treaty—in itself a dreadful commentary upon the people of the United States—in which the "Cherokee Nation" as a whole ceded to the United States all lands "owned, claimed or possessed" by them east of the Mississippi River, for the sum of $5,000,000,

* Article 8, Treaty of Fort Gibson, Feb. 14, 1833, 7 Stat. L. 414.

while the United States agreed in consideration of the sum of $500,000 to convey to the Cherokee Nation as a whole, including the eastern and western Cherokees and their descendants, by patent in fee a tract of land estimated to contain 800,000 acres adjoining the tract of 7,000,000 acres hitherto conveyed to them by the treaties of 1828 and 1833. All claims on the part of persons other than Cherokees to the land within the domain delimited by this treaty were to be extinguished and satisfied by the United States, and any Cherokee who desired to remain east of the Mississippi River and become citizen of the United States was to receive the per capita accruing under the treaty. The reunited Cherokee Nation was also accorded the right to have a delegate sit in the House of Representatives "whenever Congress shall make provision for the same," while in Article V the United States covenanted that the lands ceded to the Cherokee Nation should never, without their consent, be included within the territorial limits or jurisdiction of any state or territory.

But they shall secure to the Cherokee Nation the right by their National Councils to make and carry into effect all such laws as they may deem necessary for the government and protection of the persons and property within their own country belonging to their people *or such persons as have connected themselves with them*: provided always that they shall not be inconsistent with the constitution of the United States and such acts of Congress as have been or may be passed regulating trade and intercourse with the Indians; and also, that they *shall not be considered as extending to such citizens and army of the United States as may travel or reside in the Indian country by permission according to the laws and regulations established by the Government of the same.*

Although this treaty was signed by persons purporting to represent the eastern Cherokees, and assent to its provisions was given by the two delegates from the western Cherokees, Ross and his followers who were absent from the council that adopted the treaty and knew of the frauds perpetrated against the early immigrants in the matter of transportation costs, etc. disputed its validity. The authority of the western delegates was also denied. Nevertheless, in 1838 the Ridge or Treaty Party, numbering over two thousand, including about three hundred slaves, began their movement to the West while Ross's party, consisting of 14,757 persons, addressed a memorial to the President protesting against the treaty. Now again John Seargent intervened in their behalf with the result that the Government promised the cost of transportation would not be charged against the tribe if all the Cherokees would remove, which fully satisfied Ross, whose party now followed.

In August, 1838, while on their way to the Indian Territory, the eastern Cherokees resolved in council "that the inherent sovereignty of the Cherokee Nation, together with the Constitution, laws and usages of the same, are and by authority aforesaid, are hereby declared to be in full force and virtue, and shall continue so to be in perpetuity, subject to such modifications as the general welfare may render expedient," and refusing to submit to the government of the western Cherokees upon their arrival, offered to unite in a general council which should frame a constitution and establish a government for all. The western Cherokees declined to make this arrangement, insisting that the eastern Cherokees had entered their territory without their permission, and that their character was that of aliens or immigrants, subject to the constitution and laws existing among them. In July, 1839, however, a convention was called at which a new constitution providing that "the land of the Cherokee Nation shall remain common property" was adopted, and an act of union passed, which was ratified by a popular convention in January, 1840.

The validity of the ratification was denied by the western Cherokees and although the government set up under the Constitution of 1839 was upheld in Washington as the lawful one, grave disorders continued destroying the general peace of the territory. Finally on June 18, 1846, the western Cherokees agreed to submit their claims to a board of commissioners to be appointed by the President and Senate of the United States, with the result that on August 6, 1846 a treaty was concluded with the United States by the delegates of all the parties to the controversy.

Whereas serious difficulties have, for a considerable time past, existed between the different portions of the people constituting and recognized as the Cherokee Nation of Indians, which it is desirable should be speedily settled, so that peace and harmony may be restored among them; and whereas certain claims exist on the part of the Cherokee Nation, and portions of the Cherokee people against the United States; therefore, with a view to the final and amicable settlement of the difficulties and claims before mentioned, it is mutually agreed by the several parties to this Convention. . . .

So read the preamble. This treaty provided that the lands then occupied by the Cherokee Nation should be secured to the *whole* Cherokee people for their *common* use and benefit. All difficulties and disputes were adjusted and a general amnesty declared in favor of those who had broken the peace and violated the law. Payments on claims that had been erroneously settled with moneys belonging to the Cherokees were to be reimbursed, and it was expressly declared that that portion of the Cherokee people known as "Old Settlers," or "Western Cherokees," had no exclusive title to the 7,800,000 acres ceded to the Cherokees then east of the Mississippi. It was also agreed that inasmuch as this territory had become the common property of the whole Cherokee Nation, the Cherokees then west of the Mississippi, by the equitable operation of the treaty of 1828, had acquired a common interest in the lands then occupied by the Cherokees east of the Mississippi River, although no provision therefor had been made by the treaty of 1835. So also a common interest on the part of both groups in the funds of the Cherokee Nation was declared to exist. A scheme of distribution among the eastern and western Cherokees was set out in detail. This agreement, placing the title to the Cherokee lands in the tribe forever, not to be alienated except to the United States, converted the fee simple title originally acquired by Ross into a base, determinable, or qualified fee which was the same title as that acquired by the other Georgia tribes. They, like the Cherokees, organized for themselves governments similar to those of the States, including a governor, a bicameral popularly elected legislature, national courts, school system and treasury. The institution of slavery was recognized by them all.

Beginning in 1832 the Baptist Church established missions throughout the Indian Territory. It continued to conduct a higher school in Kentucky known as the Choctaw Academy, which was patronized in the main by the Choctaws and Creeks. In 1835 the Methodists also established missions in the territory. Both the Baptists and Methodists met with the most encouraging success.

Although the removal of the tribes was in charge of the military, there can be no doubt the most unconscionable frauds were perpetrated by the civil agents in respect of their rationing and transportation just as in the case of the first emigrants. Not only was the sum of $1,047,067 appropriated in 1838 to defray the cost of moving 17,000 Cherokees expended but $189,422.76 in addition. There is no way to determine exactly the losses in life. The first migrants undoubtedly suffered the most through lack of experience in handling such a movement. In addition to the Cherokees who had already arrived in 1838, between that year and 1853 about 15,000 more reached the Indian Territory, or a total of approximately 21,000. Of those who set out from Georgia it is said that fully a fourth perished on the way. Over three hundred were drowned at one time in a steamboat accident while crossing the Mississippi. Therefore, since a total of not less than 57,000 emigrants from the South arrived, not less than 15,000 perished on the way. No one may say how many times the pathetic romance of Evangeline reoccurred among them.

While the transfer of the Southern tribes to the Indian Territory was, perhaps, one of the most woeful and cruel as well as remarkable events in American history, apparently the Acadian tragedy in which they had already become involved in 1835 in their effort to escape destruction at the hands of American civilization made no appeal to Jackson whatsoever. Having created the year before the machinery of a system which could not be expected with reason to function effectively at once, he plumed himself on his achievements. Thus, in 1835 he assured Congress that the plan of removing the Indians who had remained in the East was approaching its "consummation"; that it should be persisted in since all experiments for the improvement of their lot had failed. "It was a proven fact," he declared, "that they could not live in contact with civilization and prosper." The only Indians remaining east of the Mississippi from whom an engagement to remove had not been obtained were two small bands in Ohio and Indiana, not exceeding 1,500. He congratulated the country on the large receipts for the past year—$11,000,000—derived from the sale of lands in which the Indian title had been extinguished!*

Though in fact this declaration was an unwitting condemnation of himsef, nothing could have been more unfair than Jackson's conclusions regarding the Indians. Unquestionably among them there were irreconcilables like Red Jacket, who at every Seneca Council cried out against the white man's civilization. Still a pagan, he hated the missionaries, or the "black rats," and believed them all to be like the hypocrite, Eleazer Williams, of whom he had such bitter experience. Yet, Red Jacket was of ill repute among his own people, and exercised only a limited influence over them. This was ignored by Jackson, who took the professions of Indians of Red Jacket's type as exemplifying Indian nature and the general attitude of the race. He shut his eyes to the fact that for every advocate of resistance among them like Pontiac there was a Little Turtle, that for every Tecumseh or Meteor, there was a Ross or a Swan, for every Black Hawk, a Quinney, for every incorrigible like Red Jacket a Hole-in-the-Day. Indeed, at this very moment, old Hole-in-the-Day said: "Immigrants are coming among us. They won't be satisfied with the land now open for entry, and we cannot resist their encroachments if we would. We should, therefore, prepare for a state of things which we can not avert, and settle down like the whites."

Be all this as it may, Jackson had counted his chickens before they were hatched. Within a few days after this message to Congress, the Seminoles of Georgia rose. A treaty of removal to the Indian Territory had been foisted upon them by General G. A. Thompson at Payne's Landing in 1832 with the corrupt aid of chief Emathla. No longer able to endure the punishment still being wreaked upon them, they repudiated the treaty on the ground it did not express the will of the tribe, and took to the warpath in defense of their rights under the leadership of the patriotic and gallant young warrior, Osceola. This they did with a unanimity and a resolution which characterized few Indian wars. Even their Negro slaves aided them with desperate courage.

Osceola was born in the Creek country on the Tallapoosa river about 1803. He was also called Powell from the fact that after his father's death his mother married a white man of that name. His paternal grandfather

* Seventh Annual Message, Dec. 7, 1835.

was a Scotchman, and it is said the Caucasian strain was noticeable in his features and complexion. Not a chief by descent, nor, so far as is known, by formal election, he became the leader and acknowledged chieftain of his tribe by reason of a very real ability and commanding personality.

Hostilities were initiated in December, 1835, with the killing of Emathla and General Thompson. Secreting the women, children, and old men of the tribe in the gloom of a great swamp where the white troops for a long time were unable to find them, Osceola set about the work of harassing the troops. On December 28th, he was attacked by a force of one hundred men under Major Dade. This he promptly defeated. Of his enemies but two or three wounded men escaped. Beginning with General Gaines, one officer after another was placed in command of the troops sent against him. To aid them in the pursuit blood-hounds were brought from Cuba and rewards of $200 offered for every dead Indian.

Osceola's skillful use of the difficult swamp country in which he elected to fight baffled his opponents until General Jesup, maddened by the public cry for more energetic action, seized the heroic young chieftain and his attendants while they were holding a conference with him under an American flag of truce. The dastardly ruse by which he was finally captured did not even win the applause of the Buffalo Party much less the approval of those who had some sense of national honor. It was a great mistake from every point of view. Jesup had not yet learned that his act alone would have sufficed to make Osceola a martyr. He, like many another American, had not discovered what the British had learned. One of the ways not to encourage native resistance is not to make native martyrs. When they must be dealt with, adequate force should be employed to crush their resistance at once, but with meticulous regard to the native idea of what is fair. Fairness and power above all things they respect.

While the most relentless persecutors of the Seminoles protested against Jesup's act, nevertheless, Osceola was imprisoned in Fort Moultrie where, broken in spirit and exhausted by his ceaseless exertions in the unequal conflict with the whites, he died in January, 1838, another martyr to the forlorn cause of Indian liberty. The circumstances of his capture more than anything else caused the Seminoles to fight on until finally starved into submission in August.

Thus ended the third Seminole War, one of the most desperate Indian wars in the history of the country. In it over 1,500 American soldiers were killed. It was the last stand of the Indians. The Pax Romana, east of the Mississippi, had been established. All now was quiet along the Atlantic. The last great Indian leader in the East had perished in the youthful Osceola whose name belongs in the roll of martyrs to Indian liberty. Yet, his sacrifices were vain. Broken utterly, bleeding, helpless, only the pitiful remnants of a once proud race of warriors remained.

Some of the Seminoles, including a number of their Negro slaves who had taken part in the strife, fled to Mexico and settled south of the Rio Grande near Eagle Pass, Texas. In 1856 about 2,000 removed to Indian Territory. They received a part of the Creek domain by voluntary cession and were reorganized as the Seminole Nation. Thus were brought together again the Five Civilized Tribes.

Meantime, what had occurred in the Indian Territory, where for the first time Indian governments were allowed to function free of interference by state governments, had utterly belied Jackson's view that the red race was incapable of adapting itself to civilized ways. For the states set up by them there was no precedent in history, nothing to guide them in putting their governments into operation. Within a few years, however, they were functioning effectively. Nor must it be forgotten that the five tribes had been transplanted as communities. Dumped down in an untamed desert after an ordeal which alone might have produced a state of anarchy among them, these essential agriculturists, habituated to the climatic and soil conditions of the South Atlantic states, had no opportunity gradually to adapt the machinery of their governments to the entirely novel economic conditions of a harsh and unfamiliar environment. Moreover, the widest differences of opinion existed among them with respect to the institution of slavery. Unlike what Jeffreys hoped the Virginia tribes would do when in 1679 he set apart a domain to them, they did not cut each other's throats. This they left to John Brown and the white slavers in Kansas.

Unquestionably they made mistakes and many of them. So, too, grave abuses occurred among them. Looking at these in a broad way, or in the light of the attending circumstances, it is astonishing that more did not occur. It is amazing that these people could have settled down to a fairly well ordered life as rapidly as they did. The prosperity which they achieved must be taken as tribute to the genius of the race. Their more serious troubles did not commence until the whites appeared among them in sufficient numbers to start the old processes.

Finally, if what they accomplished be compared with what was soon to occur among the whites in the suddenly transplanted society of the Pacific Coast, one must conclude that the fault was not with the Indians as declared by Jackson. Nor were they less capable of self-determination under civilized conditions if given reasonable freedom to adapt themselves thereto. They were certainly not less orderly and self-restrained than the whites of the West. While the aboriginal Indian system is destroyed by civilization, the Indian himself readily adapts himself to it, as declared by the late James Mooney, one of the foremost authorities on the red race. Thus, it would have been fairer had Jackson given as the basis for the removal of the Indians from the East the fact that the white man would not permit them to prosper there.

At any rate the transplanted tribes of the Indian Territory dwelt in absolute harmony during the two decades preceding the Civil War, while the whites engaged in violence among themselves and with the Indians elsewhere. Nor is there to be found among the whites such an example of generosity as that which characterized the voluntary cession by the Creeks of part of their territory to the oppressed Seminoles.

22

The Oregon Trail and the Forty-Niners

OVER THE ROLLING PRAIRIES of the midcontinent for centuries the bold cossacks of the red race had roamed undisturbed. There were the Sioux, Cheyennes, Crows and Arapahoes in the North just where Father Hennepin and La Seur had found them. The Kansas, Blackfeet, Arickarees and Osages were in Kansas as in the days of Coronado. The Utes and Shoshones were in Utah and Colorado, the Pawnees, Commanches and Kiowas in Texas. In 1826 a single covered wagon wended its toilsome way from the Missouri River across the prairies, up the arid slopes of the Rockies, and down to the shore of the Pacific. Little did these people understand the portent of that first "Arkansas traveler" as they watched him vanish beyond the utmost rim of their own world. How seldom now does a visitor to the sun-kissed vineyards of California, as he dashes in a motor over the same route followed by the first covered wagon but a century ago, contemplate the terrible tragedies of which it was the forerunner.

Strange to say the experience of over three centuries was not to profit those who soon followed. The processes which began in Vinland in 1004, in San Salvador in 1493, and in Virginia and New England early in the seventeenth century, were to begin all over again. Many of the inhumanities already perpetrated were not only repeated but outdone.

Having arrived in Texas in 1832, the following April Houston had been sent as a delegate to the Mexican Constitutional Convention of San Felipe, deputed to draft a memorial to the Mexican Congress. In this he was to ask for the separation of Texas from Coahiula where the anti-

American party was in control, and a tentative constitution for Texas as a new member of the Mexican Republic. This, of course, was but preliminary to the War of Texan Independence which, fomented by him, followed in 1835. Selected as commander-in-chief of the Revolutionists, in April, 1836, he routed the Mexican army, captured Santa Anna, its commander, and in May caused the independence of Texas to be proclaimed. Thereupon, he was elected President of the Republic of Texas. Reelected in 1841, he was, of course, urgent in the demand that immigration to the West be promoted in every way possible.*

As is usually the case, into the empire won by Jefferson, explorers, trade prospectors, and the flag, came the church. Thus, in 1831, the Baptists had established missions for the Shawnees on the upper Missouri, and the Methodists and Presbyterians in Oregon in 1836 and 1838, respectively.

The anti-slave party had long since seen through Jackson's purpose. To counterbalance the acquisition of Texas by King Cotton, it was now to engage in a race for Oregon. Accordingly, in 1841, the West was thrown open to settlement by the Preemption Act of September 4, without other provision for the protection of the Indians than that of the general land laws requiring the Indian title to be extinguished before their lands became a part of the public domain. This Act provided that every person who was the head of a family, whether a widow or a single person, if over the age of twenty-one, who was a citizen of the United States, or had applied for citizenship, might, by settling in person on the public lands, improving the tract settled upon, and erecting a dwelling thereon, purchase as much as one hundred and sixty acres. Upon its passage immediately great caravans of settlers put out from the Missouri heading for Texas and Oregon. Fortunately for the forerunners of the Five Civilized Tribes, they at least were protected by the military, so that the preemptors were compelled to pass them by. But for the Indians of the West, generally police protection was utterly lacking. For the aborigines who had welcomed Lewis and Clark but a few years before, who in 1826 had watched the first covered wagon crawl through forest and desert, undaunted by the towering Rockies, far unto the utmost rim of the universe, it was the romance of an unexampled tragedy. No man in the councils of the nation but Sam Houston, the Cherokee Blackbird, cried mercy!

The machinery created by the Act of 1834 was not only utterly incapable in its nature of coping with the situation, but as a matter of fact was to

---

* Upon the admission of Texas to the Union in 1845, he became one of its first Senators, and in 1859 its Governor. After opposing the Kansas-Nebraska Bill and the secession of Texas he was declared deposed. He died July 26, 1863, at the age of seventy. The Cherokee Blackbird was the only American who was ever governor of two states, member of both Houses, and president of a foreign state.

break down entirely. The Bureau of Indian Affairs was soon nothing but an agency of Congressional politics. The representatives it sent forth, dependent upon the favor of local politicians and interests, were but the agents of the latter. Recognizing no obligation to the Indians, they spared no opportunity in executing the will of their masters to prey upon and fatten off the helpless wards of the nation. The army posts from which they conducted their operations quickly dotted the West. They were not designed primarily to protect the Indians but merely to help the immigrants seize the land. In justice to the army, however, be it said, not it but the politicians who made use of it, were the real enemies of the Indians. This the latter recognized. Almost invariably military men were their friends, loathing like Houston the system of which the Indians were the victims. Among the soldiery an Indian agent was held in more contempt than the vagabond Indians who lived on the offal of the posts. Van Loon said:

> This chapter in our history is not very flattering to our pride. On every page we read the records of greed and cruelty, of broken treaties, and the entire episode is steeped in the contraband rum of the prairie saloon. But what are ethical principles pitted against the laws of nature? The two million or more square miles of mountain and plains which the United States had bought, swapped, or taken from the French, the Mexicans and English, and which contained some of the richest deposits of gold and silver and lead and copper and oil of which the world had never dreamed, were in the possession of a weak race which fought with bows and arrows and they were eagerly coveted by a strange race with rifles and cannon.

Yet, however hypocritical the government may have been in dealing with the rights of the Indians, it would be a grave error to charge the average preemptor or settler in the West with hypocrisy. He was undoubtedly as sincere in his belief that he was entitled to the land there as he was earnest in his demand upon Congress that the way be cleared for him. These people were no more hypocrites in their attitude toward the Indians than were their Puritan ancestors in the bigoted conviction that the natives of America were the spawn of Hell.

Let us witness them as they press westward.

Now and then the great emigrant caravans of "prairie schooners" come to a halt, the rumble of their weary wheels a moment ceasing. It is the Sabbath, perhaps—a day of rest for man and beast. In the silence that reigns, grizzled patriarchs with flowing beards like the prophets of old draw out of their vehicles thumb-stained testaments. About the wheeled tents gather firm-jawed women—women with eyes that know no fear— mothers of a conquering race. Children innumerable cluster at their knees, awed by the strange vastness of the land of milk and honey which earnest parents tell them God in his mercy is delivering to them. The readers turn to Deuteronomy or Joshua to find the sanction of God for their invasion. Does the holy testament itself not promise them these prairies, the cattle that graze upon them? Surely, they are but God's reward to those who smite the heathen!* So it is that down by the corrals where the weary, trace-stained beasts nip at the succulent prairie grasses, young men bearing great rawhide whips, muskets slung at every shoulder and huge bowie knives thrust in their belts, are planning a Sabbath recreation. How proud they will be to write back to the East the story of their "first Indian!"

There have been many ways of counting coup. Scalping was the Indian way. But scalps were never taken by the aborigines except in warfare as evidence of battle prowess—victory as they saw it over an enemy in combat. The hunting down and shooting at sight of helpless Indians, the stripping of their beads and much-prized apparel, the leaving of their maimed bodies to fester upon the ant hills, their bones to bleach on the burning sands of the prairie desert was another. Which was the more cruel?

King Cotton had seized Florida, and was sending agents into California, Mexico, Honduras, Costa Rica, Nicaragua, New Grenada and Cuba to prospect for new slave territory. No less willing were his adversaries to fight for free territory.

Various attempts had been made by Great Britain to settle the northeastern boundary of the Oregon Territory, but the Americans were determined to surrender nothing. In 1843 when the population was already 3,000, the Oregonians met and organized a territorial government, and by 1844 the slogan of "Fifty-four forty or fight," had become the rallying cry of the National Democratic Party. While there were several senators who advocated war, the majority of Congress believed that the best method of gaining possession of the disputed territory was by its actual settlement. This was proceeding apace. The immigrants from the East had rapidly outnumbered the few British fur traders who occupied the country. Thus was the Buffalo Party encouraged, aided and abetted by the sentiment in the North which demanded the creation of new nonslaveholding states. In the tremendous dispute which followed, war with Great Britain was averted only by the narrowest margin through the agreement of 1846 that the Oregon boundary should be the 49th degree of north latitude.

Oregon and Texas having been acquired, only the imperial domains known as California and New Mexico remained to be obtained to complete the present continental possessions between Canada and the Rio Grande from ocean to ocean. In 1840, Monterey had become the capital of California. A year later the Russians had abandoned the trading post at Bodega Bay which they had maintained for twenty years. Although under

* Deuteronomy, 2, 3; Joshua, 8.

the Mexican government California enjoyed virtual autonomy, almost immediately the American immigrants began to speak of its annexation, by force if need be, to the United States, in disputes which arose with others who favored a British protectorate. Long before the Mexican War, American political agents were secretly at work in California in the interest of King Cotton. When on March 5, 1845, Polk—a Tennessean who with his party desired additional slave territory—became president, the result became certain, for the great bulk of the population lay below the line established by the Missouri Compromise of 1820.

Diplomatic relations with Mexico were soon suspended. The Mexican Minister had demanded his passports a few days before from Calhoun, the Secretary of State under Tyler. He had sent a leave-taking note in which he angrily charged that the United States had assumed a hostile attitude toward his country in the resolution of Congress providing for the annexation of Texas whose independence had not been recognized by Mexico. It is plain that the republic to the south was receiving at the hands of King Cotton no more consideration than Spain had received with respect to Florida. Tyler and Calhoun, now Polk in the saddle!

Polk's plan was to acquire New Mexico and upper California by purchase, if Mexico could be forced to part with them, as Spain had been made to cede Florida. He was willing to pay as much as $40,000,000, but hoped to establish the southern boundary from the mouth of the Rio Grande along latitude 30° north to the Pacific at a cost of not more than $15,000,000. In all this the cabinet concurred.

For a while his negotiations produced no result. Texas was, therefore, annexed, and a claim quickly put forward by the United States that not the Nueces, as declared by Mexico, but the Rio Grande was its southern boundary. Taking advantage of the controversy which naturally now arose, Polk ordered General Taylor to cross the Nueces, and later advanced him to the Rio Grande, thus insuring a collision near Matamoras, resulting in the defeat of the Mexicans at Palo Alto on May 8, 1846, and at Resaca de la Palma the next day. Four days later Congress declared war.

Meantime, one Larkin, agent of King Cotton, had been actively at work in California fomenting with the support of the Administration discord between the inhabitants. Before the revolt in the South planned by him could take place, on June 14, 1846, a small party of annexationists, aided by Colonel John C. Frémont, who very conveniently happened to be in this quarter on an official exploring expedition, seized the town of Sonoma, raised the Bear Flag, and to celebrate the seventieth anniversary of the Union, declared the independence of California on July 4th!

The Pacific squadron was also conveniently near at hand. Pursuant to instructions from the Navy Department, Commodore Sloat landed at

Monterey and Yerba Buena (San Francisco), hoisted the American flag, and gave notice that he would carry it through California. Between Commodores Sloat and Stockton and Colonel Frémont, the four provinces were quickly brought under American control. Without the slightest delay California was declared on August 15, 1846, to be a territory of the United States. From all of this it is plain that again King Cotton had effected, by a most carefully worked out conspiracy, the conquest of more land. In order to reassure the inhabitants, Sloat announced that "all persons holding titles of real estate, or in quiet possession of lands under color of right, should have their titles and their rights guaranteed to them."

The conquest of New Mexico proceeded with equal celerity. General Kearney took possession without a shot in the name of the United States on August 22, 1846. Leaving Santa Fe for California, he was appointed governor of that territory.

Upon the outbreak of the Mexican War the Indians stood aloof. They looked on not without interest, but amazed by the great armies of pale-faces who had come among them. It was not their fight. Their aid was not enlisted, so they kept out of it, just as they would have done in the East, no doubt, had the French and the British colonists not driven them into their wars and exploited them politically.

The Indians of the Pacific Coast, Arizona and New Mexico, were very numerous and entirely different from those of the East and the midcontinent. As usual, however, they were at first friendly disposed. From the earliest times the Franciscans had been engaged in Christianizing them. In 1823, when the last and most northern mission was built at Sonoma, there were twenty-one religious establishments in California. Not an Indian in California or Oregon had been on the warpath for years. A more unresisting, harmless people could not be imagined. In all there were over 200,000 of them in California alone, grouped in eighteen distinct non-nomadic tribes which occupied definite domains.

When the Spaniards came to Mexico in 1521 they found the Indians possessed of a communal organization upheld by traditions that went so far back into prehistoric times that there was no memory of any other life. The basis of the political form of private organization was the common ownership of land with the parceling out of tracts to the individuals who were able to work them, a sort of temporary tenancy continuing only for life. There were no land titles and system of heritage. The natives lived in villages. They seldom had their homes on the land they worked. In theory, if any ownership existed, it was vested in the cacique, or petty chief, who in turn owed feudal fealty to the rulers of his clan and nation. No individual enterprise except exploitation by the priests and caciques was possible, and no ambition for land ownership, for homesteads, or for economic change entered into Indian psychology.

The Spaniards themselves introduced the idea of property as it was known in Europe, but the communal system of the Aztecs gave them the opportunity to fasten a serfdom not unlike the feudal system on the natives. Great tracts of land and whole villages were distributed among the conquerors, and there grew up the anomaly of Spanish possessors of vast estates within which their Indian serfs held communal titles to village lands.

These were the basic facts of the relationship of the worker to the land in Mexico. They go back to certain early conditions which antedate the present era by four centuries. Before their conquest in 1521, the Indians had no private ownership, only tribal ownership of land. The Spanish government, recognizing this system, endowed all the Indian towns with three sorts of real estate, preserving the communal idea, but giving it some legal basis. These properties, community owned, were: (1) a town site (funds legal); (2) pasturage (edigos); and (3) commons (tierras communales). The town site was the square which could be inserted in a circle with a radius of 600 yards (about 2,000 feet), the center of the circle being the center, or plaza, of the town, each family having its hut and "orchard" in the village. The pasturage or egidos, was a one-league square of grazing land, so that the natives "could feed their cattle," as the law expressed it. The commons, of varying size, were forest and farm lands owned by the community, and heads of families were assigned certain sections which they were to work, and they could neither sell, mortgage, nor lease it.

The Indian properties remained untouched, even after the independence of Mexico from Spain, until in 1856 the Juarez government ordered the allotment of all real estate belonging to the villages to the members of the community. Thus, the Mexicans could have taught the Americans much.

The origin of the great landed estates of Mexico goes back to another form of property, the royal grants of the Spanish Crown, something entirely outside the laws creating the Indian land titles just described. There were oppressions under these grants, and yet as a general rule the Indian properties, town sites, pasturages, and commons which were embraced by the larger grants were recognized both by the colonial government and by the haciendados. It was this condition that brought about the anomaly of an Indian belonging to an hacienda, and at the same time owning his own property within the confines of the hacienda, an anomaly which made it vitally important for the Mexican government to place the ownership of the Indian lands upon a modern legal basis, submission to the laws of survey being made incumbent on the haciendados, miners, and ranchers as well as upon the Indians.

In 1777 the Spanish government, which had never recognized the title to the land claimed by the missions, began the foundation of pueblos or towns. Upper California was divided into the four provinces of San Diego, Santa Barbara, Monterey, and San Francisco. By repeated laws the pueblos or Indian communities were declared to hold their lands of which they could not be deprived without their consent under the communal system, or in common, and though patents were given whites they were taken subject to the Indian title. The land where the Indian was seated was his so long as he desired to use it.

After the Mexican Revolution of 1821, the missions began to decline. The Indians were partly emancipated in 1826, and the process of secularization, which had begun in 1833, was completed by 1845, but the Indians of California were ever deemed to possess under the Mexican law an undisputed title of possession and use of the land actually occupied by them, under which it was held to be an aboriginal right which antedated the sovereignty of Spain and Mexico, not derived from either, but recognized and protected by both.

The Mexican War was ended by the Treaty of Guadalupe Hidalgo, signed February 2, 1848. In this treaty it was provided that the United States should pay Mexico $15,000,000. The United States also assumed the payment of certain claims which had long been due citizens of the United States by Mexico, amounting to $3,250,000. Mexico waived all claim to Texas, and the Rio Grande was established as the southwestern boundary of the United States. This boundary admitted that the provinces of New Mexico and California had become a part of the territory of the United States. The treaty provided, in articles VIII and IX, full and complete protection for all property rights of Mexicans, whether residing in the ceded territories or elsewhere.

Article VIII, in part, reads:

Mexicans now established in territories previously belonging to Mexico, and which remain for the future within the limits of the United States as defined by the present treaty, shall be free to continue where they now reside, or to remove at any time to the Mexican Republic, retaining their property which they possess in the said territories, or disposing thereof, and removing the proceeds wherever they please, and without their being subjected, on this account, to any contribution, tax or charge whatever.

In a protocol dated May 26, 1848, suppressing certain articles of the Treaty of Guadalupe Hidalgo, the following explanatory words were employed by the contracting parties:

1st. The American Government by suppressing the IXth article of the Treaty of Guadalupe Hidalgo and substituting the IIIrd article of the Treaty of Louisiana, did not intend to diminish in any way what was agreed upon by the aforesaid article IXth in favor of the inhabitants of the territories ceded by Mexico. Its understanding is that all that agreement is contained in the IIIrd article of the Treaty of Louisiana. *In consequence all the privileges and guarantees, civil, political, and religious,* which would have been possessed by the *inhabitants* of the ceded territories, if the IXth article of the treaty had been retained, will be enjoyed by them, without any difference under the article which has been substituted.

Inasmuch as the Mexican Constitution of 1836 stipulated that "Indians" were "Mexicans," the last clause of Article IX necessarily protected the Indian property right of occupancy.

Moreover, the concluding sentence of article XI of the Treaty of Guadalupe Hidalgo was as follows:

And, finally, the sacredness of this obligation shall never be lost sight of by the said Government, when providing for the removal of the Indians from any portion of the said territories or for its being settled by citizens of the United States; but, on the contrary, special care shall then be taken not to place its Indian occupants under the necessity of seeking new homes, by committing those invasions which the United States have solemnly obliged themselves to restrain.

Nothing could be more clear from the foregoing language than that both the United States and Mexico, the high contracting parties in the Treaty of Guadalupe Hidalgo, deemed the tribal Indians to possess an inviolable right of occupancy in their tribal domains, and that the United States assumed a solemn obligation to protect them in that right.

Yet, the first whisper that went out about the nugget found at Sutter's Mill, January 24, 1848, set the world awheel for California. In endless streams the "Argonauts" already pouring southwestward to Texas and across the continent to Oregon, abandoning their original quest for mere land, swept onward to the Eldorado of California. In such unexampled numbers did the "Forty-niners" cross the intervening deserts and mountains, or arrive by ship around the Horn, that by the end of the year 1849 the white population of California numbered 100,000! The world had never before known such an adventurous migration, or perhaps so many utterly reckless persons brought together in one frenzied quest. A large percentage of them were young, unmarried men with but one thought—to make a "strike."

If the lilting refrain which carried the pioneers of the West across the continent was for them a song of joy, for the Indians of the Pacific Coast it was a dirge. No longer was there to be warfare upon "incorrigible" savages but just plain, cold-blooded murder.

The enormous accession of public lands resulting from the acquisition of Oregon, California and New Mexico compelled the creation of a Department of the Interior in 1849 to administer the public domain. In order to coordinate the work of the land office with the Department of Indian Affairs created in 1834, the latter was now transferred to the Department of the Interior of which it became a bureau. As far as any protection it afforded the Indians it might as well not have existed. The peaceful tribes along the Pitt River in California, forced to the wall, stood at bay. Against them, to satisfy the demands of the Californians, federal troops were despatched in April. On June 5, 1850, responsive to the Buffalo Party, Congress enacted a law authorizing the President to appoint one or more commissioners to negotiate treaties with the several tribes in Oregon for the extinguishment of their claims to lands lying west of the Cascade Mountains, and if found expedient and practicable, for their removal East of the mountains. The old cry had risen in California. It was unbelievable that Indians should be permitted to retain "gold lands." Congress must remove them, and quickly.

In consequence of all this, thousands of Indians had fled across the Rio Grande so that the Mexican government was protesting loudly. Therefore, since the United States was bound by the guarantees of Indian rights contained in the Treaty of 1846 with Great Britain as well as that with Mexico, it was compelled to take some action that would at least be a semblence of fulfilling those guarantees. So it was that on September 9, 1850, the day California was admitted as a state to the Union, it passed the Territories Act, creating the territories of New Mexico and Utah, and containing certain provisions relating to all the territories then existing or to be erected. In this Act it was expressly declared that nothing therein contained should be construed to impair the rights of person or property pertaining to the Indians in any territory, so long as such rights remained unextinguished by treaty between the United States and the Indians, or to include any territory which, by treaty with any Indian tribe, was not, without the consent of such tribe, embraced within the territorial limits or jurisdiction of any state or territory. And it was expressly provided that "all such territory shall be excluded out of the boundaries, and constitute no part of any Territory now or hereafter organized until such tribe signifies its assent to the President to be embraced within a particular Territory."

Having done this Congress was ready to proceed with its old treaty policy. Accordingly, the Act of September 28, 1850 provided for the appointment of not more than three agents for the Indian tribes of California, while the Indian Appropriation Act of September 30, 1850, carried an item of $30,000 with which to defray the "expenses of procuring information and collecting statistics necessary to the Indian Bureau and for making treaties with, and presents to, the various tribes of Indians residing within the limits of the United States upon the borders of Mexico." There was also an item of $25,000 "to enable the President to hold treaties with the various tribes in the state of California." Thus, to bring about the removal of the Oregon, California and New Mexican Indians from the path of the clamorous whites, and to satisfy Mexico and Great Britain as well, a total of $95,000 was appropriated by Congress at this time. Nowhere was there any suggestion that the welfare of the Indians was to be considered. They were merely to be removed for the convenience of the gold-seekers.

Alexander H. H. Stuart, the Secretary of the Interior, now directed the Commissioner of Indian Affairs to issue the necessary instructions to Messrs. McKee and Barbour, former Indian agents. They were expected to negotiate the treaties, and to ascertain the most suitable places to establish "reservations" for the tribes. In the instructions issued the commissioners it was declared that "the Department is in possession of little or no information respecting the Indian of California." Yet, they were charged "by all possible means to conciliate the good feelings of the Indians, and to get them to ratify those feelings by entering into written treaties binding on them towards the government and each other."

Here an old acquaintance is recognized. Still the hidden policy of which Indian treaties were born was silently at work. Again governmental diplomacy was preparing the Indians by "every possible means" to dispose of their lands for the benefit of the whites—to accept the benefits described by the Reverend Jedediah Morse in 1821 as the equivalents which "humanity and expediency" had substituted for the land to which the whites were deemed by their forefathers to be entitled. For them the benefits of an agricultural life on a reservation were still deemed a sufficient consideration for the lands they might be "disposed" to part with by negotiations with experienced, adroit, clever agents of the government backed by the power of gold and bayonets. Most of them had been murdered by an unrestrained mob of cutthroats, in aid of whom they had been driven to bay by the armed forces of the United States. The exigencies of the survivors would be availed of, for surely they were in an amenable state of mind to negotiate not merely for their lands, but for their lives, now that their backs were to the Pacific!

Congress was under a treaty obligation to Mexico not only to protect the property rights of the Indian tribes but the titles of all persons holding grants from the Spanish or Mexican governments as well. Therefore, pursuant to instructions from the Secretary of State, the military governor of California had rendered a report on March 1, 1849, setting forth among other things the laws and regulations of Mexico with respect to the Missions. This was followed on April 10, 1850 by the report of a confidential agent on the existing private land claims. Thereupon the Act of March 3, 1851 was passed, creating a commission to examine all private land claims and to confirm those which were valid. Inasmuch as commissioners had already been appointed to negotiate treaties with the Indian tribes, whose rights had been guaranteed by the Treaty of Guadalupe Hidalgo, and their claims were in no sense private claims, Section 16 of the Act contained the following provision:

Sec. 16. And it is further enacted, That it shall be the duty of the commissioners herein provided for to ascertain and report to the Secretary of the Interior the tenure by which the mission lands are held, and those held by civilized Indians, and those who are engaged in agriculture or labor of any kind, and also those which are occupied and cultivated by Pueblos or Rancheros Indians.

Upon arriving in California the Indian treaty commissioners set to work industriously and between May 13 and November 4, 1851, negotiated eighteen separate agreements with the eighteen tribes of California.

In the letter of the Commissioner of Indian Affairs transmitting these tentative agreements to the Secretary of the Interior, they were fully described. Some of the stipulations contained in them were characterized as new, especially those requiring the Indians to relinquish all claims to their tribal domains and accept in lieu thereof not reservations proper but small tracts not previously occupied by them, a method designed, of course, to remove them entirely from the gold country and thus make way for the miners. Yet, no annuities whatever were provided to enable the Indians to subsist while converting the land to be assigned them in the vast wilderness of California to agriculture. All disputes henceforth were to be adjusted by "local agents," and all legal controversies were to be settled by the civil tribunals of California.

How the "head men" of the tribes were induced to sign these outrageous, confiscating instruments may well be imagined. According to Captain H. W. Wessels, commanding the military escort of one of the commissioners, they had not the least idea of what they were doing.

The Klamath and Trinity Indians, with few exceptions came freely into camp . . . exhibiting an appearance of often, cheerful frankness entirely different from that of any tribes heretofore met with. . . . Their dwellings are simple in construction but far superior to those of any tribe in upper California. An excavation some ten or twelve feet square being made in the ground, a house of boards placed upright against its sides is formed, with sloping roofs and entirely closed, except a circular opening in front, which permits ingress and egress to the occupants. . . . Their fishing dams exhibit considerable skill and great perseverance, being formed of timbers laid horizontally, securely fastened together, and strengthened by uprights; the whole extending across the river, and necessarily possessing great solidity, in order to resist so rapid a torrent as the Klamath. Salmon are taken in large quantities and in some seasons constitute their chief article of food. These fishing dams, however, prove a constant source of trouble and complaint, the upper tribes being often dissatisfied, because the salmon are obstructed in their passage by the dams below them. Their females are virtuous, and as far as we could observe during our stay among them, behaved with propriety; in fact the whole people manifested much satisfaction in their intercourse with the command, it being the first detachment of soldiers they had ever seen. . . . Our means of communicating with these tribes were so imperfect that it was impossible to gain an accurate knowledge of their customs, ceremonies, etc. or their notions as to their own origin or a future state. I was also unable to ascertain their numbers, and it is extremely doubtful if any reliable information on that point can at present be had.

After much delay, arising from misunderstandings and jealousies between different tribes, a treaty of peace and friendship was formally signed on the 8th by such chiefs as were willing to do so, and a reservation marked out, embracing the point where the two rivers join and extending along the crest of the mountains for some miles above and below on each of them. It is impossible, at this time, to find any one sufficiently familiar with their language to explain abstract ideas, or to teach them according to our notions the obligations of a bargain or a contract; and it is hardly probable that they had a very clear conception of the character of the instrument to which their marks are affixed, though I believe them to be sincere in their promise to remain good friends in the future. Mr. Durkee, an enterprising adventurer from New England, has resided on these rivers for several months, and by good management has acquired a considerable influence among the tribes near the junction; he has also some knowledge of their language, and in due time may qualify himself so as to render valuable service as an interpreter.[*]

Meantime, California had appointed a Superintendent of Indian Affairs —the Honorable Edward F. Beale. The United States commissioners had gotten into touch with him at once. Learning of the eighteen tribal treaties effected by the United States commissioners, the Legislature of California violently opposed the granting of any land to the Indians. The debates in this august body show that the popular demand was for the absolute extinguishment by the United States of the tribal claims. The lands of California were to be freely at the disposal of the state. Therefore, Beale was at once dispatched to Washington where the President and the Secretary of the Interior heard his side of the question, for there were many views as to what should be done with the Indians of California.

Beale was prompt to put himself on record. In his opinion the treaties were wise. The system of tribal areas established by them was "but the natural result and consequence of the policy pursued throughout." It was evident, he declared, that if allowed to roam at pleasure their complete extinction would be inevitable. He would not believe that the government would deny them the occupancy of "a small portion" of the vast country to be taken from them from which "such extraordinary benefits would

_____

[*] Ex. Doc., 3rd Sess., 34th Congress, Vol. 9, 1856-1857.

undoubtedly accrue to the whites," who were already the recipients of many. But to remove them east of the Sierra Nevada mountains as urged by the Californians in order to separate them from the whites and release their lands to settlement was impractical in his opinion, for the reason that there was no knowledge of the country beyond—no more "than the interior of Africa." It was reported to be a vast desert, almost unrelieved by verdure, with all cultivable areas already occupied by other Indians. It was not to be expected that they could be induced to remove to such a region voluntarily or be forced in that direction. To move them north would be but to add the 100,000 Indians remaining of over 200,000 less than three years before, to the already overflowing Indian population of Oregon where the same demand on the part of the settlers existed as in California. To move them south would be but to place them directly in the path of immigration. "The results to the weary immigrants would be disastrous." Moreover, the treaty stipulations with Mexico forbade their establishment on the southern border. Again, there they would be in the way of the projected railroad to the Southern Pacific. He had not examined all the eighteen reservations but he had seen some of them in the southern part of the state. Even if some of them might contain gold lands yet unprospected by the whites, the same was true of any location. At least the tracts selected by the commissioners had the advantage as a rule that they embraced "only such lands as are unfit for mining or agricultural pursuits."

The reservations made in the south, he declared, were "undoubtedly composed of the most barren and sterile lands to be found in California." Those who opposed their being set apart to the Indians in no instance had been able "to point out other locations less objectionable or valuable than those already selected." He did not believe that the land in any of the proposed reservations compared favorably "with the agricultural and valuable portions of the State" from which the Indians were to be removed. Moreover, it was necessary to protect the Indians, for only recently two or three whole villages, men, women and children, had been massacred by the whites. The saving of the Indians from such butchery justified the proposed system which, therefore, he approved. But he objected seriously to the provisions contained in the treaties looking to the supply of the Indians with agricultural implements and the establishment of schools among them! He did not believe they were disposed toward agricultural pursuits; such implements as were furnished them would soon be taken by the whites. As for education—it would not "subserve their interests," "their present state of civilization and advancement" precluding "the possibility of their appreciating the benefits to be derived from such instruction." Since no money consideration was to be passed to them, the promise to provide them with breeding stock and a temporary supply of beef cattle might be considered as a trifling price for their title if the provisions as to agricultural implements and education were stricken out! The preceding spring whole mining districts had been idle by reason of Indian disturbances, resulting in a great decrease in the "exportation of gold." If this were the result of a war with a few tribes the consequences of a war with the whole Indian population might readily be imagined. Beef and flour, he argued, were but substitutes for annuities, and even the supply of them was to cease at the end of two or three years. Consequently, the treasury was in fact effecting a great saving. Reassured by the Secretary of the Interior that "economy" might be "ill timed in the present case" by reason of the possible cost to California of a great war, he urged the ratification of the treaties without delay, minus their "objectionable features."

Need more be said? When it is considered that Beale's ideas were as much in advance of those of the public, as modern ideas of justice are of his, the case of the Californian Indian is made out. The Superintendent of Indian Affairs of California though opposing the Legislature and the popular views of California, was himself endorsing the complete confiscation of Indian lands, the removal of the Indians to domains which he believed to be utterly incapable of supporting life by agriculture, the withholding from them of agricultural implements, the denial to them of the education which they craved, and the furnishing of breeding stock to them only by reason of the fact that it would increase and multiply in their hands and render it unnecessary to furnish beef and flour to them for more than two or three years. He must have known that large numbers of them had been Christianized by the Spaniards, that education had long been available to them at the missions, that thousands of them for years had subsisted by agriculture, olive, grape and citron culture, and the breeding of cattle on their ranches in which their rights for two centuries had been established under the Spanish law. He himself told of the destruction of their villages and approved of the waste places selected on which to confine them for the reason that agricultural pursuits on their part would be impossible in such places to the exclusion of the whites. Plainly it was the purpose of Beale and those behind him to convert the Indian population of California back to the nomadic hunter state from which they had emerged even before the coming of the Spaniards.

Despite the outcry in California and the great lobby in Washington, on May 14, 1852, immediately after the receipt of Beale's letter, the Commissioner of Indian Affairs transmitted the eighteen treaties to the Secretary of the Interior, declaring in his letter of transmittal that they had been withheld till then by reason of Congressional opposition. The same day the Secretary of the Interior transmitted them to the President with Beale's report and other documents attached, the President in turn transmitting them to the Senate. July 8, 1852, the Senate resolved not to ratify the eighteen treaties. Thus, it is seen that the Senate simply did not have the courage to set apart any land in California to the Indians in exchange for their tribal domains.* Six months had passed since the tribes signed the treaties. Seeing that they would be destroyed if they remained on their lands while the treaties were passing through the tortuous channels of the government, the commissioners had urged them to remove without delay to the reservations set apart to them. Abandoning their ancient homes in obedience to this advice, in November, 1851 the migration of the tribes commenced, continuing for seven months.

What occurred during this grand trek is all but unbelievable. The migrations of the Georgian tribes was a pleasure journey in comparison. Once uprooted the migrants were harried on their way with indescribable cruelty. Moreover, the general movement gave rise to the rumor that they had taken to the warpath so that when small groups appeared they were often slaughtered at sight, questions being asked afterwards. Consequently their losses were appalling.

But this was not all. On reaching the waste places erroneously designated in the treaties as reservations, they found them already occupied by miners and gold prospectors who had no idea of yielding up the claims they had staked. There too, the Indians were met with arms and butchered mercilessly.

Nevertheless, no provision whatever was made for them either by the state or federal government. Left to disperse, to become mere vagrants upon the earth, the few survivors were to exist henceforth upon the mere tolerance of the usurpers. Treated at best as squatters on their own lands by the state government, thousands of them were forced to become mendicant hangers-on in the vile alleyways of a mushroom civilization. Thus forced into contact with the dregs of human society, abused, landless, penniless, infected with diseases they had never known before, a population of over 200,000 semi-civilized, semi-Christian aborigines was all but wiped out in a decade of so-called white civilization to the eternal disgrace of California, the nation, and the United States Senate.

Politics having achieved their infamous purpose, and the California Indians having been left to rot for the benefit of the land-grabbers of California, the United States and the state proceeded as usual to appropriate their lands, contrary to the then existing land laws of the United States forbidding the issuance of patents to land in which the Indian title had not

* When the eighteen treaties were dug out of the pigeon holes of the Senate in 1906, the following characteristic endorsements appeared on them: "June 7, 1852. Read and with the documents and treaties referred to the Committee on Indian Affairs and offered to be printed in confidence for the use of the Senate. January 18, 1905. Injunction of Secrecy removed."

been extinguished. Some of this land today constitutes the great national parks and forest reservations in California, the title to which, the United States or the guardian, could not have acquired by adverse possession or laches.*

* Pursuant to the Acts of 1891 and 1906 the government has expended about $200,000 on education and welfare work among the California tribes. By the Act of May 18, 1928 (45 Stat. 602), it authorized the institution of a suit against the United States in the Court of Claims, by and in behalf of the California Indians, such suit or suits to be brought by the Attorney General of the state, without expense to the Indians for legal services. Pursuant to this Act suit was instituted on their behalf on August 4, 1929, to recover the sum of $12,800,000. The amount claimed on their behalf is not, however, the value of the property of which they were divested, but merely the alleged value of the property which was promised them in the unratified treaties in exchange for the land yielded by them.

In 1910 it was estimated by the Bureau of Ethnology that there remained about 15,000 California Indians for whom the only land provision ever made by the government up to the present is the purchase in 1925 at a total cost of $7,650 of the three small tracts of 190 acres each which were capable of affording homes for about 250 Indians. Strenuous efforts have been made in recent years by the Indian Board of Cooperation of California, a private philanthropy, to succor the California Indians, but despite the appalling evidence adduced before Congress, so far they have failed. For full record see Hearing Before Sub-Committee of the Committee on Indian Affairs, H.R., 66th Cong., 2nd Sess., March 23, 1920; Ibid, 66th Cong., 2nd Sess., April 28, 29, 1922; Ibid, 69th Cong., 1st Sess., 1926.

The per curiam decision (1926) of the Supreme Court in the test case of Super et al. v. Work, et al. brought on behalf of the Indians by the Indian Board of Cooperation, defeated the hope which the Indians had entertained of having their status as well as their legal rights clearly defined. The Supreme Court simply affirmed the decision of the lower court upon the authority of Barker v. Harvey, U. S. Title Ins. & Trust Co., and Lone Wolf v. Hitchcock.

23

The Civil War, Railroads and the Minnesota Sioux

Fortunately for the republic, there was no Tecumseh to organize the Indians generally during the period of the Civil War when the government was compelled to withdraw every regiment of troops but one from the West, for during this period the greatest unrest existed among the tribes.

Some of the Indians of the Southwest were deeply agitated by Colonel Albert Pike, the Confederate Commissioner of Indian Affairs, who was deputed to negotiate treaties of alliance with them. But like the western Sioux and the unwarlike Snakes, Yakimas, Coeur d'Alenes, and Paloos of Oregon and Washington, they did not take to the warpath out of sympathy for the South. It was because they saw in the war an opportunity to strike a blow in their own cause. The same is true of the Sioux of Minnesota, among whom, due to the stupidity of the whites, and the resentment of the Sioux over the violation of the treaty of 1858 respecting the Pipe Stone Quarry, an uprising occurred in 1862 under Little Crow. This is popularly known as the Minnesota Massacre. In it about eight hundred whites including a hundred or more soldiers were killed.

The Five Civilized Tribes in the Indian Territory, however, engaged in the war, for by reason of their ownership of slaves, many of whom they had brought with them from Georgia, Alabama and Florida, they were deeply interested in the question of slavery. Yet, like all other border people, such as the Marylanders, Virginians, Kentuckians and Missourians, they were also on the economic borderline and consequently found their interest divided.

A part of the people of each tribe, the slave-holding planters among the Indians, or the Indian aristocracy, desired to maintain the institution of slavery at least until they were compensated for the surrender of their property. The nonslaveholders having no material stake in it were in sympathy with the abolition movement. The slaveholding element in the main being the richer and, therefore, the more enlightened faction, was naturally in control of the tribal governments just as the same faction in the Southern states was in the saddle. Thus, the Indians were governed in their sympathies and attitudes by very much the same influences as those which dictated the alignments of the whites.

At the close of Buchanan's Administration nearly all the Indian agents in the Indian Territory were secessionists, and the moment the Southern states commenced passing ordinances of secession, these men exerted their influence to get the five tribes committed to the Confederate cause. Occupying territory south of the Arkansas River, and having the secessionists of Arkansas on the east and those of Texas on the south for neighbors, the Choctaws and Chickasaws offered no decided opposition to the scheme. With the Cherokees, the most powerful and most civilized of the tribes, it was different. Their veteran chief, John Ross, who was opposed to hasty action, in the spring of 1861 issued a proclamation, enjoining his people to observe a strictly neutral attitude during the war. A majority of the Cherokees at this time were full bloods, called Pin Indians, and not being slaveholders, sympathized with the Union. Therefore, when Albert Pike, the Indian Commissioner of the Confederacy arrived with General Ben McCulloch, commanding the Confederate forces in western Arkansas and the Department of Indian Affairs, to negotiate a treaty with the Cherokee Nation, Ross was still able to hold the Cherokees neutral.

After the battle of Wilson's Creek in which the Union forces under General Lyon were defeated and forced to fall back from Springfield, Missouri, to Rolla, believing that the South would probably prevail, Ross deemed it wise to change his policy. He now permitted John Drew, a staunch secessionist, to raise a regiment of Pin Indians of which he was commissioned Colonel and William P. Ross Lieutenant-Colonel. Colonel Stand Watie, the leader of the secession party, also commenced to raise a regiment for McCulloch's division. Therefore, upon his return from the far West in September where he had been engaged in stirring up the plains tribes, Pike revisited Ross, finally negotiating the treaty of Park Hill of October 7, 1861, with the Cherokee Nation. Treaties with the other tribes were also secured in each of which it was provided as in the Cherokee treaty that the troops raised should not be employed outside the Indian Territory.

The Creek treaty was obtained by Pike through the influence of ex-Chief D. N. McIntosh who was appointed Colonel of their regiment. Like his brother Chitty, he was a leader of the secession faction. Nevertheless, at least two thirds of the Creeks were behind Chief Hopoeithleyohola who, Pike declared, was determined to remain neutral, not out of loyalty to the United States but as the best means of self-preservation. The truth is, there was a feud between him and the McIntoshs, whose father's death warrant he had signed in 1825, for bribery in connection with the fraudulent Creek treaty of Indian Springs. At any rate, after making several abortive efforts to obtain a conference with the old chief, Colonel Douglas H. Cooper, commanding the Department of Indian Operations of the Confederacy, determined to compel his submission or drive him out of the Creek country. Accordingly he set out to collect a force of white troops for that purpose. Seeing this, in November the Chief at the head of his faction started for Kansas. Overtaken by Colonel Cooper at the head of the Confederate Creeks, Choctaws, Chickasaws, Cherokees, and some Texans, a fight ensued. During the skirmishing of the night the Pin Indians of Drew's Cherokee Regiment which had been raised by Chief Ross, went over to Hopoeithleyohola. The next day they fought on his side in the desperate action known as the battle of Chusto Talasah in which five hundred of the Union Indians were reported by Cooper to have been killed and wounded.

The Confederate Indians of Stand Watie's regiment, and those of Drew's regiment, after returning from Kansas, participated under General Pike in the battle of Pea Ridge in March, 1862. They were charged with scalping and mutilating the federal dead on the field. Hearing of this, General Pike called on the Surgeon and Assistant-Surgeon of his field hospital for reports in which it was stated that one of the federal dead had been scalped. Pike then issued an order, denouncing the outrage in the strongest language, and sent a copy of it to General Curtis. He claimed that part of the Indians were in McCulloch's corps in the first day's battle, and that the scalping was done at night in a quarter of the field not occupied by the Indian troops under his immediate command.

After the battle of Pea Ridge the federal government sent an expedition of five thousand men under Colonel William Weer, 10th Kansas Infantry, into the Indian Territory to drive out the Confederate forces of Pike and Cooper, and to restore the refugee Indians to their homes. Following a short action at Locust Grove, near Grand Saline, on July 2, 1862, Colonel Weer's cavalry captured Colonel Clarkson and part of his regiment of Missourians. On July 16 Captain Greeno, 6th Kansas Cavalry, captured Tahlequah, the capital of the Cherokee Nation, and on July 19, Colonel Jewell, 6th Kansas Cavalry, captured Fort Gibson, the most important point in the Indian Territory.

The Confederate forces were now driven out of all that part of the Indian country north of the Arkansas River. Thereupon the loyal Indians of the Cherokee, Creek, and Seminole tribes were organized, by authority of the United States government, into three regiments, each fully a thousand strong, for the defense of their country. The colonel and part of the field and line officers of each regiment were white men. Most of the captains of companies were Indians. Colonel William A. Phillips, of Kansas, who was active in organizing these Indian regiments, commanded the Indian brigade from its organization to the close of the war. He took part with his Indian troops in the action at Locust Grove, and in the battles of Newtonia, Maysville, Prairie Grove, Honey Springs, and Perryville, besides many minor engagements. In all the operations in which they participated the Indians acquitted themselves creditably, and to the satisfaction of the federal commander.

To the Confederacy the Indians of the Territory contributed four regiments and a battalion of infantry, in all about 3,500 men.

Ross having been deposed as head chief, on February 18, 1863, the Cherokee National Council, then composed of a majority of Union sympathizers, repudiated the treaty with the Confederate States of America. At the same time slavery was abolished among the Cherokees. The other tribes soon followed this example. Thus was peace finally reestablished in the Indian Territory.

Here let it be said that the American Indians were represented on the personal staff of General Grant, the Generalissimo of the federal armies, in the person of Ely Samuel Parker (Do-ne-ho-ga-wa), a Seneca chief. As a captain and assistant adjutant general, entrusted with all of Grant's personal correspondence, he was Grant's military secretary at Appomattox and wrote with his own hand the articles of Lee's surrender. An engineer by education he was a cultured man with a wide knowledge of his race. He was a direct descendent of Sayenguhaughta, of Revolutionary fame, and of Djigonasch who, with Hiawatha and Dekanawida, is supposed to have founded the Iroquois Hodenosaunnee. In 1865 he was appointed Brigadier-General of Volunteers, and in 1866, First-Lieutenant of Cavalry in the Regular Army. On March 2, 1867, he was promoted to brigadier-general.

In 1860 the forces of disunion were not arrayed along the Potomac alone. They existed elsewhere than in the South. To Lincoln's broad vision it was apparent that in the South they were reacting in but one form. The coercion of the secessionists back into the Union would not make less artificial and slender the ties which bound the far West to a reunited East. Within a decade, men on the Pacific Coast had favored a British protectorate in preference to an alliance of their country with the Union. As late as 1861 it had been necessary to resort to the expedient of exempting California from the duty of furnishing troops to the federal army in order to hold the state in the Union. Between the far off regions of the Pacific Coast and the East, there existed a void unspanned by national sentiment. He did not fail to see that the West was an empire differentiating more and more along the lines of its distinct social and economic problems from the East. If these two empires were to be welded into a single body politic there must be a common mind, a common heart, common veins for the flow of its life blood. Washington must be the common mind—a thing which until now it never had been, standing as it did between Boston and Charleston. One flag, bestarred by every state, must flutter in each. The veins? Transcontinental railroads nourishing the nerve centers of every member!

No longer were the vast expanses of the western prairies, the broad sweeping rivers coursing through them, the soaring altitudes of the Rockies and Sierras to be traversed by a pony express. The "Iron Horse" was to take its place. The midcontinent must be peopled quickly to bind the West and East together.

First, he proposed to encourage a more rapid distribution of the public domain by granting land free to those who would actually settle upon and improve it. Second, to make vast grants to the states for the purpose of encouraging education and public works. Third, to bring about the construction of the needed transcontinental railroad systems without further delay by granting land to the corporate interests that would undertake to build them, sufficient in value to insure volunteers for the task. Accordingly in 1862-63, he secured the passage of the Morrell Act providing grants aggregating 1,000,000 acres to the states, particularly designed to insure military instruction in the so-called "land grant" colleges—one in each state; the Homestead Act providing that actual settlers by a continuous residence of five years and improvement of the land should be entitled to a free patent to a quarter section of the public land, or 160 acres; and the great Railroad Act enfranchising the Union and the Central Pacific Railroad companies.

What Lincoln's inward convictions were concerning the people inhabiting the great midcontinental regions of America history does not disclose. It is said that when he was informed of the Minnesota Massacre of 1862, he exclaimed in soulful anguish: "If I survive I will reform the Indian system." But the fact is, his humanitarianism failed utterly to envisage the human rights involved in the execution of his grandly wise plan of national development, and to provide for those it must of necessity displace. Just as the scheme for the settlement of the Northwest Territory, the Southwest Territory, the Louisiana Territory, the Pacific Seaboard and the great Middle West by the sale of public lands had failed to anticipate the needs of the Indians, so did the Homestead Law and the plan for the erection of the transcontinental railroads. For the Iron Horse "Poor Lo" was to furnish the grazing and the highways!

When Lincoln signed the Act authorizing the construction of the Union and Central Pacific railroads, an army of laborers poured into the West while along the rights of way upon which they were employed a chain of mushroom towns, each a center of disturbance, with incredible rapidity spanned the continent. As the war drew to a close thousands of unemployed soldiers from both North and South sought employment at the hands of the railroad contractors, adding their members to those of the homesteaders under the Act of 1862 who had preceded them. To feed the enormous new population of the West, meat was required. Buffalo contractors were at once busily employed in meeting the demand. Scouring the plains without regard to the necessities of the Indians who depended upon the buffalo, like the Indians they were slaughtered promiscuously. To prevent interference by the Indians with the work of construction and destruction that went on apace, the fierce Pawnee cossacks of Oklahoma and Northern Texas were employed and organized in mounted bands under fearless white veterans "to protect" the laborers. From this innumerable fictitious homesteaders and other intruders benefited greatly. All along the railroad rights of way they were more or less free from the first to "squat," gradually pushing farther and farther out into the Indian country. So long as land was not actually occupied by Indians, the homesteaders assumed it was open to preemption. It was the old story, all over again, of the Forty-niners and the gold days in California. Of what had occurred along the Oregon trail, Cruze has given a very vivid picture in his "Covered Wagon." Mr. Fox's Iron Horse shows that the nation rode rough shod in its economic progress over the Indians of the plains in the 60s. Historically these spectacles would have been far more accurate and valuable had they stressed the ouster of the Indians rather than representing them as intractable "bushwhackers" in the path of civilization.

With what importance to the Indians the economic policy of Lincoln was fraught is readily seen from the magnitude of the federal grants of the next decade. To the states, including the grants to the land grant colleges, 71,000,000 acres of the public domain were granted before 1873. Under the Homestead Act 85,000,000 acres were preempted within an amazingly brief time. To the corporate interests which undertook to finance the construction of the transcontinental railroads, 155,000,000 acres including the rights of way and alternate sections of non-mineral-bearing lands were granted outright.

The part taken by the Indians in the war, and particularly the so-called Sioux Massacre in Minnesota, were made much of by those who were determined to suffer no restraint upon the taking of their lands. It was an outrage, it was declared, that a people so savage, so disloyal should be allowed to block the economic development of the country. To permit them to do so passed the bounds of reason. Accordingly, hardly had the

Indians of Oklahoma declared peace, when on the floor of Congress it was proposed to confiscate the property of the Five Civilized Tribes.

It was, of course, beyond the power of Congress to pass a bill of attainder. Yet, the Indians were to be penalized in another way. In 1854 Congress had created the Court of Claims as a tribunal in which the United States, or the Sovereign, should voluntarily submit to the determination of its contractual obligations, pursuant to which innumerable claims arising out of their treaties had been filed by the Indians. Now, by the Act of March 3, 1863, it withdrew from the Court of Claims jurisdiction over all claims based on an Indian treaty that was not pending on December 30, 1862, and at the same time provided that no interest should be allowed on any claim up to the time of the recordation of judgment thereon where the payment of interest had not been expressly stipulated. In other words, unblushingly, it deprived the Indian tribes of the only legal means they had of enforcing their constitutional rights and recovering the value of the property which was being ruthlessly taken from them in violation of the solemn treaty obligations of the United States.

Thus encouraged by Congress, it was not unnatural that the States should now undertake to subject the property of the Indians to taxation. In this situation, again the Supreme Court came to their relief, declaring in two exhaustive decisions that since the tribes were dependent domestic communities of the United States, even though they may have "outlived many things, they had not outlived the protection afforded by the Constitution, treaties and laws of Congress."* The same year the clause of the Constitution conferring upon Congress the power to regulate commerce with the Indian tribes, was construed by the Supreme Court to include the exclusive power to administer Indian property.† These two decisions compelled Congress expressly to exclude the Indians from the enumerations in the Fourteenth Amendment upon which direct taxes and representation in Congress are based.‡

It was just at the time that Longfellow was writing his great poem that a tremendous intellectual force came into being among the Sioux. In 1540 de Soto had found them in East Arkansas, and in 1557 another Spaniard had encountered them in the highlands of South Carolina. John Smith had found them represented by the Manahoacs of Virginia, while in 1640 the Jesuits of Canada knew much about them. Father Hennepin had paid them a long visit in 1680 and in 1708 Le Seur had also visited them. After this they were allies of the French in 1755 and of Pontiac, and later of the

---

* *The Kansas Indians*, 5 Wallace, 755.
The New York Indians, 5 Wallace, 767.
† *U. S. v. Holliday*, 3 Wallace, 407.
‡ Art. I, sec. 2, par. 3 and Amendment XIV, sec. 2. The Amendment was proposed June 13, 1866 and adopted July 28, 1868.

British. Lewis and Clark had spent the winter of 1804-05 with the Yankton Sioux. Upon the birth of their chief's son they had wrapped the future Struck-by-the-Ree in the American flag. Treaty relations with the Sioux generally had commenced in 1826. Soon after this they had begun to feel more and more the pressure of white immigration to the Northwest and were constantly being importuned to yield their lands.

At times it almost seems as if in the distressing ordeal of the struggle by the Indians to maintain themselves in face of an advancing hostile civilization that some incorrigibly evil spirit dictated the course of the government. Thus, when in 1851 the Wahpeton and Sisseton Sioux were induced to cede their lands in the territory of Minnesota, at the instance of the Evangelical missionaries who had long since discovered the power exercised over the red race by the shamans at the Red Pipestone Quarry, the government deliberately made the Sissetons, or the traditional guardians of the Quarry, include the same within their grant.

Among the Hunkpapa Teton Sioux at this time was a young chief named Tataka Yotaka, or Jumping Badger, the son of the subchief Four Horns. The latter as Sitting Bull, which implied that he had become a councillor, was undoubtedly one of the Pipe Stone Quarry shamans. Having distinguished himself in battle against the Crows by counting coup when he was fourteen, the boy was proclaimed a great hero at a tribal feast and given his father's name of Four Horns. This meant that he was twice as bold as a buffalo bull. Carefully schooled to succeed his father as Councillor, he was just seventeen years of age in 1851, when the government, following the gold rush to the West, demanded that the Wahpeton, Sisseton, Yankton, and Yanktonais Sioux of Minnesota yield their lands and remove beyond the Missouri River to join the other three tribes.

It was now seen by the Sioux statesmen that the missionaries were determined to break up the religious organization of the Dakota, or League of Friends. At this the whole Sioux Nation was enraged. Seeing the danger that the settlers were in, even some of the whites protested. Accordingly, in 1855 Longfellow published the Song of Hiawatha, pointing out clearly what the Quarry really meant to the red race and showing that it should be regarded as a distinct asset instead of a danger.

An intelligent government would have recognized that the real danger lay in interference with the tribal religious institutions. As a matter of fact its course led to an inevitable result. When Four Horns succeeded his father as Sitting Bull in 1857, already a definite program had been formulated to unite the Sioux more firmly after the Iroquois manner to prevent the further alienation of their lands. Indeed, of an Indian nationality such as that Tecumseh had proposed, the new Councillor was already the avowed exponent. Accordingly, when this same year the government undertook to extinguish the rights of the Yankton Sioux and concentrate them upon their present reservation in South Dakota, they firmly refused to enter into any treaty that did not restore to them, as the Levites of the Dakota, the exclusive guardianship of the quarter section embracing the precincts of the sacred Quarry. In vain the chieftains were taken to Washington where agents of the government, including an Indian bribed to do the work, tried to argue, then scare them into betraying the tribe. In the end some of them were bribed. Seeing that the Sioux were about to lose the Quarry, Struck-by-the-Ree, unquestionably at Sitting Bull's instance, now drove a shrewd deal with the government. In other words the chieftains agreed to sign the treaty of 1858 in which the Yanktons released their claims in approximately 11,000,000 acres of land and reserved to themselves 340,000 acres in South Dakota, upon the express stipulation that the sacred precinct of the Quarry should be surveyed, marked, set apart to the Yanktons, and protected forever by the government against trespass.

His real object being unknown, upon his return to Minnesota the patriot who had thus recovered the shrine for the whole Sioux Nation by a sacrifice on the part of the Yanktons, was charged with having betrayed his tribe so that a factional conflict was narrowly averted by Sitting Bull. But having obtained the desired removal, the government at the instance of the missionaries now refused to carry out the express stipulations of the treaty with respect to the survey of the Quarry precinct. The result was a state of mind among the Sioux which produced a train of dire events, directly connected with the flagrant violation of their religious liberty.* Nor is it possible to understand these events without bearing in mind the government's continuing trespass upon the sacred precinct of the oracle whose retention by Sitting Bull was essential to his scheme of Indian nationalization; that long before Sitting Bull was heard of in the East as the great Sioux shaman he was already dominating Indian tribal politics throughout the country and was engaged in a contest not only with the gold seekers but with the missionaries whose unreasoning zeal to civilize the Sioux was to cost thousands of innocent lives. Plainly, too, he was trying to protect the gold which the medicine men generally knew was in California and in the realm of the Dakota.

In 1866, having greatly enhanced his prestige by defending the Sioux rights against the Crows and Blackfeet, Sitting Bull assailed Fort Buford. His influence was now tremendous. This same year treaties of amity were negotiated with the Five Civilized Tribes in which the former expressly repudiated their treaties with the Confederate States of America and abolished slavery within their domains. The Negroes but not their former masters became citizens of the United States! In these treaties it was stipulated that the freedmen should be subject to the tribal laws, and that

---

* The so-called Minnesota Massacre of 1862; Sitting Bull's attack on Fort Buford in 1866; the Sioux Uprising of 1874; the death of Custer; the refusal of the Sioux to cede more land in 1888; the Ghost Dance of 1890; Wounded Knee.

a general council should be established composed of delegates from each tribe, while the Choctaw and Chickasaw treaties provided for the election of a delegate to Congress whenever Congress should authorize the admission into its body of a delegate from Indian Territory.

Now it was that John Ross, the old "Swan," who for just half a century had striven for his people, while still laboring for them in Washington, passed away—one of the greatest statesmen ever produced by his race.*

With the Sioux an amity was also declared, but no sooner was it signed than about one hundred of those engaged in the massacre of 1862 from whom secret agents of the government had obtained confessions, were executed.

Reaction against the Indians was to take still another form. In the treaty of 1867 by which Russia ceded Alaska to the United States, contrary to the established practice of international law, not only were no guarantees of Indian rights contained, but an express guarantee was given by Russia against the assertion in the future of any claim of right on their behalf.

This year the Sioux were required to yield more land and to agree to concentrate upon a specified reservation when called upon to do so. All manner of benefits were promised them while the land given up by them was to be sold for their account.†

---

* August 1, 1866.

† The Buffalo Party never had any idea of fulfilling this treaty which required the Sioux to concentrate on or before January 1, 1876. At this time the bankers and the politicians were employing General George A. Custer, of Civil War fame, a perfectly ruthless man, to crush Satanta and Lone Wolf, at the head of the Cheyennes, Kiowas, Arapahoes, Comanches, and Apaches in the Indian Territory. Their real object has been brazenly told.

"Now," says Whittaker, Custer's biographer, "it was necessary to overrun the northwest. When Custer pacified the Kiowas, Arapahoes, and Cheyennes by force, physical and moral, the Sioux of the northwest had fared very differently. They had frightened the Government into a treaty, the treaty of 1868, by which the United States had promised to give up to them forever a large expanse of country, and not to trespass thereon. Now that the danger was over, and the Pacific Railroad safely completed to the south, thanks to Custer, the treaty with the Northern Indians became irksome. It was all well enough to *promise* a lot of naked savages to give them up so much land, but it could not be expected that such a promise should be *kept* a moment longer that was necessary to secure a quiet building of the railroad. It was now time to break the treaty. A northern Pacific road had become necessary, and its route was to lie right through the very midst of the territory solemnly promised the Indians by the treaty of 1868. As a practical measure to provoke an Indian war, there is nothing so certain as the commencement of a railroad."

Here then is the real reason for the Indian wars of what was perhaps the most disgraceful period in American history. Today it seems almost unbelievable that a people who called themselves civilized could have treated a helpless, dependent race in this way.

24

Driven to the Reservations

THE HISTORIAN IS WONT to represent Ulysses S. Grant in the successive role of a rugged, hard-hitting, relentless warrior, a soldier President relatively of no great proportions, an unsuccessful and embittered adventurer upon the field of finance. Thus, from posterity, the true magnitude of his character has been obscured. The truth is Grant, like Washington, was not only a great soldier, but one of the greatest humanitarians of his age.

Immediately upon becoming President he appointed General Ely Samuel Parker, the Seneca chief, and his former military secretary, Commissioner of Indian Affairs.* But while Parker's influence over Grant was undoubtedly great, the latter's experience of the Indians and Indian country, first as an officer in the army, and then as Secretary of War, was alone sufficient to make him sympathetic with the idea of drastic Indian reforms.

A ceaseless strife was still being waged against the tribes of the West and this with the aid of the Government. Since 1849 twenty-two campaigns, officially designated wars, had been conducted against them. Even now the Kiowas in the Indian Territory were being assailed. The popular outcry about the cruelty of the Indians did not deceive him. As between white man and red man he knew only too well who was the most cruel. Coming into power with a soul unembittered by the narrow prejudices of Jackson, conscious of the great wrongs that had been done a dependent race, and like Washington with nothing more to ask of the frontier voter,

---

* Parker now resigned his commission in the army. He served as Commissioner of Indian Affairs from 1869 to 1871.

he was determined to arouse the nation to a high sense of its moral duty. Among his first acts was the replacement with army officers of the dishonest Indian agents whom he knew had long been mere exploiters. Inasmuch as these offices were rich ones, long and loud was the cry which went up in the lobbies of Congress. Parker's appointment was bitterly assailed as evidence of too much Indian influence. In answer Grant compelled Congress to appoint an able commission of philanthropists to render without delay a report to Congress that would fortify him in the fight which he saw was impending. Following the rendition of this report he laid down the gauge of battle in his first message.

From the foundation of the Government to the present, management of the original inhabitants of this continent—the Indians—has been a subject of embarrassment and expense, and has been attended with continuous robberies, murders, and wars. From my own experience upon the frontiers and in the Indian countries, I do not hold either legislation or the conduct of the Whites who come most in contact with the Indian blameless for these hostilities. The past, however, cannot be undone, and the question must be met as we now find it. I have attempted a new policy towards these wards of the nation (they cannot be regarded in any other light than as wards), with fair results so far as tried, and which I hope will be ultimately attended with great success.

A system which looks to the extinction of a race is too horrible for a nation to adopt without entailing upon itself the wrath of all Christendom and engendering in the citizen a disregard for human life and rights of others, dangerous to society. I see no substitute for such a system, except in placing all the Indians on large reservations, as rapidly as it can be done, and giving them absolute protection there. As soon as they are fitted for it, they should be induced to take their lands in severalty and to set up territorial governments for their own protection.*

In this language, like that of Washington, there was no compromising with present nor palliation of past wrongs, no moral quibbling. It was the same Grant, magnificent in his magnanimity, who returned Lee's sword at Appomattox, who sent home the Confederate soldiery with their mounts to do the spring planting.

The reply of Congress was immediate. A statute was enacted in 1870 prohibiting the assignment of army officers to civil duties! So great was the opposition to Parker that he resigned his office in 1871 to accept another public office in New York and thereby relieve Grant from the attacks of the Buffalo Party on his account. Armed with the information gathered by his commission, and with the support of the public interest aroused by it, Grant next proceeded to put into effect the reforms it recommended.

The decisions of Marshall that the tribes were not independent nations but dependent communities, had made them necessarily a part of the body politic. So, in 1866, the Supreme Court had declared that all the guaran-

* First Message, Dec. 6, 1869.

tees of the Constitution applied to them. How could such a people be dealt with by treaties?

It was justice above all else which the solution of the Indian problem required. Yet, so long as the tribes were free to negotiate, it was only natural each should seek peculiar advantages. To the extent their demands were recognized injustice necessarily must result to the other tribes. In the second place the policy of dealing with them as if they were not subject to ordinary law tended to encourage the popular impression that there was something peculiar about Indian property rights, that the principles of the common law pertaining to ordinary communities did not apply to them. This was a view which was often reflected even by the courts.

Grant knew that where an obligation is fixed by a sovereign state upon a dependent people which the latter are required to perform in accordance with the interpretation of that obligation by the sovereign, the act amounts to no less than legislation by the superior over the will of the inferior. Such is the case of a proclamation, resolution, or act in which, upon the fulfillment of the conditions mentioned therein, the state holds itself out in advance as willing to be bound. The accepted practice at international law, where an understanding between a sovereign state and an aboriginal tribe or dependent community was necessary for the purpose of arranging the terms of the guardianship of the state over the tribe, was a charter determining the manner of the administration of the tribe as a dependent community, or an act of legislation. Where it was necessary to obtain the assent of the tribe to insure acceptance of the act, that assent might be manifested before or after the act. It was by such legislation that the British Crown dealt with its dependent Indian communities after the establishment of peace between Great Britain and France, as evidenced by the Royal Proclamation of October 7, 1763, which gave notice of the sovereignty of the British Crown over the communities occupying the territory ceded by France, and the basis of their obligation and of its own. Therefore, pursuant to the recommendation of the Board of Indian Commissioners in its report of 1869, Grant caused to be passed the Act of March 3, 1871. This act provided that thereafter

no Indian nation or tribe within the territory of the United States shall be acknowledged or recognized as an independent nation, tribe, or power with whom the United States may deal by treaty: Provided, further, that nothing herein contained shall be construed to invalidate or impair the obligation of any treaty heretofore lawfully made and ratified with any such Indian nation or tribe.

Having done this, in 1873 he addressed Congress as follows:

My efforts in the future will be directed to the restoration of good feeling between the different sections of our country . . . and, by a humane course, to bring the aborigines of the country under the benign influences of education and civilization. It is either this, or war of extermination. Wars of extermination, engaged in by people pursuing commerce and all industrial pursuits, are expensive, even against the weakest people, and are demoralizing and wicked. Our superiority of strength and advantages of civilization should make us lenient toward the Indian. The wrong inflicted upon him should be taken into account, and the balance placed to his credit. The moral view of the question should be considered and the question asked, cannot the Indian be made a useful and productive member of society by proper teaching and treatment? If the effort is made in good faith, we will stand better before the civilized nations of the earth and in our own consciences for having made it.*

If the Indians were to be civilized they must be encouraged to abandon their tribal relations by offering them individual estates in lieu of mere communal privileges. Accordingly, the Commission recommended that Indians like the whites be accorded the right to patent homesteads. Therefore, Grant secured the passage of the Indian Homestead Act of March 3, 1875, which provided that any Indian born in the United States, who was the head of a family, or who had arrived at the age of twenty-one years, and who had abandoned, or might thereafter abandon, his tribal relations, should be entitled to all the benefits of the Homestead Act of 1862 with the wise restriction that the land patented by him might not be alienated or incumbered for a period of five years from the date of the patent. It also left the Indian homesteader free to participate as a distributee of any tribal annuities, funds, lands, or other property so that he was under the necessity of making no sacrifice in the acquisition of an individual estate. But revolutionary as was this Act it was designed to provide for individual Indians only. In the nature of things it could contribute nothing to the economic welfare of the tribal Indians. They too must be saved.

Following upon the financial panic of 1873, Congress enacted various preemption laws that still further encouraged the agricultural development of the West. In many instances the courts had confirmed tribal rights in lands vastly in excess of present needs. Since the future welfare of the Indians required that they transform their mode of life to that of the economic order of the nation by acquiring the habits of an agricultural people, their own interests demanded that they concentrate upon lands suitable for agriculture. This being so, there could be nothing unjust in a policy designed to induce them to surrender their surplus lands upon just compensation. Moreover, considerations of safety dictated such a policy. The experience of the past showed that it was futile for them to expect to retard the advance of civilization. Whatever the courts might rule, however numerous the national police, they could not hold those domains indefinitely. It were far better for them that they surrender lands surplus to their necessities in the process of economic transformation, provided they

* Second Inaugural Address, March 4, 1873.

receive ample consideration for them, and concentrate upon reservations adequate to their present needs. There they might be in part, at least, removed from the deeper channels of the onrushing tide of civilization which they were not prepared to breast, more easily be protected by a limited police, and be tutored in agriculture while receiving a rudimentary education designed to emancipate them eventually from the communal system. Meantime the proceeds from the surplus lands disposed of by them were to be invested for the communal good.

This, in brief, was the Grant-Parker plan to emancipate the Indians from the economic slavery of the aboriginal communal system. Having abolished the baneful system of Indian treaties by the Act of 1871, under which the Indians had been made the shuttlecock of local politics in Congress, Grant set to work with the utmost energy to negotiate with all the tribes executive agreements pursuant to which he might secure to them definite reservations. The power of the President to do this was subsequently upheld.* Penalties were also provided for settling on Indian reservations, and the United States attorneys were charged with the duty of representing tribes whose rights in them were infringed. Thus, it is seen that God at last had sent to the red race another "great oak" among men, one who like Washington combined the knowledge, power and moral courage to do justice. Grant's Indian policy alone would mark him for posterity as a great man.

Notwithstanding the valiant efforts made by Grant and Parker to protect the tribes, it must not be thought that the Indians were at first conscious of any improvement in their lot, nor was it easy to convince them that the new system was not merely another ruse to despoil them—another "one-sided treaty."

Time was necessary for the acquisition of sufficient experience to enable the reservations to be administered efficiently. The ceaseless encroachments upon them resulting from the lax and at times dishonest administration of the land laws, bootlegging, the activities incident to the construction of the great transcontinental railways, the illegal taking of rights of way through their domains, were constant sources of irritation. Moreover, with the appearance of the host of settlers and railroad laborers in the West, the supply of buffalo meat and big game upon which the Indians had relied for their sustenance had been enormously reduced, so that often sheer want of food was the cause of friction, while the radical change, confinement upon reservations involved in their mode of life increased sickness among them to an appalling extent, and with it unrest. The result was sixteen more officially styled Indian wars, besides innumerable disturbances, occurred before violence came to an end. It

* *U. S. v. Leathers*, 6 Sawyer 17, 26 Fed. Cas. No. 15, 581; *U. S. v. Clapox*, 35 Fed. 575.

would be fruitless to narrate the story of all these wars, since the facts of two of the more noteworthy will disclose the causes of them all.

The treaty of 1863 with the Nez Percé created one of the great conflicts in the Indian wars. The government claimed that the Nez Percé had sold them the Wallowa Valley in Oregon which was the home of the Joseph band of that tribe. In fact the treaty had been fraudulently signed by people from another band of the tribe and there had been no relinquishment by the Oregon Nez Percé of the land. Under the leadership of Chief Joseph and Looking Glass, two stalwart chiefs of the tribe, the Nez Percé refused to move from their homes. In one of the great speeches of Indian history Joseph eloquently described the fraudulent negotiations which had been held:

Suppose a white man should come to me and say, "Joseph, I like your horses, and I want to buy them," he said. I say to him, "No, my horses suit me, I will not sell them." Then he goes to my neighbor, and says to him: "Joseph has some good horses. I want to buy them, but he refuses to sell." My neighbor answers, "Pay me the money, and I will sell you Joseph's horses." The white man returns to me and says, "Joseph, I have bought your horses, and you must let me have them." If we sold our lands to the government, this is the way they were bought.*

Chief Joseph was a man of extraordinary presence and appearance, with forehead, brow and chin suggestive of a Webster. In fact he was one of the most remarkable leaders produced by the red race. In vain a commission dispatched in 1875 to negotiate with his tribe attempted to induce its removal to the Lapwai reservation in Idaho. The following year while he was actually engaged in gaining consent to this the impatient whites fell upon the Nez Percés to hasten their decision. Driven to the wall, at this they declared war—a war of self-preservation, in which Joseph was to display an amazing capacity as a tactical leader in a retreat after several defeats worthy to be remembered with that of Xenophon's Ten Thousand. In spite of the fact that General Miles assailed him in front, Howard from the rear, and Sturges with federal Indian scouts from both flanks, Joseph managed to fight his way over one thousand miles to within fifty miles of the Canadian border, carrying with infinite care the women, children, aged and infirm of his tribe. Surrounded at last by reason of large reenforcements to his enemies, on October 5, 1877 he surrendered conditionally. The promises made by his military adversaries, who had only praise and intense admiration for him, to their utter disgust were ignored. The little band of four hundred and thirty-one souls was removed to Fort Leavenworth, Kansas, there to be placed under narrow confinement as prisoners of war!

* Josephy, Alvin, *The Nez Perce Indians and the Opening of the Northwest*, Yale Univ. Press, 1965, pp. 488-489.

The Nez Percé suffered greatly during their stay in Kansas. Most of Chief Joseph's family died of disease and the conditions got so bad that the little group was on the verge of breaking out in a manner comparable to the flight of the Cheyennes of Dull Knife. Finally the government relented and allowed Joseph to return to the Northwest, but not to his lands. They had long since been stolen by the white settlers. Instead his band was unceremoniously placed on the newly created Colville Reservation in northeastern Washington state. Lingering with a broken heart, his family long since perished as a result of the white man's treacherous betrayal, Joseph died there in 1904, a lonely and pathetic monument to the white man's inability to deal justly with the red man.

The other incident will serve to illustrate the situation in which the tribes found themselves. Since their migration west to Minnesota and the Dakotas, the Sioux had been a nation of warriors, yet they had that curious integrity common to few nations in history, in that they were bound together by their religion more firmly than by any other cultural factor. The White Buffalo Woman who had delivered the sacred pipe dominated the sense of peoplehood by which the Seven Council Fires lived.

In a sense the Yankton treaty of 1858 had made that tribe the Levites of the Sioux Nation among whom the shamans at the sacred Quarry were seeking to maintain the old organization of the Dakota or League of Friends just as the Iroquois had struggled to maintain the Hodenosaunnee. But a stupid government had learned nothing from the Sioux uprising of 1862 in Minnesota. On the other hand it had made matters worse with its gross violation of good faith by executing participants in the alleged massacre after promising them amity. Now in 1871, at the instance of the missionaries the local representative of the Land Office actually gave one August Cluensen a permit to settle within the sacred precinct of the shrine obviously for the purpose of watching the Sioux shamans and breaking up the immemorial mystic rites which Catlin had discovered in 1830, despite Longfellow's appeal for reason! At this the whole Sioux Nation was so enraged that the government became greatly alarmed and in 1872 hastened to survey and mark the Quarry tract as expressly provided in the treaty of 1858. A quarter section of approximately 648 acres embracing the whole deposit of catlinite, or red pipestone, and the peculiar wooded "prairie mountain" where, beside the Winnewassa Falls the sacred shrine was located, was now marked and set apart to the Yanktons and a plat thereof filed in the Land Office. But another trick had been perpetuated. The three great boulders which had been borne to this prairie region by the glaciers, known as the Three Maidens from which the present red race was supposed to have descended after the destruction of the original tribes, were left out of the tract as a concession to the missionaries

Moreover, at their instance, in 1874 the Secretary of the Interior actually confirmed the permit previously given to Cluensen, by a patent in direct violation of the treaty of 1858, and soon one Carpenter and others settled upon the tract.* All this, of course, in the minds not merely of the Sioux but of the Indians throughout the United States, was but further evidence that the government was utterly untrustworthy, that its solemn written pledges meant nothing. Apparently it was utterly reckless of the fact that in trying to destroy the most sacred institution of the race it was rousing the hostility of every native shaman in America and thereby multiplying the difficulties of the Indian problem immeasurably. Again the Sioux were bitter. Such was the situation when in 1876 a Sioux appeared in a bar room with a huge lump of gold which he admitted he had picked up in the Black Hills where all these centuries the white man had been searching for it. At last Eleusis had disclosed her mysteries! The great secret of the mystic shrine was out! So this was why the Sioux—"the worst enemies of all"—had clung so tenaciously to their country! Within an incredibly short time the whole Sioux country like California in 1849 was overrun. In vain Grant tried to control the situation. It would have required a larger army than that of the United States to eject the trespassers, to say nothing of checking this new "gold rush."

Unfortunately for the Sioux they were not like the placid, nonresisting tribes of California. Far from it. A fearless race of warriors, they were as proud and inflexible in the assertion of their rights as the Highland clans of Scotland whom they resembled strikingly. Besides, they possessed a number of superb fighting chieftains in Red Cloud, Crazy Horse, Spotted Tail, Rain-in-the-Face, American Horse, Gall and Crow King. Indian warriors of the most skillful and intrepid type, they would have done any race of soldiers credit.

Although Red Cloud, Makhpéya-luta, or Scarlet Cloud, Makpia-sha, head chief of the Oglala Teton Sioux, a very great man, often called the Lord Chesterfield of his race, was far and away the ablest of these chieftains, Sitting Bull was still their spiritual leader who though a medicine man had greatly enhanced his prestige by frequently taking to the warpath in recent years against the Cheyennes and Crows. When, in 1876 in order to clear the gold country for the whites, the government demanded that the Sioux go upon the reservation defined in the treaty of 1867, Sitting Bull was determined to save the Black Hills for his people, if possible to do this. Accordingly, he insisted that this treaty had already been violated by the whites; that it was useless to expect it to be respected any more in the future than it had been in the past. Therefore, in the spring of 1876 Sheridan planned a campaign against them which he hoped

---

* *United States v. Carpenter,* III U. S. 340.

would cause them to submit without bloodshed. Three expeditic
set under way against Sitting Bull, who was supposed to have m
6,000 Sioux somewhere near the juncture of the Rosebud and the Yellowstone. General Crook with 2,500 troops was to advance from the east;
General Terry with an equal force from the south; and General Gibbon
with a smaller command was to follow along the Yellowstone from the
west. With Terry's force was the 7th U. S. Cavalry under Colonel George
Armstrong Custer, Brevet-Major General, U.S.V., who had been ordered
to the Sioux country from Kentucky in 1873. A veteran of the Civil War,
he had also seen service against the Indians in Indian Territory where he
had decisively defeated the Cheyennes at Wichita in 1868.

Feeling his way along the Yellowstone, Crook came upon a band of
about one hundred Sioux under Crazy Horse on the Rosebud and was
defeated on June 17, but not decisively. Meanwhile Terry and Gibbon had
effected a junction in Yellowstone at the mouth of the Big Horn without
encountering opposition in force. After the encounter with Crook, Crazy
Horse led his warriors southwesterly toward the land of the Greasy Grass,
called by the whites the Little Big Horn River. It was a favorite hunting
spot of the Sioux and one which they had bitterly defended a decade
earlier against Fetterman. In 1868, Red Cloud had forced the United
States to withdraw its forces from the Big Horn country—the only occasion on which the United States was thoroughly humbled by an Indian
tribe.

Combining with Sitting Bull's Hunkpapas and the Cheyennes of Two
Moons, the Indians settled down for some hunting in the traditional summer fashion. Realizing that they were being hunted by superior forces,
they had chosen to make their stand in their ancient hunting grounds
where they had previously had such good luck against the United States
forces. The camp was placed on the western bank of the river and contained somewhat over 2,000 Indians, the majority women and children,
but still having a substantial number of warriors present. Upon discovering the presence of the Indian camp, Terry decided to march upon them at
once, sending Custer with his command ahead to prevent the camp from
dispersing and heading eastward back into the Black Hills and Badlands
country of South Dakota. He prepared to follow by boat up the Big Horn
with the remainder of his force reenforced by Gibbon's command. It was
understood that he would meet Custer at the junction of the Big and Little
Big Horn rivers on June 26.

Riding day and night and pushing his troops to the limit of their endurance, Custer reached the appointed place early the morning of June 25, a
whole day before Terry's command was due to arrive. Years before in
Oklahoma Custer had found a helpless Cheyenne village and had torn it to
bits in what was later called the "Battle of the Washita." Thus his only

experience in fighting Indians had been when he had been able to attack a virtually defenseless village and he assumed that his luck would hold with the combination of Sioux and Cheyenne on this occasion. But Custer was facing a desperate camp of Sioux whom he had treacherously betrayed years before when he led the expedition into the Paha Sapa, the Black Hills, the religious sanctuary of the tribe. In the camp there were also relatives of the Cheyennes with whom he had smoked the pipe in Oklahoma years before. And Custer had violated the pipe ceremony of the Cheyennes.

The night before the battle, the standard of the 7th Cavalry which had been planted firmly in the ground outside Custer's tent had fallen to the earth with a resounding thud, an omen of what was to come. Sitting Bull had done the Sun Dance earlier and had received the vision of many soldiers falling into his camp helpless and the Indians were encouraged by the omens while an air of pessimism spread through the soldier camp.

Custer made a dreadful blunder. Not the slightest opportunity was afforded for a parley. What Custer desired was a fight, not an understanding with the Indians who had as yet not initiated a single overt act of aggression. Learning of the arrival of large forces and of the threats of such men as Custer, they were naturally expecting trouble.

With Crazy Horse and Gall there remained not over a thousand warriors, the majority of them young boys in their early teens. Dividing his command into three groups, Custer sent Major Reno with three troops to attack the village from the south, Captain Benteen two miles southwest to attack the village, and with six troops he marched toward the north to complete the pincers movement against the Indian village. Thus, fatally divided, the 7th Cavalry assailed the Sioux almost at sight. Reno was swiftly driven back across the river to refuge among the bluffs on the east side of the Little Big Horn. The substantial body of Indian warriors attacked Reno and it can be said with some certainty that he faced the worst onslaught of fighting all through the battle.

Benteen was also driven back and soon joined Reno's command in safety on the eastern bluffs where they were pinned down by sallies and snipers.

In the meantime Custer had worked his way completely around the Indian village and charged down upon it from the northeast. He caught the people completely by surprise. At his approach only four Cheyenne warriors were present in camp and, like doomed Four Horsemen, they gallantly rode to their deaths in a desperate effort to oppose the troops massing across the river. But Custer did not get much farther than the river. When word came of a larger force attacking the village from the north a substantial number of warriors broke off from the attack on Reno and swarmed northward against the new threat.

Custer was driven back across the river and up the slope toward the northwest by Gall and the Hunkpapas. Meanwhile the great military genius of the Sioux, the charismatic Crazy Horse swept through the ranks of the Indians calling them to follow him. Gathering up a large force of men who were eager for the chance to fight along side the undefeated Sioux chieftain, Crazy Horse led his forces farther north along a ridge hidden from the view of Custer's men. Just as the soldiers were dismounting to make their "stand" Crazy Horse and his men hit them from behind and swept down the slope nearly destroying the formation that the soldiers had attempted to make. In less than half an hour Crazy Horse had turned a certain defeat into the most stunning victory of the Sioux nation. When Custer saw the situation he shot himself; he could not face being a captive of the Indians. Sitting Bull was distressed by Custer's death and went upon the field to cover Custer's face with a silk handkerchief which the General had given him. "Why, Yellow Hair, did you do this dreadful thing?" he asked as he stood over Custer's body. Nor did he fail to see what it meant to his own people.

The Cheyennes remembered that Custer had violated their pipe ceremony by lying to them of his intentions toward them prior to the Washita fight. At that time the medicine men of the tribe predicted that they would meet Custer on another occasion and that the pipe would provide their victory. So winning against their hated foe was sweeter for them than it was for the Sioux who had only wished to keep their hunting lands.

The next day Terry arrived and relieved Reno and Benteen who themselves had been hard pressed. Crazy Horse withdrew into the western parts of Dakota. There is not much doubt that both Reno and Benteen could have been destroyed had Sitting Bull permitted it. But he had had enough conflict for the time being and wisely understood that there would be no quarter given the people who had destroyed Yellow Hair. But he was content, by his medicine he had defended his people.

25

The Reform Movement

WITH THE REFORMS INSTITUTED by Grant, Congress showed little sympathy. The corporate interests affected by them were particularly powerful, controlling to a large extent not only the legislatures of the western states but Congress. The federal government was, therefore, not disposed to go out of its way to protect the Indians in the face of the political pressure that was constantly brought to bear upon it. Moreover, it was often itself the aggressor. After 1863 the Indians could not sue the government upon a treaty without a special act conferring jurisdiction on the Court of Claims. Therefore, they were rapidly robbed of much of their property.

Nevertheless, during the Administration of Hayes the good work commenced by Grant was continued. Acting under the influence of Carl Schurz, his Secretary of the Interior and a powerful advocate of Indian rights, Hayes proposed the legislation providing for the bringing of fifty Indian boys and an equal number of girls annually to the Hampton Normal School, and also the organization of auxiliary Indian cavalry units for the Army which would not only provide a peculiarly effective frontier police but serve as a medium of education for the Indians as well. And it was in 1879 that the Bureau of American Ethnology was established by Congress and placed under the Smithsonian Institution, being directed to collect all the archives, records and materials relating to the Indian tribes which had been gathered by the Department of the Interior.

Fortunately, a very great man—Major J. W. Powell—of vast experience in Indian research work, was assigned to the organization and conduct of the Bureau and whose monumental services as Director continued until his death in 1902. A devoted friend of the red race, and taking a broad view of the opportunities presented him, he set out to accomplish many practical ends such as a study of the relations, location and numbers of the tribes, and their classification into groups and families; a study of the numerous sociological, religious and industrial problems involved in adjusting the race to civilization; a history of the relations of the red and white races; an investigation of the psychology, medical practices, and sanitary conditions among the tribes, and their hygienic needs; the preparation of bibliographies of valuable literature relating to the Indians; a study of their economic resources; the preservation of their antiquities and a record of their traditions; and a handbook embodying in condensed form essential information concerning them. Nor was the strictly scientific aspect to be ignored. Thus, all departments of anthropological research— physical, psychological, linguistic, sociological, religious, technical, and esthetic—were to be embodied in numerous bulletins and reports for the guidance of the government and the education of the world generally. It was a marvelous work he undertook, of the first importance to the government, and it was to be done well with the aid of a corps of devoted scientists whose extraordinary achievements only went to show what Congress could have accomplished for the Indian had their welfare been at heart.

Upon retiring from the cabinet, Schurz, as editor of the *New York Evening Post*, continued to labor in the cause of reform to which he devoted the full influence of a pen dipped in the well of a broad experience with Indian affairs. The records of the land office showed that down to 1880 nearly a billion acres of public land to which the Indians' title had originally attached, had been disposed of with total receipts to the government of $233,000,000 and to the Indians of $9,000,000. In the hands of the government remained another billion acres of public land valued at $1,000,000,000, while the partly indifferent land reserved to the use of the tribes out of their original holdings aggregated but a little over 100,-000,000 acres. Pointing to these figures, which spoke eloquently for themselves, he ridiculed the hackneyed fable of the government's generosity in providing reservations for the Indians.

Surely, he insisted, the preemptors, homesteaders, gold-diggers, and the railways at last must be sated with Indian lands. So too, the country must have had enough of war. The Indians were not buffalo, nor the savage, unreasonable people the frontiersmen in self-interest had represented them to be. Long declared by the courts to be the wards of the nation, they were in fact children for whom it was responsible to the Creator of mankind. The Indian problem was no longer a military one to be solved by the rough-and-ready methods of a frontier that was gone. They were not on the borderline of civilization—on the edge of it—but were within the nation, a part of it, the flesh and bone of an empire, merging their blood rapidly with that of their conquerors. Their problem was a sociological one to be solved along sound economic lines, not to be left to prejudice and greed as in the past, but demanding the utmost patience, foresight and reason.

If they were ever to be emancipated from the aboriginal order, the reservation system at best was to be viewed as a temporary expedient only. That the tribal Indians should be isolated, forever incarcerated within the bounds of an aboriginal social and economic order, was unthinkable to anyone with the slightest conception of morality. As a permanent system it was fundamentally faulty since it encouraged a spirit of dependence on the part of the Indians. This, the Indian Homestead Act of 1875 had not tended to counteract. The trust period upon Indian homesteads was too short. It created but a new estate for the whites. Instead of being charged with ineptitude for a civilized life, with being lazy and shiftless, they must be helped to transform their nature to fit the white man's society, which was not possible with the limited education being furnished them on the reservations. At the same time they must be encouraged to abandon the communal system by being protected while adapting themselves to a new order of life.

Meantime Sitting Bull had gotten into touch with Washington and with the aid of the Canadian government negotiated an amnesty agreement. The refugees were to return and go upon the Sioux Reservation. To satisfy public opinion he himself was to be confined for two years at Fort Randall. But he did not agree to this until the government instituted legal proceedings to annul the patent granted in the Pipestone Quarry Tract.* Accordingly, in 1881 Sitting Bull was confined.

To the irrefutable arguments of Schurz the President now responded by placing before Congress in 1882 a comprehensive plan for the creation and safeguarding of individual Indian trust estates, and on December 15, 1882 there was formed at the instance of John Welsh of Philadelphia the Indian Rights Association of which a representative was to be kept in Washington to scrutinize legislation and exercise a general surveillance over the administration of Indian Affairs. It was this same year that the Carlisle Indian School was established in an old military barracks with a few Indian prisoners at the instance of Lieutenant, afterwards General, R. H. Pratt, their custodian in Florida.

While many forces were now cooperating to bring about better understanding of the Indians, none contributed more, perhaps, than the activities of a great showman—Colonel William F. Cody.

* *U. S. v. Carpenter*, 111 U. S. 350

Born in Scott County, Iowa, in 1845, he enlisted at fifteen years of age, in the perilous service of a pony express rider through the Indian country before the construction of the railroads was undertaken, serving during the Civil War as a government scout and guide in the Indian warfare of the West. Enlisting in the 7th Kansas Cavalry in 1863, at the close of the war he contracted with the Kansas Pacific Railroad to furnish buffalo meat to its laborers from which fact he derived the world-famous sobriquet of "Buffalo Bill." Serving from 1868 to 1872 with the Army as a colonel of scouts in the Sioux country, the latter year he was elected to the Nebraska Legislature. While with the 5th Cavalry in the Little Big Horn Country in 1876, with his own hands he slew Chief Yellow Hand, and upon the suppression of the Sioux uprising there was, perhaps, not a man in the United States possessing a wider or more varied experience of the plains and the Indians.

A man of extraordinary appearance and address, and a born politician with an intuitive understanding of the red man, he conceived the plan of introducing to the world beyond the frontier the veteran warriors of the plains with mutual benefit to them and himself. Accordingly, in association with the celebrated Colonel Thomas Ochiltree, of Texas, and others, in 1883 he organized the Wild West Show, a spectacular representation of actual life on the plains, which at once began to tour the country. The several hundred Sioux, Blackfeet and Crow warriors whom he assembled, including such famous chieftains and veterans of Custer's Massacre, as Red Cloud, Spotted Tail and Red Shirt, took the East by storm and for a time Barnum sank into insignificance. Everywhere Cody was entertained, mingling in the most distinguished society and losing no opportunity to advertise the red man. The civilized East gazed agape upon the Indians, the cowboys, the bucking bronchos, the buffalo, whose real acquaintance they now made for the first time. As an exhibitor of the genuine character of the "Wild West" and its people, and the author of numerous fascinating books of frontier adventure, including his own autobiography thrilling with romance, Cody carried on during the next ten years a propaganda of incalculable value to the Indians. Among them his power and influence was almost unlimited. The experiences of the Indians themselves had a pacifying effect. They carried back to the West amazing stories of the friendship of the white man beyond the frontier.*

The year Cody and the Wild West Show first visited the East—1883— there was established by the philanthropist, Alfred K. Smiley, the Indian Conference at Lake Mohonk where all manner of intellectual gatherings were wont to take place. Of its Council, the Board of Indian Commission-

* In 1888 he was invited to visit Europe where he and his people also secured an ovation. There he was honored by kings and emperors.

ers which included Smiley, was made a part. Every year thereafter it was to meet during four days to discuss needed reforms, to formulate legislation, and advise the government. In a large measure it was to be responsible in the future for the President's Indian policy. This same year Sitting Bull was released from confinement and rejoined his people.

On March 3, 1885, a statute was enacted by Congress making Indians criminally liable before the federal courts for certain offenses committed against each other as well as against other persons, whether committed within or without Indian territory. While the constitutionality of this legislation was attacked, the Supreme Court was unanimous in upholding it.

Cleveland, like Washington and Grant, was under no delusion as to the cause of most of the trouble with the Indians. Not hesitating to order intruders out of their country, and to employ the military to enforce his proclamations, "it is useless," he declared, "to speak of settling the Indian problem by their destruction and neglect." Notwithstanding the unpopularity and adverse political effect of his methods, on March 13, 1885, he boldly addressed to Congress the following message which left no more doubt as to the duty of the nation than as to the true condition of its Indian wards:

The report of the Secretary of the Interior, containing an account of the operations of this important Department and much interesting information, will be submitted for your consideration.

The most intricate and difficult subject in charge of this Department is the treatment and management of the Indians. I am satisfied that some progress may be noted in their condition as a result of a prudent administration of the present laws and regulations for their control.

But it is submitted that there is lack of a fixed purpose or policy on this subject, which should be supplied. It is useless to dilate upon the wrongs of the Indians, and as useless to indulge in the heartless belief that because their wrongs are avenged in their own atrocious manner, therefore they should be exterminated.

They are within the care of our Government, and their rights are, or should be, protected from invasion by the most solemn obligations. They are properly enough called the wards of the Government; and it should be borne in mind that this guardianship involves on our part efforts for the improvement of their condition and the enforcement of their rights. There seems to be general concurrence in the proposition that the ultimate object of their treatment should be their civilization and citizenship. Fitted by these to keep pace in the march of progress with the advanced civilization about them, they will readily assimilate with the mass of our population, assuming the responsibilities and receiving the protection incident to this condition.

Our Indian population, exclusive of those in Alaska, is reported as numbering 260,000, nearly all being located on lands set apart for their use and occupation, aggregating over 134,000,000 acres. These lands are included in the boundaries of one hundred seventy-one reservations of different dimensions, scattered in twenty-one States and Territories, presenting great variations in climate and in the kind and quality of their soils. Among the Indians upon these several reservations there exist the most marked differences in natural traits and disposition and in their progress toward civilization. While some are lazy, vicious, and stupid, others are industrious, peaceful, and intelligent; while a portion of them are self-supporting and independent, and have so far advanced in civilization that they make their own laws, administered through officers of their own choice, and educate their children in schools of their own establishment and maintenance, others still retain, in squalor and dependence, almost the savagery of their natural state.

In dealing with this question the desires manifested by the Indians should not be ignored. Here again we find a great diversity. With some the tribal relation is cherished with the utmost tenacity, while its hold upon others is considerably relaxed; the love of home is strong with all, and yet there are those whose attachment to a particular locality is by no means unyielding; the ownership of their lands in severalty is much desired by some, while by others, and sometimes among the most civilized, such a distribution would be bitterly opposed.

The variation of their wants, growing out of and connected with the character of their several locations, should be regarded. Some are upon reservations most fit for grazing, but without flocks or herds; and some on arable land, have no agricultural implements. While some of the reservations are double the size necessary to maintain the number of Indians now upon them, in a few cases, perhaps, they should be enlarged.

Add to all this the difference in the administration of the agencies. While the same duties are devolved upon all, the disposition of the agents and the manner of their contact with the Indians have much to do with their condition and welfare. The agent who perfunctorily performs his duty and slothfully neglects all opportunity to advance their moral and physical improvement and fails to inspire them with a desire for better things will accomplish nothing in the direction of their civilization, while he who feels the burden of an important trust and has an interest in his work will, by consistent example, firm yet considerate treatment, and well-directed aid and encouragement, constantly lead those under his charge toward the light of their enfranchisement.

The history of all the progress which has been made in the civilization of the Indian, I think, will disclose the fact that the beginning has been religious teaching, followed by or accompanying secular education. While the self-sacrificing and pious men and women who have aided in this good work by their independent endeavor have for their reward the beneficent results of their labor and the consciousness of Christian duty well performed, their valuable services should be fully acknowledged by all who under the law are charged with the control and management of our Indian wards.

What has been said indicates that in the present condition of the Indians no attempt should be made to apply a fixed and unyielding plan of action to their varied and varying needs and circumstances.

The Indian Bureau, burdened as it is with their general oversight and with the details of the establishment, can hardly possess itself of the minute phases of the particular cases needing treatment; and thus the propriety of creating an instrumentality auxiliary to those already established for the care of the Indians suggests itself.

I recommend the passage of a law authorizing the appointment of six commissioners, three of whom shall be detailed from the Army, to be charged with the duty of a careful inspection from time to time of all the Indians upon our reservations or subject to the care and control of the Government, with a view of discovering their exact condition and needs and determining what steps shall be taken on behalf of the Government to improve their situation in the direction of their self-support and complete civilization; that they ascertain from such inspection what, if any, of the reservations may be reduced in area, and in such cases what part not needed for Indian occupation may be purchased by the Government from the Indians and disposed of for their benefit; what, if any, Indians may, with their consent, be removed to other reservations, with a view of their concentration and the sale on their behalf of their abandoned reservations; what Indian lands now held in common should be allotted in severalty; in what manner and to what extent the Indians upon the reservations can be placed under the protection of our laws and subjected to their penalties, and which, if any, Indians should be invested with the right of citizenship. The powers and functions of the commissioners in regard to these subjects should be clearly defined, though they should, in conjunction with the Secretary of the Interior, be given all the authority to deal definitely with the questions presented deemed safe and consistent.

They should be also charged with the duty of ascertaining the Indians who might properly be furnished with implements of agriculture, and of what kind; in what cases the support of the Government should be withdrawn; where the present plan of distributing Indian supplies should be changed; where schools may be established and where discontinued; the conduct, methods, and fitness of agents in charge of reservations; the extent to which such reservations are occupied or intruded upon by unauthorized persons, and generally all matters related to the welfare and improvement of the Indian.

They should advise with the Secretary of the Interior concerning these matters of detail in management, and he should be given power to deal with them fully, if he is not now invested with such power.

This plan contemplates the selection of persons for commissioners who are interested in the Indian question and who have practical ideas upon the subject of their treatment.

In order to carry out the policy of allotment of Indian lands in severalty, when deemed expedient, it will be necessary to have surveys completed of the reservations, and I hope that provision will be made for the prosecution of this work.

The threatening and disorderly conduct of the Cheyennes in the Indian Territory early last summer caused considerable alarm and uneasiness. Investigation proved that their threatening attitude was due in a great measure to the occupation of the land of their reservation by immense herds of cattle, which their owners claimed were rightfully there under certain leases made by the

Indians. Such occupation appearing upon examination to be unlawful notwithstanding these leases, the intruders were ordered to remove with their cattle from the lands of the Indians by Executive proclamation. The enforcement of this proclamation had the effect of restoring peace and order among the Indians, and they are now quiet and well behaved.

Still one of the greatest obstacles in the way of the preparation of the Indians to take part in the economic life of the country on a basis of equality with the white man was their natural disinclination to abandon tribal relations and the communal system of the aboriginal order.

Unquestionably the Indians had a constitutional as well as a moral right to elect between the old order and the new, a fact which, building on the foundation laid by Marshall, the courts had gradually recognized. Thus the political, dependent communities of the United States known as Indian tribes were judicially defined to be communities of persons of the same or similar blood united under one leadership or government, and inhabiting a particular though sometimes ill-defined territory.* These communities were likened to municipal corporations,† possessing the right to regulate to some extent the domestic intercourse of their members.‡ The descent of the communal land was held to cast according to tribal laws of which the courts must take judicial notice.§ The title in the land was held to be vested in the community, not in its members severally, or as tenants in common,|| while the right of each tribesman to participate in the enjoyment of the tribal property was said to have been derived from tribal membership.¶ This right terminated with the loss of membership in the tribe.** It was neither alienable nor descendable and was, therefore, not an estate of inheritance or a right that might be transmitted to an heir.†† The children of a tribesman acquired the common rights not by inheritance, but by virtue of the tribal laws making them members of the tribe.‡‡

Such being the law, it was impossible of course, to compel a tribe possessing an estate in its land, such as that of the Five Civilized Tribes, to change its system of land tenure for the benefit of its members. But as to the tribes possessing a mere right of occupancy the situation was different. In 1886 the Supreme Court reaffirmed the guardianship of Congress over

* *Montoya v. United States* (1900), 180 U. S. 266.
† *U. S. v. Kagama* (1886), 118 U.S. 375. *Jones v. Meehan,* 175 U. S. 1.
‡ *Gray v. Goffman,* 3 Dill (U. S.), 393.
§ *Old Settlers v. U. S.* (1893), 148 U. S. 427. *U. S. v. Cherokee Nation* (1906), 203 U. S. 76.
|| *Stephens v. Cherokee Nation,* 174 U. S. 445; *Duke v. Goodall,* 5 Ind. T. 145, 82 S. W. 702; *Conroy v. Ballinger,* 216 U. S. 84.
¶ *Sully v. United States,* 195 Fed. 113; *Oakes v. United States,* 172 Fed. 305.
** *Sizemore v. Brady,* and *Gritts v. Fisher,* supra.
†† *Mullens v. Pickens,* 250 U. S. 590; *Haynes v. Barringer,* 168 Fed. 221; *Simpkins v. Wer,* 45 Okla. 327; *Sizemore v. Brady, Gritts v. Fisher,* supra.
‡‡ *Gritts v. Fisher,* supra.

them, holding that its authority might be delegated to the Executive, thus making it possible for the government to assume a more or less arbitrary control over their property.*

As a result of the groundwork done by Washington, Monroe, Grant, Schurz and Arthur, and the present efforts of Cleveland, finally, at the instance of the Lake Mohonk Conference there occurred in 1887 an epochal piece of legislation known as the General Allotment Act. This provided

that in all cases where any tribe or band of Indians have been, or shall hereafter be, located upon any reservation, created for their use, either by treaty stipulation or by an act of Congress or Executive order setting apart the same for their use, the President of the United States shall be, and he hereby is, authorized, whenever in his opinion any reservation or any part thereof of such Indians is advantageous for agricultural and grazing purposes, to cause said reservation, or any part thereof, to be surveyed, or resurveyed if necessary, and to allot the lands in said reservation in severalty to any Indian located thereon in quantities as follows:

To each head of a family, one quarter of a section;

To each single person over eighteen years of age, one-eighth of a section;

To each orphan under eighteen years of age, one-eighth of a section; and

To each other single person under eighteen years now living, or who may be born prior to the date of the order of the President directing an allotment of the lands embraced in any reservation, one-sixteenth of a section. . . .

Having learned the necessity of safeguarding Indian allottees against fraud and their own improvidence, Congress was again careful to avoid investing them with an absolute title in the first instance. What was inaptly called an allotment patent was, therefore, no more than an allotment certificate under which for a period of twenty-five years, subject to extension by the President in his discretion, the United States was to hold the land allotted in severalty in trust for the sole use and benefit of the allottee, or in case of his death, of his heirs. As in the case of Indian homesteads a patent in fee was to issue at the expiration of the trust. Meantime any attempt by the allottee to incumber the land was to be of no force and effect.† Detailed administrative provisions were made for carrying out the provisions of the Act without fraud or injustice.

At last it was possible for Indians not belonging to one of the Five Civilized Tribes to acquire an estate of inheritance in the tribal domain, and to encourage them to do so it was provided in the Act of 1887 that upon taking an allotment in severalty the allottee was to become a citizen of the United States.

Soon special allotment agreements were negotiated with each of the

---

* *United States v. Kagama* (1886), 118 U. S. 375.

† *Monson v. Simonson* (1913), 231 U. S. 341.

Five Civilized Tribes under which tribal relations might be retained by allottees who were not to become citizens of the states in which lay the land patented to them unless that were expressly provided in their tribal agreement.

The result of these provisions was that a new form of citizenship came into being for which there was no precedent in the law of the United States—a national citizenship unaccompanied by state citizenship, so that there were now three classes of Indians—ordinary citizens, or citizens of both a state and the United States, citizens of the United States only, and noncitizens.

26

Geronimo and Sitting Bull: the Last Indians

DURING CLEVELAND'S FIRST ADMINISTRATION new Indian disturbances occurred which, like Cody's Wild West Show, served the purpose of arousing fresh interest in the Indian problem. In 1878 Mexico had insisted that the Treaty of Guadalupe Hidalgo was being violated by the United States through depredations by the Chiricahua Apaches. These people had long since been ousted of their lands in the territory of New Mexico, the western portion of which was now Arizona. Upon an attempt to remove them to the San Carlos reservation on the southern frontier of Arizona, under the leadership of Geronimo they fled into Mexico.

Born about 1834 near Fort Talerosa in New Mexico, Geronimo was not a man of any particular strength of character like Chief Joseph, nor did he even possess the really patriotic motives of the admirable Black Hawk. Yet the Apache warrior was perhaps the finest fighting man ever produced by the American Indian race. His uncanny generalship and devilish brand of psychological warfare made his capture a literal impossibility without the use of modern communications devices and the pursuit of thousands of soldiers. Geronimo was an astute man. Originally a minor chieftain he rapidly rose to prominence as the Apaches were systematically being mistreated by their agents, and discontent with existing conditions made it imperative that some form of resistance to slow starvation and destruction be made by the miserable Apache captives on the San Carlos reservation in Arizona.

Geronimo and his band were originally arrested when they came back from Mexico on a raiding expedition in the early 1880s. They were confined on the San Carlos Reservation in Arizona for a number of years and made honest attempts to become farmers. But the Apaches were by nature hunters and desert wanderers. Confinement to the reservation was destroying their spirit. Their agents systematically cheated them and the government refused to provide them with any water for their farming operations. The water instead went to white settlers who casually took the Indian water rights while maintaining that the Indians could never make a go of civilized existence.

Finally in 1882 Geronimo, at the head of one of the Apache bands, made a raid into Sonora, probably for no other purpose than calling attention to their needs, for they surrendered quickly to General Crook when overtaken in the Sierra Madre. After that he was assigned one of the best farms in the reservation where he settled down to making "tiswin," a native intoxicant. This enterprise was soon stopped. Gathering together a large band of malcontents, during the years 1884-85 he terrorized Arizona, New Mexico, and Sonora and Chihuahua in Mexico. Again Crook was sent against them, very properly, with orders to capture or destroy Geronimo and his followers.

In March, 1886, after a warfare in which the Apaches displayed all the wiles of the most devilish cunning, a truce was made, followed by a conference at which terms of surrender were agreed upon. Geronimo and his followers, however, were not impressed by the terms offered and eluding Crook, who thought he had them cornered, again fled to the Sierra Madre across the Mexican border where General Miles finally overtook them the following August. The entire band numbering about 340, including Geronimo and Nachi, the hereditary chief, were deported as prisoners of war to Florida, later to Alabama, finally being settled at Fort Sill, Oklahoma, where the tribe still resides under military supervision.

In order to finally capture Geronimo General Miles had to use thousands of troops stationed on every high point in Arizona territory. These troops, using the new heliographs, watched the desert below for signs of dust raised by the fugitive Apaches. By signaling from mountain to mountain Miles was able to bring his troops to bear on the Apache band with swift pursuit instead of attempting to follow the band with one troop of men as had been the usual method of fighting against Indians.

Even with the use of these startling devices Miles was forced to recruit thousands of friendly Apaches to follow Geronimo and guide the white troops under his command. With their women and children and fatigued horses the Apaches were still able to make better than fifty miles a day in the Arizona and Mexican deserts. Thus they frequently outran the United States troops sent against them simply because of their incredible capacity to endure hardship.

A most tragic incident closed Geronimo's career. During his days as marauder he punished the friendly Apaches who refused to join him on the warpath. So many of them gathered at Fort Apache in the White Mountains to seek Army protection against his vengeance. After his capture he and his band were returned to Fort Apache shortly before they were sent to Florida. When the orders came to send his band to Florida, every Apache at the Fort was sent with him. Thus the friendly Apaches who had sought the protection of the United States and even the Apache scouts who had served Miles in capturing Geronimo were sent off in exile to share his fate. Of the 750 Apaches sent to prison in Florida only forty were of Geronimo's band. The remainder were friendly Apaches and scouts. The scouts protested that they were employed by the United States and should not become prisoners of war. But to no avail.

When the band was moved from Florida only Geronimo and a couple of others remained from the original fugitive band. The rest were the scouts and friendly Apaches who had trusted the government. The band was placed at Fort Sill, Oklahoma where it resides today. For decades the band was considered dangerous and kept in prisoner-of-war status so that a Congressional investigating committee in the 1920s were startled to learn that some grandchildren of the original scouts were classified as prisoners of war even though they had been born two generations after the Geronimo escapade. In such ways has the United States rewarded those Indians who served it.

The situation of the Indians, particularly in the Indian Territory, was becoming more and more confused as time went on by reason of the unwillingness of Congress to make a study of the Indian problem and deal with it intelligently in its entirety.

Among the Sioux the situation was even worse. The fact that Sitting Bull had predicted exactly what happened on the Little Big Horn and his subsequent imprisonment, after voluntarily surrendering himself to the government, had tended to enhance his influence. The fact that he had provided the medicine for Custer's defeat and his determination to continue the old Sioux religious rites stood him in good stead among the restless reservation people. His years with Buffalo Bill's Show only served to enhance his national reputation so that his influence continued to grow and his image as a potential troublemaker (as he was viewed by suspicious Army and government people) was increased.

In 1884 the Supreme Court held, in *United States v. Carpenter*, that the traditional shrine of the Indian race, the Red Pipe Stone Quarry at Pipestone, Minnesota, was exclusively owned by the Yankton Sioux, and that they had the exclusive title to the quarter section embracing the Quarry as surveyed and marked in 1872. This was a tremendous victory for the traditionalists of which Sitting Bull was the foremost proponent. He compelled the removal of the squatters who had moved into the Quarry in

1887. In 1888 he flatly refused to allow the Sioux to cede any more of their lands so that the whole Sioux nation now looked to him as its leader in a time when the utmost pressure was being placed on them to cede their lands in the Dakotas.

Everywhere Sitting Bull's reputation spread so that many of the scattered tribes of the West had heard of him and looked to him for leadership in the stubborn resistance to efforts to civilize them. Everywhere people said that Sitting Bull made "good medicine." Such was the situation while Cody was in the East with the Wild West Show arousing the country to some sense of Indian nature and while the government was doing everything it could to discredit Sitting Bull and destroy his influence over the Indians.

But the government itself was still incorrigibly stupid so that once more it changed its policy. Accordingly, Congress resorted to the device of the Act of March 2, 1889 providing for the condemnation of a right of way for a railroad straight through the Quarry tract, and its survey and appraisal. By this means the politicians proposed to turn the light of civilization upon the "mystic shrine" of Winnewassa while the government was ostensibly exercising its right of eminent domain. With the prospect of compensation dangled before them, it was hoped that the Yanktons who were poor, in need of money, and now widely separated from the Sioux Nation as a whole, would betray the latter and sell the Quarry tract.

The consequences were inevitable. When the surveyors and contractors went upon the tract and routed out the shamans, at Sitting Bull's direction Kicking Bear instituted the first Messiah or so-called "Ghost Dance" on the Standing Rock Reservation. This was but a form of spiritual emotionalism induced directly by the destruction of the ancient Sioux shrine.

Actually the Ghost Dance religion stemmed directly from the deserts of Nevada. There a frightened group of Paiutes had been gathered at the Walker River reservation during the frequent roundups of Indians in Nevada. In this nebulous situation it was, perhaps, inevitable that the Indians would look to religious salvation since all attempts to defeat the white man by military means had failed. In addition negotiations with the United States had always resulted in further deprivation of Indian rights since the treaties never said what the Indians were told they said.

One day a man named Jack Wilson (Wovoka) began preaching the destruction of the white man and the return of the buffalo. By doing a special dance until the point of exhaustion he showed people that they could bridge the gap between this world and the next and contact their dead relatives. By some miraculous means the dead warriors of the past would return in the spring of 1890, the millions of buffalo that had once roamed the west like the nomad Indians would once again cover the plains and the white man would be driven from the continent. To ensure the success of the Indians special ghost shirts were worn which would protect the warriors against the bullets of the white man. This gospel of salvation spread throughout the West, arriving quickly at the Oklahoma reservations on which thousands of discontented Indians lived.

Hearing of the new religion, Kicking Bear and some other Sioux went to Nevada to visit the new prophet and returned with solemn affirmations of the validity of the new religion. Indian country was in the throes of religious revolution. Many thought that it was better to perish than go on suffering at the hands of their heartless oppressors. In other words, the Ghost Dance was but a recurrence of the spiritual unrest which was promoted by the Indian prophets under Pontiac and Tecumseh in 1762 and 1805, respectively.

The demand for Sitting Bull's arrest was immediate. Now, Red Cloud, long the friend of peace, leaving Cody in the East, hastened home to protect his people from what he saw was in store for them. But the situation had gone too far for either Red Cloud or Sitting Bull to control. The local settlers, the politicians, Congress, and the Superintendent of the Reservation had all lost their heads, particularly the last. His attempted arrest of Sitting Bull at a moment of the utmost religious fervor not unnaturally led to an effort to prevent this, resulting in the unnecessary killing of Sitting Bull himself by Sergeants Red Tomahawk and Bullhead of the Indian Police on December 15, 1890.* Thus, the great medicine man of the red race became even more of a martyr in the eyes of his people.

In retrospect the uselessness as well as the stupidity of such things seems plain. Yet the killing of Sitting Bull was but the prelude to another tragedy. Some of the followers of the Ghost Dance religion were chased from the Standing Rock Reservation southward to Cheyenne River Agency. There they were given refuge in a remote valley on the extreme western edge of the reservation by Chief Big Foot. Soon the Army demanded that the whole band surrender themselves at the agency for confinement. Big Foot gathered his frightened people and headed south toward the Pine Ridge Reservation where he thought that he could receive the assistance and protection of his old friend Red Cloud who had now become the leading spokesman for the disorganized and disheartened Sioux Nation.

It was the dead of winter and the little band had few possessions of worth. Hunting was now a past way of life. The tepees they used were thin cloth and canvas tents that did little to keep out the cold. On the whole it was a very ragged, starving, frightened band that Big Foot led south.

The agent at Pine Ridge became frightened at the thought of more Sioux arriving at his agency. It had been little over a decade since Crazy

* Many Indians believed this had been prearranged by the politicians.

Horse had been assassinated at Fort Robinson, Nebraska, and tempers continued to run high between the followers of the great Sioux war chief and the followers of Red Cloud and American Horse, chiefs who had not been at the Little Big Horn. Learning that Big Foot had brought his band to the northern edges of the reservation in the Badlands country, and could be intercepted at Wounded Knee creek, General Miles rushed Colonel Forsyth there with two squadrons of the 7th Cavalry—Custer's old regiment and the worst that could possibly have been selected—with eight Hotchkiss guns, to round them up. Surrounding the camp, after training the deadly machine guns upon it, Forsyth called upon the band to assemble and surrender their arms. Over a hundred of them came out of the tepees peacefully and seated themselves in a group upon the snow while the village was being searched. Then firing commenced and instantly the machine guns began to spit. That night 146 Indian corpses lay upon the snow-covered plain.* The incident was, of course, reported as unavoidable but it is an established fact that Forsyth, having finally gained control of his men, stopped the slaughter with the plea: "We did not come here to butcher them." Although the troops also suffered casualties and claimed they had been fired upon first, the Sioux insist to this day that Big Foot's band was deliberately massacred in retaliation for the killing of Custer and his men, and that the troops believed they would be saved a hard winter's campaign in the Badlands if the Sioux generally were terrorized into returning to the reservation. The reader may draw his own conclusions. Soon, a large force of regular troops was assembled by General Miles and after an impressive review proceeded to round up the survivors without trouble.

In 1891 Congress passed an act authorizing an Indian school to be erected on the Quarry tract but was advised that it need expect no peace with the Sioux despite the killing of Sitting Bull, until it had restored the property to its owners. Upon the advice of the Yankton Superintendent and Major McLaughlin, the latter one of the ablest men in the Indian service, in 1891 the money obtained by the government for the right of way through the Quarry tract was distributed to the tribe. So also, by Act of 1892, Congress directed that a treaty be negotiated by McLaughlin designed to end the dispute over the possession of the Indian shrine. Long and adroitly McLaughlin labored. In the agreement finally obtained by him it was provided that

if the Government of the United States questions the ownership of the Pipestone Reservation by the Yankton Tribe of Sioux Indians under the Treaty of April 19, 1858, including the fee of the land, as well as the right to work the quarries, the Secretary of the Interior shall as speedily as possible refer the

* This appears from unpublished photographs in possession of the Smithsonian Institution and of the author. Mooney had no doubt that the band was deliberately massacred.

matter to the Supreme Court of the United States to be decided by that tribunal. And the United States shall furnish, without cost to the Yankton Indians, at least one competent attorney to represent the interest of the tribe before the court. If the Secretary of the Interior shall not, within one year after the ratification of this agreement by Congress, refer the question of ownership of the said Pipestone Reservation to the Supreme Court as provided for above, such failure on his part shall be construed as and shall be a waiver by the United States of all its rights to the ownership of the said Pipestone Reservation, and the same shall thereafter be solely the property of the Yankton tribe of Sioux Indians, including the fee to the land.

This agreement was duly ratified by an Act of Congress in 1894. Nevertheless, it had merely been utilized to quiet down the Sioux Nation as a whole. Nor did the government ever have any idea of carrying out the agreement. After the Sioux had been so thoroughly disorganized that they were powerless to resort to hostilities, Congress passed another Act (June 7, 1897) at the instance of the missionaries, directing the Secretary of the Interior to negotiate an agreement with the Yanktons looking to the purchase of the Quarry.*

Disorganized, discouraged, broken and bleeding, robbed of its sacred shrine and its gold alike, its warriors no more than prisoners, now at last the proud Dakotas bowed to the inevitable. The great Red Cloud like Little Turtle and Rift-in-the-Clouds before him, saw the uselessness of further resistance. It could only mean more Wounded Knees. The thumb of Fate was turned downward upon the native warrior in the arena of modern warfare. He could not face machine guns. Had not the great oracle in the days of Dekanawida and Hiawatha counselled peace?

Then upon the ground the warriors
Threw their cloaks and shirts of deer-skin,
Threw their weapons and their war gear,
Buried all their warlike weapons.

Then a voice was heard, a whisper:
Broken are the spells that bound you,
All the charms of old magicians,
All the magic powers of evil.

Yes, at last the missionaries were happy. Now they undertook to silence the voice of Winnewassa by causing the lonely cascade to be destroyed with dynamite and the hallowed grove of the shamans to be laid low while the ancient graves of the red men buried at the shrine were desecrated by the curious. Today, in the thrifty little town of Pipestone where the Indian traditions are cherished, the pipestone finds a ready sale among the tour-

---

* An agreement was negotiated by McLaughlin Oct. 2, 1899, in which the United States was obligated to pay $100,000 for the Quarry but Congress refused to ratify it.

ists who little suspect the great human tragedy that centered in this spot.*

Although Indian disturbances occurred in 1895 among the Bannock tribes, in 1898 among the Chippewas at Leech Lake, in 1913 among the Navajo, and in 1915 among the Paiutes, the passing of Sitting Bull virtually marked the end of organized Indian resistance. Surely, the nation may not point with pride to the unenviable record of having waged fifty-two separate official wars and an almost ceaseless strife against its aboriginal wards during the first hundred years of its existence.†

With the continued development of the natural wealth in the West, and particularly oil, the Buffalo Party girded itself for a new assault upon the red man and took to the field of politics again in all its pristine strength. The far West had been very largely appropriated by homesteaders and with the exception of the Indian reservations brought under the control of the rapidly forming states. But in the aggregate there was still a vast amount of public land scattered here and there, particularly in Oklahoma. After the hard times of Cleveland's administration and under the pressure of a tremendous European emigration to America, for this land a great demand arose. Pressing upon the Indian Territory from all sides—now the frontier of civilization if one may be said to have remained—the settlers looked longingly upon the tribal domains which during the past half-century, with the aid of their slave labor and then the freedman, the Indians as cattle rangers had brought to a relatively high state of development.

Under the pressure of the government and the popular demand for these lands, their allotment in severalty followed rapidly the legislation designed to accomplish it. This naturally disrupted the old control of the tribal governments not only over the communal property but the persons of the allottees notwithstanding their efforts to maintain their tribal organization. The two simply did not go together. Coincidentally it gave to the Indian agents a new power, clothing them with an importance they had not hitherto possessed, since the situation was ripe with opportunities for favoritism and graft. Being human beings, it was not to be expected they could resist the ceaseless assaults upon their integrity. The effect of all this upon the Indians was disastrous for they too were human.

Under the treaties of 1866 the several tribes had obligated themselves to admit the freedmen who remained to the enjoyment of the communal land rights. Oil was now discovered in the Territory and with the enhanced value of the tribal domains the freedmen began to clamor loudly

* In 1910 Congress finally passed a Jurisdictional Act conferring on the Court of Claims jurisdiction to hear the claim of the Yankton Sioux. In a strangely confused opinion the Court simply declined to render any judgment.
† For official list of Indian Wars see Appendix.

for a proportional share in the allotment that went on apace. So, too, they demanded equal school facilities and other privileges, claiming that the tribal governments in their discriminations against them were violating the Fourteenth Amendment.

Thus, at the end of half a century, with the increase of white population, conditions developed in the Indian Territory for which the Indians were in no wise responsible and which were to put a period to the possibility of the ultimate success of their efforts at self-determination since the utmost capital of their hopeless situation was to be made by the revivified Buffalo Party. Was it not obvious, Congress was asked, that the experiment of the tribal governments was a complete failure? Even the Lake Mohonk Conferences now urged the abolition of the tribal governments along with other reforms, particularly the withdrawal of rations from able-bodied Indians on the reservations. So it was that in Harrison's administration by the Act of May 2, 1890 Congress made the law of Arkansas applicable, in so far as nontribal property rights were concerned, to the Indian Territory until other provision could be made. In consequence now operating within the Territory were six separate bodies of law!

The demand for the abolition of the Indian Territory and Oklahoma and the erection out of them of a state grew louder and louder. It was absurd, the whites declared, that the Five Civilized Tribes should be allowed to bar this, unthinkable that citizens of the United States should be subjected to their laws. Consequently, upon the reelection of Cleveland and the return to power of the Democratic Party, which was all-powerful in the West, the Territory was doomed. In the Indian Appropriation Act of March 3, 1893, the President was authorized to appoint, by and with the advice and consent of the Senate, three commissioners to enter into negotiations with the Indians for the extinguishment of the national or tribal title to any lands within the Indian Territory then held by any and all such nations or tribes, either by cession of the same or some part thereof to the United States, or by allotment and division of the same in severalty among the Indians in order that a new state might be created out of those lands. At the same time the office of Indian agent was abolished. This, it would seem, was not primarily out of consideration for the Indians but because the office was the medium of federal control over allotments.

General Hugh L. Scott, as a lieutenant, had been placed in charge of the Apaches after their removal to Oklahoma by General Miles, largely at his own suggestion. Having saved them from destruction, he has described the working of the system by which the Indians generally were still being despoiled:

In 1892 the Cherokee Commission came to the Indian Territory to arrange for the purchase of the Indian lands for settlement by white people. Its members were charged with all sorts of irregularities in obtaining agreements. The Kiowas, Comanches, and Kiowa Apaches elected me to take a delegation of their tribes to Washington to prevent the ratification of the treaty by Congress. Some of the influential Indians were said to have sold out the interests of their people to the Cherokee Commission. Quanah Parker, Chief of the Comanches, was in favor of ratification. He got permission to visit his children in Carlisle and ran over to Washington without the knowledge of his agent, and arranged for a hearing before the Committee on Indian Affairs, of which Mr. Holman of Indiana, often referred to as the Watch-Dog of the Treasury, was chairman.

Quanah's appointment for a hearing was on the day we arrived in Washington, and my delegation attended the hearing. It was Quanah's appointment, and he had it all his own way at first, and held the sympathy of the committee, who wanted to open the land for settlement by the white man. The committee was about to close the hearing and go to lunch when I asked for my day in court. The chairman asked, "Who are you, and what are you, a white man, doing here?" I handed him my card, and he exclaimed, "Why, you are a soldier; how do you come here?" I told him I was there by order of the commanding general of the army. Quanah jumped up in a great rage and said he wouldn't have any white man speak for him or his people.

I said: "Quanah objects to my speaking here for the Kiowa and Comanche people, but he is speaking only for himself and not for his people, who have not sent him here, and he does not represent their sentiment. If he has any credentials, as I know he has not, let him produce them. Here are my credentials, signed by the agent of the Kiowa and Comanche people, certifying to my election with this delegation to represent them in open council, and I would like to be heard." Whereupon the committee agreed to hear me for an hour at 1 P.M.

We met and quarreled from 1 until 5 P.M., the delegate from Oklahoma the most conspicuous in the opposition in support of Quanah. We metaphorically kicked shins, pulled hair, gouged, bit, and scratched, catch-as-catch-can, no holds barred, all the afternoon. Quanah announced his intention of killing me before I could get back to Fort Sill, and the committee reserved decision.

The Southwest from St. Louis down was determined to open the country, fraud or no fraud. In those days I used to be the enemy of the Indian Department and everybody in it, but three men I knew to be honest, James McLaughlin, J. George Wright, and Major Larrabee in the Indian Office; and I set out to arrange an interview with President Cleveland through the War Department. I knew nothing of the tangled mazes in those days to be found in Washington—wheels within wheels, and deep pits for simple people like me—but some of the wiser men in the War Department advised me to arrange my interview through the Commissioner of Indian Affairs, Mr. Browning.

Ahpiatom, a Kiowa of my delegation, made a great impression among the senators and congressmen in the President's anteroom. He was beautifully dressed in soft yellow buckskin with long fringes, and with his silver medal on his breast. He looked off down the Potomac like an eagle off a crag, paying no more attention to the senators handling his ornaments than if they had been ants crawling about his feet, and despising their effeminate curiosity about his trinkets.

We went in with Commissioner Browning and when I had finished stating my case to the President, he jumped up from behind his desk, striking one hand into the other in emphatic indignation, and exclaimed: "I will not permit it. I will see justice done to those Indians as long as I am in power!" And he did. Through the influence of President Cleveland and of Senator Matt Quay of Pennsylvania, who had a romantic interest in the Indian as well as a wide knowledge of them and their history that filled me with amazement, the ratification of that agreement was prevented, against all the power of the Southwest, for seven years; but I later picked up a newspaper in Havana and read that it had been ratified, fraud and all.

I never applied in vain to President Cleveland or Senator Matt Quay for help in getting justice for the Indian. We caught a soldier with a twenty-five gallon jug of whiskey selling drinks to the Apache Indians in violation of the law. I wanted to try the man by court martial as the only way I could get him punished, but the judge-advocate in Chicago decided that he would have to be tried in one of the civil courts in Oklahoma. On my way to Washington, I stopped in Chicago to see the judge-advocate about it; I was only a first lieutenant in those days, a football for everybody. I told the judge-advocate that it was very necessary to stop whiskey selling to Indians, who as said before were homicidal monomaniacs under its influence, and I could not get the man punished in the civil courts. He ruffled up like an angry owl, snapping his eyes at me, and said, "Do you have the audacity to come here and say to me, sir, that the courts of the United States will not do justice?" I told him that was just what I had come to tell him, since he did not seem to know about it; but he would not recede.

I explained the matter to President Cleveland, who took me into the old cabinet room upstairs in the White House, where the Cabinet was then assembling. He called the attorney-general, Judson Harmon, to whom he introduced me, saying, "This is Lieutenant Scott of the Seventh Cavalry, who will explain a case to you which I want you to prosecute without mercy." The case was called at El Reno, but the prosecuting attorney failed to subpoena the principal witness, and the case was thrown out of court. I did not have money enough to pursue it further, but it went far enough to frighten people from selling any more whiskey to Apaches in my time.

Before going home I met a friend on F Street late one Saturday night, who told me that R. V. Belt had just been nominated for commissioner of Indian affairs, which gave me a severe jolt; and I considered how I could prevent his confirmation, concluding to try Senator Matt Quay first.

I found that the senator was sick in bed. He sent for me to come upstairs and asked how I was getting on, he asked what my trouble was, and I said, "R. V. Belt has been nominated commissioner of Indian affairs." He broke out, "What damned business is that of yours?" I told him that Belt had once been assistant commissioner, had learned the inside of the Indian Office, and had become counsel in claim cases against the Comanches aggregating two million dollars, with fees amounting to some three hundred thousand dollars to be got out of those Indians, and the Senate was going to confirm him in a position to adjudicate on those cases himself.

The senator seemed disgusted with my meddling in political appointments. He turned over with a snort with his face to the wall, as if to say, "Get out; I am tired of you." I got out, believing that my shot missed the mark and I would have to try somewhere else.

There was a man from Oklahoma, originally from Pennsylvania, who wanted to be assistant commissioner of Indian affairs. He thought I had some influence with Senator Quay from Pennsylvania which I could be induced to exert for him, and he dogged me wherever I went. He asked my delegation to sign a petition for his appointment, and when they asked what they should do about it I told them to forget how to write their names, and they forgot. The man followed me into the room of General Miles's aides in the War Department on Tuesday and told me that R. V. Belt had been nominated as commissioner of Indian affairs on Saturday. On the following Monday Senator Quay appeared in the office of the Secretary of the Interior and told him that if he did not withdraw the nomination he would defeat its confirmation in the Senate. The Secretary withdrew it. This caused great joy in the Indian Office, where Belt was much disliked.

When one of the new agents was appointed at Anadarko for the Kiowa and Comanche Indians, Captain Schuyler and I, passing through, stopped to call on him and make his acquaintance. He occupied most of our call by trying to find out from us the different ways an agent could obtain graft, which disgusted us extremely. His grafting during the next two years went beyond bounds, and his actions were a demoralizing influence on his Indians, who were all enraged against him on account of the treatment received from him, and he became intolerable to everybody. I wrote out a request for the Indians to present to the Indian Office for his dismissal, gave it to my cousin with a pen and bottle of ink, and told him to go in to the buggy with I-see-o to every Kiowa and Comanche Camp and witness himself the signature of all those who wanted to sign. They drove down two teams and followed some Kiowas clear up into the Cheyenne country to get their signatures. The agent had saved enough in two years on his salary of eighteen hundred dollars, as he said, to buy a new farm for himself and set his son up in a store and his son-in-law in a canning factory. The agent was subsequently discharged.

Shortly after President McKinley was inaugurated, General Miles introduced me to a Mr. Tonner, who lived next door to the President at Canton, Ohio, and had come on to Washington with him to receive some vacancy as yet undetermined. In the meantime he had a desk in Secretary Cornelius Bliss's private office. He later became the assistant commissioner of Indian affairs, and told me that they were going to send that discharged agent back to Anadarko. I told him that it would be only over my dead body and raised a fearful racket in Washington. After some two weeks, Tonner told me that the administration felt that the ex-agent was too strong for them and they would have to do something for him, but if I would shut up and go home they would not send him to Anadarko, merely appointing his son-in-law agent for the Poncas. I answered that I was not holding any brief then for the Poncas, for whom I was very sorry, but that if they would really promise not to send him to the Comanches I would shut up and go home. The son-in-law was sent to the Poncas and went to the penitentiary in a year and a half.

Dark days were these for Poor Lo! It has been shown that early in the eighteenth century, Amherst had recommended the use of blankets infected with smallpox as a good means of exterminating Indians. Scott relates how measles was made use of a century and a half later for the same purpose.

One reason why I was so averse to the return of the ex-agent to the Comanches was his action when the children in the Kiowa school began having measles. Instead of keeping them quarantined in the school and treating them rationally, he turned them all out, carrying the infection into every family, and shortly afterward he brought the Comanches as well as the Kiowas into the same camp at the agency for a payment.

The parents had no knowledge of the proper treatment of measles and put the children in water to allay their fever, with the consequence that the Kiowas lost three hundred children of measles in one month. The sight of so many mourning parents in one camp was heartrending.

Scott then goes on to describe how easy it was to prevent the exterminating mortality among the Indians upon their reservations by the simplest instruction in hygiene and the provision of a few milk-cows for the children whose mothers were in the habit of weaning them with dry jerked beef.

Above all else Scott was a truthful, straight-thinking man. One cannot read his testimony without a flush of shame rising to the cheek. Yet, in every generation equally reliable evidence of "the Indian system" is to be found. In the face of it what becomes of American history as written for public consumption? Is it not because the so-called historians have lacked the moral courage to tell the truth that the "system" has persisted, that the voices of those who like Roger Williams, William Penn, Sir William Johnson, Sam Houston, Prescott, Parkman and Scott who in each age have dared tell it, have been drowned?

But the facts cannot be erased. They must forever remain indelible. Determined to abolish the Indian Territory as a means of getting rid of the Indian governments therein, on March 29, 1894, the Senate adopted the following resolution:

Resolved, That the Committee of The Five Civilized Tribes of Indians (Cherokee, Choctaw, Chickasaw, Muscogee or Creek, and Seminole Nations), or any sub-committee thereof appointed by its chairman, is hereby instructed to inquire into the present condition of the Five Civilized Tribes of Indians, and of the white citizens dwelling among them, and the legislation required and appropriate to meet the needs and welfare of such Indians; and for that purpose to visit Indian Territory, to take testimony, have power to send for persons and papers, to administer oaths, and examine witnesses under oaths; and shall report the result of such inquiry, with recommendations for legislation. . . .

whereupon Cleveland appointed the comission which took its name from Henry L. Dawes, of Massachusetts, its chairman.

Proceeding with its mission at once, on May 7 and November 9, 1894, the Commission rendered preliminary reports, and after being enlarged to five members by the Act of March 2, 1895, on November 18 following rendered an exhaustive report disclosing the deplorable existing conditions. Its work not having been completed, it was continued by the Act of June 10, 1896.

The labors of Cleveland, the reformer, were at an end. He had struggled against the Buffalo Party over a period of twelve years. While his purpose was high, it cannot be said he had succeeded in reforming the Indian system. McKinley did nothing material for the Indians generally. By the Act of June 28, 1898, Congress denied the claims of the freedmen on the Indian Territory to a right of participation in the trust funds which accrued in the hands of the government from the lease of the unallotted communal lands of the tribes, and in the school funds arising out of royalties. By the Act of March 3, 1901 all the inhabitants of Indian Territory were declared to be citizens of the United States. Now it was impossible for the tribal governments to discriminate against the freedmen or to enforce their laws against the whites. Moreover, by that Act the fate of those governments was sealed, containing as it did provisions for their abolition and that of the Indian Territory and the admission to the Union of Oklahoma as a state on March 4, 1906.

Thus may be said to have come to an end, after a trial of seven decades, the Indian governments set up during Jackson's administration in Indian Territory. It was through no fault of the Indians but by reason of the fate that had pursued them even into the haven which had been set apart to them in the heart of a desert. How clearly Monroe had foreseen this eventuality!*

---

* For a good picture of conditions in the Indian Territory at this time see Edna Ferber's *Cimmaron.*

27

The Early Indian System Analyzed

THE ABOLITION OF INDIAN TERRITORY and the tribal governments long operative there may be said to have marked the passing of the old and the beginning of the new order in Indian affairs.

As pointed out by Cleveland there were now 260,000 Indians, living on 171 reservations embracing 134,000,000 acres, located in twenty-one different states and territories.

But why were these people at the end of a century in the condition described by Cleveland?

Of the reservation system under which this condition of affairs had been brought about, in 1898 Lyman Abbott, one of America's foremost scholars and humanitarians, and a profound student of political science said:

It assumes that the Federal executive can administer a paternal government over widely scattered local communities. For such a function it is peculiarly unfitted. The attempt to engraft a Russian bureaucracy on American Democracy is a fore-doomed failure. . . . Our government is founded on the principle of local self-government; that is, on the principle that each locality is better able to take care of its own affairs than any central and paternal authority is to take care of them. The moment we depart from this principle we introduce a method wholly unworkable by a Democratic nation.

In this opinion the Lake Mohonk Indian Conferences concurred. The "Russian bureaucracy" was, of course, the Bureau of Indian Affairs.

Since its establishment in 1834 it had been presided over by more than a score of commissioners with an average tenure of office of less than three years. The statutes and regulations relating to the administration of

Indian affairs had come to fill many hundreds of pages although the entire law of Great Britain concerning the Indian subjects of the Crown were embodied in fifty-four. Each session Congress added to the bewildering mass, without making any effort to transform in fundamental respects a system designed for the governance of savage and semi-civilized tribes. Naturally, these laws and regulations were incomprehensible to the Indians as well as to many of the officials set over them. Confused, and often conflicting, they received varying interpretations by the executive and judiciary departments, while almost invariably the Department of Justice gave them the one most favorable to the government.

One statute forbade the employment of counsel by a tribe except with the approval of the Secretary of the Interior and Commissioner of Indian Affairs, whose practice it was to require attorneys to bear the risk of all costs and expenses. Since Congress usually limited attorneys' fees in the jurisdictional bills to ten per cent of the recovery and the lawyers were compelled to compete before the tribes for contracts, the system compelled that which is ordinarily deemed champerty, promoted barratry, involved the tribes in politics over the selection of their attorneys, and discouraged professional men of high attainments from accepting Indian cases against the government.

All Indian bills had to be referred to the committees on Indian Affairs of Congress which in turn referred them to the Secretary of the Interior who, of course, relied upon the Bureau of Indian Affairs for advice. In consequence it reflected the views of the Bureau upon which both the executive and legislative departments of the government cast the ultimate responsibility for their acts. One hundred years of such responsibility, with little outside supervision, had crystallized the methods of the Bureau in an undefeatable system, the so-called experts of which looked with contemptuous impatience upon those who "meddled" with Indian affairs.

Yet, if there was one thing for which Congress might have looked with reasonable assurance to its Indian Bureau, it was expert guidance with respect to the sociological needs of the red race. But here too the system was faulty for the ethnological experts were to be found, not in the Bureau of Indian Affairs, but in the Smithsonian Institution. However, their superior knowledge was bitterly resented by the politically appointed clerks in charge of Indian Affairs. In other words, great ethnologists like Powell and Morgan and Mooney, whose lives were devoted to the study of the nature and peculiar needs of the red race, and whose researches formed the basis of great historical works like those of John Fiske, were jealously excluded from consideration by the committees of Congress charged with passing on the policies and measures framed in the Bureau of Indian Affairs. "Impractical dreamers," they were called. Nor did a practical

Congress put much stock in anything coming within the scope of such a formidable term as ethonology!

This was unfortunate both for Congress and the Indians, since the latter, as pointed out by Cleveland, despite a common color, were by no means the same. They varied widely in blood, instincts, mental capacity, intellectual development, religious and historical backgrounds, and consequently in their needs and aspirations. Indeed, even among the Five Civilized Tribes the ethnological differences were probably as wide as among the three great families of Negroes—Senegambian, Guinea and Congo. The variations, not alone in ethnological respects, between the sedentary agricultural Iroquois and Cherokees, the Mission Indians of California and the pueblo tribes of New Mexico, the river Indians of Washington and Oregon, and the roaming tribes of the plains, were no less marked than those between French peasants, German villagers, Scandinavian seamen and the Cossacks of Russia. These differences, the policy of dealing with the tribes at first as semi-independent, and later as domestic communities by treaties, had served to intensify and perpetuate.

Here one is constrained to inquire would the ancient Assyrian, Chaldean, and Babylonian empires have been possible had more in common been demanded by their despotic rulers of the diverse peoples constituting them than allegiance and support in accordance with their kind? Had uniformity of social organization and religion been demanded of the constituent peoples of the Persian Empire, would the crossing of the Hellespont have been achieved by Xerxes? If the Romans had not had a very liberal regard for local conditions and prejudices, could there have been a Pax Romana?

There can be but one answer to these questions. As fugitive as the ancient empires appear in retrospect to have been, they endured longer than any modern state has existed. This would not have been possible in the absence of the recognition of the fundamental sociological principle that an aggregation of different peoples cannot be brought into a successful union under an inflexible system that recognizes no racial differences. The British Empire probably exists today only by reason of the flexibility of the imperial system with respect to its diverse dependent peoples.

Nevertheless to the Indians an inflexible uniform system was applied just as if they came from a common mold, while Presidents and Congresses anathematized them because the many red peoples subject to this unreasonable system were not content with their bitter lots.

Moreover, it ignored utterly in a fundamental respect that which was common to Indian nature.

The Right Reverend Hugh L. Burleson, an eminent authority who for years had been a spiritual worker among them, said:

I contend that in the soul of the Indians in their native state are deep principles of character, tremendous possibilities of life and service that very few of us understand because we have approached life from a different angle. The angle is this. The Indian is a natural communist. By which I mean that the Indian thinks in terms of his group. The white man always thinks of himself first and his group last. We approach things from the viewpoint of the individual. The Indian's point of view is that of the group; his relation to and his responsibility for the group. He thinks in group terms. He has a socialized concept of life. Society has been a definite thing to which he was responsible. The family life and the tribe life have an immediate bearing upon all his actions.

Nothing could be more true if the ethnological experts are to be taken as credible witnesses. In the soul of the red man coursed crosscurrents still unfathomed by those who were making bold to preside over his destiny. Thus, the temporary governmental agents more often than not were apt like other whites to mistake the unbending dignity of the "long-haired" Indian as they called those in their native state, for stolidity, their patient, uncomplaining nature for sheer stupidity, but also more often than not the white man who formed such a judgment was the more stupid of the two.

During World War I, when the psychiatric test was applied to thousands of Indians, white men, Asiatics and Negroes, that which had long been believed by those who knew the red man was found by science to be a fact—he is not only not inferior in intellectual capacity to the white man but unlike the African and Asiatic is his superior in mental resistance to strain.

This mental quality on the part of the red race had undoubtedly resulted from the spiritual poise born of a philosophy of life that makes God not an anthropomorphic deity but a universal, omnipresent, benignant force in nature; that philosophy which has given to the Indian his ability to stand fast; that integrity, that fundamental something that can be trusted which lies at the roots of his race—that something in human character to which faith may be pinned. More than one authority believes it is this philosophy that also gives to the Indians not only their staunchness, but their self-respect, their dignity, and their tremendous strength of mind. A system which discounted this quality, confusing it with inherent stolidity and mental ineptitude, was grossly faulty.

At any rate a wide gulf of misunderstanding existed between the Indians and their governors which nothing in the Indian system was designed to bridge, and this was due as already suggested, to their radically different points of view, the divergence between which may be well illustrated by the translated nomenclature applied to the Indians.

Young-Man-Afraid-Of-His-Horse? What does such a name mean to the white man? Surely an absurd name for a warrior! To the Indian, however, it signified a valor so great, a courage so dauntless, that even the young soldiers of the enemy feared the prowess which the spirit of the knight had imparted to his steed. Thus, to the Indian mind, this gallant chieftain partook not of the nature of a soldier clown, but of a Cid or a Bayard. Yet the import of the name was missed entirely by the translator.

Rain-In-the-Face? What a comical name—to the white man! But to an Indian's mind it is not comical in the least, since it expresses the idea of one with a confidence, a faith, a courage so sublime that he can face without the slightest misgiving the storm of life and without flinching confront the wintry gales of adversity.

Sitting Bull? In no sense ludicrous to the Indian. As a young warrior he was Four Horns, a name signifying the power of two bull buffalo. As he became older and his judgment matured he was elevated to the post of a medicine man or sage. Then the bull with four horns took his seat as a councillor—sat down—became Sitting Bull. Standing Bull was a warrior who never rested, and so on.

Another case of misunderstanding may be cited. A gallant Chippewa chief, killed in battle, was borne home by his victorious warriors to a widowed bride. Soon a son was born to her. On the little mother's mind there was the impression of darkened days, cloud and rain, sudden shadows and sobbing trees. Just as the sun of her life seemed to have set forever, all joy departed, its rays had broken through the clouds and like the rainbow of a new hope, had shone out across the saddened plain of her thoughts—a rift in the clouds of her sorrow. Of this beautiful conception the Indian language—literally translated—conveys no other meaning than Hole-in-the-Day. Such a name it was that the whites, seeking to do him honor, placed on the tomb of Rift-in-the-Clouds!

Again, a little girl, blue-eyed and golden-haired, won the hearts of the Mohawks who had seen no blond person before. They gave her the name Gajajawox—a picture name suggestive of the breeze listing through a field of flowers wafting their perfume as it came—the mingled odor of the eglantine and honeysuckle borne on a zephyr! Yet, the English mind translated the name to Smell-in-the-Air which might be stink as well as perfume.

So it was that the prejudice of the white man for the red made it just as impossible for the former to interpret the latter's inherent sense of beauty as to translate his words, to understand that what the ordinary white man knew only in literature, art and music, in the Indian mind was a sensibility of which he himself was unaware, omnipresent, a living, everyday force, untainted by the artificial emotions of the former.

It has been shown that in every decade of American history, there was at least one outstanding Indian figure of towering mental and moral stature—Massasoit, Canonicus, Philip, Tamenend, Pontiac, Cornplanter,

Cornstalk, Brant, Little Turtle, Tecumseh, Red Eagle, Ross, Quinney, Sequoya, Rift-in-the-Clouds, Black Hawk, Osceola, Joseph, Sitting Bull—too outstanding, too numerous to be accidental. Yet, in one instance only—Parker, the Seneca who had soon been forced out of office by Grant's enemies—had the Indian system of the United States sought to avail itself of the counsel and aid even of those who were the staunch advocates of submission to it. Surely among a race that shed so much of its blood in the defense of the English colonies and the republic, which but recently sent thousands of its members overseas to fight for their fatherland, there must have been many who could have rendered priceless services to the white race as well as to their own.

Finally, the conclusion that the Indian problem would eventually be solved by the amalgamation of the white and red races was impelled by all the evidence available. Left alone in 1784, entirely surrounded by whites, the Iroquois tribes known as the Six Nations, possessing as they did from the first tribal estates, or the basis of individual wealth, had all but vanished as a distinct race in the process of absorption by the whites. The same absorption had occurred in the case of thousands of the Indians of the Five Civilized Tribes of Oklahoma. The great majority of Indians even of the one-eighth blood had to claim it to be distinguishable from the whites.* Nor was there to be found anything that justly could be taken to substantiate the view that the Algonquin, Siouan and other great Indian families, long accustomed to self-government as they had been prior to 1776, would not have mixed with the whites as readily as the Six Nations and the Five Civilized Tribes, had they been afforded the opportunity. The fact is, in the entire range of history a greater congeniality had never, perhaps, existed on the part of a primitive people toward an advanced civilization than had been disclosed by the red race. Within the life of those in being representatives of this race had been able to span in one great leap, as it were, the vast chasm of time that lay between the appearance of the aboriginal social forms on the American continents and modern civilization. Their actual proven ability to do this was unparalleled in history. It was not equalled by the Gauls and the Goths between

---

* Subsequent figures were to bear out this conclusion. In 1910 the ethnological experts declared that there were then 403,000 Indians of all degrees of admixture in the United States, of which number 180,000 were full bloods. In 1920, a decade later, the Census showed but 265,683 Indians, or people recognizable as such, while in 1924 the Indian Office reported 320,497 legal Indians, including 162,602 full bloods. Yet no abnormal mortality had occurred among the Indians during this period. Therefore, it is apparent that in fourteen years the full bloods decreased by 20,000, or by one-ninth of the entire number in 1910, while of the 403,000 Indians in 1910, 137,317, or approximately one-third had been absorbed to the point that they were unrecognizable to the Census as Indians in 1920. Moreover, in 1924, there were at least 54,814 persons in the United States who for legal reasons alone claimed to be Indians, that were not distinguishable as such in 1920.

whom and the civilizations developed in Greece, Rome and Alexandria, little congeniality existed for centuries.

At any rate, certain it is that in the problem of the Indian there was no complicating element of social prejudice, which had been the case with other aboriginal peoples. Some of the most illustrious white families of the nation claimed with pride an infusion of Indian blood.

If then, in fact, absorption by amalgamation was the destiny of the red race, common sense dictated that no effort be spared to prevent its being infused with Negro blood since the zambo, or red-black cross is inferior to both the mulatto and to the mestizo, or the red-white cross, and has been proven to be unabsorbable by the white race without serious injury to it. So, too, it should have been evident that no Indian should be denied the highest form of free education of which he is capable, no matter what expenditure in comparison with that for the education of others it might entail, and that rudimentary education, at least, should have been enforced upon all Indians; that medical aid, hospitals, asylums for orphans, the blind, aged, infirm, homeless and insane, should be freely placed at their disposal and no want or inconvenience on their part be suffered to exist in these respects, since these things were demanded not merely out of consideration for the Indians but in the interest of the whites as well.

It was then in no sense mere national charity that the solution of the Indian problem required. For the lame and the halt—the injured ones among them, yes—there should have been charity as for all others of their kind, but charity for the red race as such could only debase it and the white race as well. The higher philanthropy of common sense was demanded, a philanthropy designed not only to protect the Indian, but the entire nation, and it was just that in which, despite all the Bureau of Ethnology could do, the Indian system was wholly lacking.

28

The "Cowboy President" and the Buffalo Party in the Saddle Again

O_{F THE REAL NATURE} of the Indian System of the United States, Roosevelt apparently took no note. The Dawes Commission was allowed to die a natural death by him after the Act of June 1, 1902, making further provision for erection of Indian Territory and Oklahoma into a state. For the "Cowboy President" the West presented no moral question of present importance as shown by the highly romantic story of its winning in which it was portrayed as a field of grand adventure.* In fact he looked upon it through the lens of a materialism that recked not of the human problem inseparably a part of it. As president too he was thinking of his old playfellows merely as befeathered savages or as hunting companions. Of course he said a lot about their wrongs but did nothing to correct them.

The rapid development of the West had resulted in a prodigal waste of forests and water powers. To conserve those that remained, Congress had empowered the president, by Act of March 3, 1891, at his discretion to set aside as forest reserves in any state or territory the timber-bearing public lands. Of this Act Roosevelt was to take the fullest advantage. As he looked at the map and saw the vast wooded regions reserved to the Indians, he was impatient of their unextinguished rights in lands which on the map seemed entirely surplus to their needs. He did not visualize them as they now were, bending over the hoe in a life and death struggle to win a livelihood from the arid wastes whence the game had vanished with the feathers of the past. They must be made to yield the natural resources that

* *The Winning of the West*, Harold McCracken, Doubleday.

were vital to the economic welfare of the nation as a whole, nor could the government, so long as it maintained reservations for them, afford to be meticulous in the extinguishment of their rights.

His policy was not to prevail, however, without a legal contest with the Indians. Thus, with the aid of the Indian Rights Association, the Kiowa chieftain, Lone Wolf, undertook on behalf of the Kiowa-Comanche-Apache Confederacy of Oklahoma to restrain the Secretary of the Interior from appropriating to public use the lands in which the rights of these tribes had not been extinguished. When, in 1902, the Supreme Court held that Congress had "plenary" power to dispose of their property, that its power to administer Indian affairs might be delegated to the Executive, and that where lands were appropriated by Congress, the act being a political one no remedy was available in the courts,* the government construed the decision to mean that under the power delegated to it such Indian lands as it saw fit to take might be appropriated and the owners left to obtain such redress as Congress might be willing to make.

This, of course, was a preposterous position since by its own Constitution, as well as by the law of nations, the powers of the United States over all persons subject to its sovereignty, whatever their color or previous condition, and whether citizens or noncitizens, are limited to the objects for which it was created—the extension of democracy, republicanism, and equality of opportunity. It was not created for the purpose of denying those objects. It was to guard against the danger of the federal government failing to recognize and adhere strictly to its fundamental purposes that the states imposed upon it, with great particularity by separate amendments, the express limitations contained in the Bill of Rights that is included in the Constitution. Since all of the provisions of this Bill of Rights are universal in their application, it was not to be inferred, therefore, that the supreme organs of the United States, whether its Congress, President, or its Supreme Court, acting for the United States in fulfilling its fiduciary relationships under the law of nations respecting its dependent peoples, had more power than the United States, since the authority of an agent cannot rise higher than its source.

Nor is the government bound by the terms of the Constitution alone. In a report rendered by the British Parliamentary Committee on Aboriginal Tribes in 1837, though there was no exact definition given to the relationship of a civilized state to an aboriginal tribe under its sovereignty, the Committee insisted that the duties incident to such a relationship were precisely those of guardian and ward. This principle of international law has been universally adopted by the nations possessing dependencies, in-

* *Lone Wolf v. Hitchcock* (1902), 187 U. S. 553.

cluding the United States, and they have without exception accepted their political guardianships as a trust.

There could be no justification for a contrary conception on the part of the United States since even under its laws a guardian sustains in the fullest sense a trust relation to his ward. In all his dealings with the ward's property he is bound to look only to the ward's interest, and he is not permitted to occupy a position hostile to, nor to pursue personal interests antagonistic to those of the ward. Any property which he may acquire with the ward's funds or in the course of the conduct of the ward's business is charged with a resulting trust in favor of the ward, and where the guardian buys property out of the estate of the ward in order to get possession thereof and postpone payment until the ward's majority, he commits a fraudulent breach of trust. Any commingling by the guardian of the ward's funds with his own, by which their identity is lost, or any use of such funds for his own benefit is conversion. His acts are construed strictly against him whenever a question of right arises between himself and his ward. It is the duty of every guardian upon the termination of the guardianship to account for the property entrusted to him, and the right to periodic accounting cannot be denied to the ward. No laches on the part of the ward, nor any agreement or act of his or any other person, will act as a bar to a demand for an accounting, unless it be shown that the ward has received full justice, and that injustice will be done the guardian by requiring an accounting from him. If the guardian has himself used the ward's property, he will be charged with the fair value of the use, and if he has used the ward's funds in his own business, he will be charged with all profits made therefrom, or with legal interest, as the ward may elect.

Such being the principles governing the conduct of an ordinary guardian, it was nothing less than unconscionable for the government of a powerful nation which had overthrown the social and economic orders of an aboriginal people, to plead a sovereign immunity from liability upon its trust.

Yet, the policy of the government not only denied the constitutional rights of the Indians, but left them without a legal remedy, since after 1863 they could not sue the United States upon the treaties through which they derived their rights, or upon any contract unless the suit were brought within six years after their claim accrued, in the absence of an act conferring jurisdiction upon the Court of Claims to hear their claims.

Such was the situation when in 1904 the Supreme Court reversed itself, now holding that the guardianship of Congress over the persons and property of the Indians was incompatible with national citizenship on their part.*

By this time no one could determine with certainty exactly where an

* *Matter of Heff*, 1904, 197 U. S. 488.

Indian stood, and when the Oklahoma Enabling Act of 1906 made all the Indians of Oklahoma citizens of the state, but one thing was certain—they had been delivered over lock, stock, and barrel to the Buffalo Party.

Meantime, having lost all confidence in the Dawes Commission they had appealed to have the abolition of their tribal government postponed a year, which was done by the Act of February 20, 1906.

The ultimate termination of the tribal governments like the decision abolishing the guardianship of Congress over the tribal property was, of course, hailed with delight by the predatory interests which now found themselves free to obtain oil and mineral leases even from the restricted Indian allottees all over the West.

Nevertheless, the Indians continued to fight desperately for their rights,* winning in 1910 what may be said to have been an epochal legal victory over the government.

Among the tribes occupying the territory acquired from Mexico was the great Ute tribe which gave its name to the present state of Utah. The circumstances attending the occupancy of these Indians were no different in any respect from those which had existed in the case of all the tribes in the territory ceded by Spain. The first treaty between the Utes and the United States, one of peace and amity, was concluded December 30, 1842. By executive order of October 3, 1861 Uintah Valley was set apart to the Uinta tribe, a division of the Utes, and the remainder of the land claimed by the tribe was taken by the United States without compensation to its occupants. In a treaty made in 1863 the United States agreed to sell the lands which it had taken as public lands, and pay the Utes $1.25 per acre out of the proceeds. Instead of selling all the land covered by the treaty as public land, the government created out of a part of it a forest reserve pursuant to the authority vested in the President by the Act of March 3, 1891.

After making due demand the Utes filed their claims in the Court of Claims, the Court saying in its opinion:

While it may be true that the Indian title of the plaintiffs to any territory prior to the treaty of 1863 was not such a title of the plaintiffs as the defendants would recognize yet the plaintiffs were located within this territory and had the usual claim of occupancy of other Indians. Their claim was considered of such importance that the defendants, during the year following the Guadalupe Hidalgo treaty entered into a treaty with them and secured from them a concession for the right of free passage through their territory. (9 Stats. 984.) By the treaty of 1863 (13 Stats. 673.), defendants considered these claims to territorial occupancy of sufficient importance to obtain from them a cession of all "claim, title, etc. to lands within the territory of the United States," excepting certain lands which were set apart to them as their hunting grounds. By the

* See *Cherokee Nation v. Hitchcock*, 187 U. S. 294, and similar cases.

treaty of 1868 (15 Stats. 619.) the reservation in question was set apart to the plaintiffs, and by the third article of the treaty the plaintiffs relinquished "all claims and rights in and to any portion of the United States or territories except" such reservation. Even if we may admit that they had no valid title to any lands, yet they claimed some title and honestly claimed it, and the yielding of such a claim to a party who wishes to purchase it is a good consideration.*

Two years later the Supreme Court again declared, just as it had done in 1866, that all the provisions of the Constitution applied to the Indians, and could not be set aside by legislative Act.†

Woodrow Wilson, the first Democratic President since Cleveland, had hardly entered the White House with Franklin K. Lane of California and Cato Sells of Texas as his Secretary of the Interior and Commissioner of Indian Affairs, respectively, when the so-called Navajo War occurred. In the good year 1913 still the old processes were going on. As recorded in the press and according to the politicians, the Indians, of course, were at fault. But the truth is the trouble was the result of the Superintendent of the Reservation allowing his power to be misused by the Navajo police. A policeman had arrested a Navajo woman during the absence of her family and had taken her into the agency where she was locked up. Her husband returned and found his enemy had carried off his wife. He gathered his sons and went into the agency to see the Superintendent and recover her. The Superintendent was away at Farmington, but the Indian met his wife outside and took her up on his horse behind him. A policeman caught his bridle in order to stop him and some of the agency employees ran out foɪ the same purpose. He struck the policeman over the head with his quirt, broke away and took his wife home, fortified himself at Beautiful Mountain in a very difficult position on a high terrace, which only one man could climb up to at a time. Scott said:

He secured great Navajo support and they all had food and water and plenty of high-powered rifles with ammunition. The leaders had declared their intention to sell their lives as dearly as possible.

The situation was very dangerous since there were thirty thousand Navajos on the reservation. The mountains were covered with snow and a winter campaign might involve the whole tribe in one of the most difficult and little known sections of the country, among those great canyons that run down into the Grand Canyon of the Colorado, and no one could predict when such a campaign would end. The cost would run far into the millions, to say nothing about the suffering and bloodshed on both sides.

In this situation Scott, now a brigadier-general in command of a section of the border, was called upon in April, 1913 to take charge at the head of a squadron of the 12th Cavalry, and at once set out for the Navajo country. He himself has graphically described what followed.

* 45 Ct. Cl. 440, 465, decided May 23, 1910.
† *Choate v. Trapp* (1912), 224 U. S. 665.

We reached a little Navajo trading post near Beautiful Mountain late that afternoon. It was kept by a Mormon named Noel whose premises were so full of women and children we had to sleep in a tent in the yard. He was in great fear of losing his store and family in case of hostilities. The hostiles, as they were called, saw the arrival of the machine from their lookout and sent their scouts that night to prowl around outside. We saw them listening at windows and their shadows pass over the tent walls.

A courier was sent out to the hostiles next day for them to come in to the store for a conference. This was refused at first, but further messages brought them in the second day, seventy-five of them, heavily armed and they crowded in and filled the store so it was difficult to get a gangway for me to enter. The white men at the store urged me to make them leave their arms at home, but it was hard enough to get them to come, as it was. A demand for them to disarm would cause deep suspicion of a trap and probably defeat my whole plan, as they would probably refuse to parley. My son went in behind the counter with the trader, while I had a gangway made and three chairs and a bench placed in the center, my back against the counter.

Old Bizoshe, the father, was old and stiff in his ways. Those that knew him said nothing could be done to influence him, that everything had been tried in vain, that the old man was hard as iron, and the eldest son was of such mentality no one could deal with him at any time; it was just running unnecessary risk, the white men said, to try to effect the impossible.

I sat the old man near me with some ceremony. The large crowd of armed Indians jammed into the store, listened with perfect silence after they had once settled down. I said:

"My brother, I hear that you are in trouble out here. Troops have been sent for you, but I have told them not to come any nearer than the Two Gray Hills unless sent for, for I thought we could come to an agreement here among ourselves. You and I are too friendly for anything of that sort, and now I want you to tell me what is the matter."

He poured out his troubles in a perfect stream and I let him go until he was through, listening to him without interruption; then I told them I was too tired to talk any more that day. They would find a *hogan* (Navajo hut) out in the yard and plenty of food so they could rest, too, and I would talk to them again in the morning. An Indian cannot be hurried. He needs time to gather and assort his impressions and allow them to soften him. They could have the rest of the day and all night for the purpose of consideration. I had telegraphed the agent to have some food ready for me but he had failed to do so. When questioned, he said he did not believe in giving Indians food. I had to make myself personally responsible for the food at the store, rather than let the Indians get under the influence of anybody else; hunting for food might have resulted in ruining the whole project. The account was paid afterward by the department. Father Weber came in with Chee Dodge and stayed with us.

The nights were quite cold and when old Bizoshe came in after dinner to learn from the trader what we were talking about, I established him at the fire in an easy chair with tobacco and began to hunt for his weak spot. I found he was an historian and asked him, "Where did the Navajos come from first?"

That was enough to launch him on the history of the trials and tribulations of the Navajo. He told us they had come up out of the underworld through a hole in the La Plata Mountains of Colorado. The Pueblos of the Rio Grande came up out of the same hole but on the opposite side, which argues for a long stay of the Navajos in that country. When he went out later, I told Hunter, "It's all right, we've got him bagged. He is coming our way when the time comes."

They all crowded into the store next morning with their arms and Chee Dodge acted as interpreter. I felt them out a little from time to time to see how soft they were and decided to risk a direct demand and asked the old man, "What are you going to do?"

He answered, "I am going to do just what you tell me to do." Each in turn said the same thing.

I then said: "Very well, the marshal has warrants for you (Bizoshe) and three sons. The wagons will be here this afternoon and I want you to get in them and go with me to Gallup. You will have to answer for your conduct to the judge at Sante Fe. We will start tomorrow morning."

They said they would like to go home to make arrangements for the care of their stock during their absence. They were told they could go on one condition, that they must promise to return in time to leave on the wagons a little after sunrise. I was urged not to let them go lest they run clear away and start trouble among the other Navajos, but I allowed them to leave. The wagons did not get in when expected and it was obvious they must have a night for rest after that long drive, so our start was put off another day.

The eldest son said his child was sick, and he would like to go back and spend the night with it. He made the same agreement as the day before. Having left all their horses at home they had to do all their traveling back and forth on foot, six miles in and out. He was back as promised but when the others got into the wagon, he refused to go, saying his child was too sick to leave. We looked each other in the eye for a minute and I told him, "You go and get into that wagon." He got into the wagon.

We drove up to the camp and were soon surrounded by the soldiers who had come all the way from Nebraska to fight "those wild Navajos" and wanted to see what they looked like. The Indians, not understanding the language, became uneasy lest they intended some damage and I told them to go and get their dinner in the store and stay there until I came for them. The rear-guard of the troops had to stay back with some belated wagons and I got back before their arrival at this camp, making good my word. The marshal, hearing the glad tidings of our arrival, hurried up to see his prisoners but finding them all gone thought I had allowed them to escape. He became wrathy but he was soon pacified.

We changed teams here and started on a long cold drive for the Navajo School where we spent the night and went to Gallup next day. The old man was very frail; he had no idea how to ride in a side-seated ambulance. The weather was very raw and cold and I had to hold a blanket around him at the ninety miles into Gallup lest he get pneumonia. They were given a good supper and a warm place to sleep in the jail and it was explained that we would have to separate here. Father Weber and Chee Dodge would go to Santa Fe with them. I turned them over to the marshal and bade them be of good cheer. The old man hugged me to his breast, making a rather affecting scene when we parted. I sent a letter to the judge by Father Weber, saying he would probably find the four Navajos had been as much sinned against as sinning, if not more so; that if he intended to imprison them at Santa Fe, the expiration of their sentence would find them there with no money, unable to speak the language and without transportation home. Thus, they might be thrown on the charity of the town; whereas if their sentences expired at Gallup, on the edge of their reservation, their people could come for them on horseback and take them home safely. The judge gave them thirty days in Gallup.

Thus by the common sense of the old soldier was brought to an end the menace of another Indian war. Scott's story is of the utmost historical value. It shows that the conditions prevailing among the tribal Indians in 1913 were virtually identical with those of a century or more before; that there had been an unpardonable failure in the governmental process applied to them. But what seems most extraordinary is that the administrative system of the United States had not developed in a century and a half a single individual deemed to possess the confidence of the Indians sufficiently to accomplish what Scott was able to do. This fact alone is sufficient to condemn it.

Within a year after the 1912 decision of the Supreme Court, there came to issue before the American-British Claims Arbitration the great international case of the Cayuga Indians in which the British government undertook to recover just compensation from the United States under the provisions of the Treaty of Utrecht for the lands of the Cayugas appropriated after the War of 1812 by the State of New York. Though the decision of the Umpire was not rendered until January, 1926, making an award to Great Britain, in the appendix to the answer filed in 1913 by Robert Lansing, Secretary of State, to the Memorial of His Brittanic Majesty, after reference to the Indian System of the United States, the following unequivocal declaration was made:

Under that system the Indians residing within the United States are so far independent that they live under their own customs and not under the laws of the United States; that their rights upon the lands which they inhabit or hunt are secured to them by boundaries defined in amicable treaties between the United States and themselves; and that whenever those boundaries are varied, it is also by amicable and voluntary treaties, by which they receive from the United States ample compensation for every right they have to the lands ceded by them.

Nothing could better illustrate the callow sophistry of the government in dealing with its Indian wards than this, the most solemn declaration which one government could make to another. Not only had Indian treaties been abolished, as shown, in 1871, but even Indian citizens were being divested more and more of all control over their property at the very hour this declaration was made. Indeed, the discovery of oil throughout the West in large quantities had stimulated the Buffalo Party to such an extent that the situation of the Indians had become a public scandal in 1913, when President Wilson was finally compelled to institute an investigation of conditions in Oklahoma.

Although the facts uncovered by this investigation were so shocking that the President himself forbade their official publication, he continued to do nothing to reform the Indian System. The Bureau of Indian Affairs went on widely advertising the sudden wealth accidently acquired by the Osages and a few hundred individuals among the Five Civilized Tribes, as evidence of its liberality with respect to the Indians generally and its efficiency in administering their estates. Yet, it had in its hands trust property aggregating over four billion dollars in value that was yielding the beneficiaries less than two per cent in dividends!

In 1885, in his *Congressional Government*, Wilson had pointed out the defects in the American system of government. The following year Carnegie published his *Democracy Triumphant* and in 1896 Lecky published *Liberty and Democracy*. They took no cognizance of the Indian. In *Despairing Democracy* (1897), William Thomas Stead of England told the truth.

Wilson was as cold to the Indians as Jefferson, McKinley, Roosevelt, and Taft had been. Nevertheless, there was one who to his everlasting credit could not be silenced. No one knew better than Franklin K. Lane, Secretary of the Interior, what was going on. Making the facts known semi-officially, in his official report for 1914, he put himself on record in the following bold words:

That the Indian is confused in mind as to his status and very much at sea as to our ultimate purpose toward him, is not surprising. For a hundred years he has been spun around like a blindfolded child in a game of blindman's bluff. Treated as an enemy at first, overcome, driven from his lands, negotiated with most formally as an independent nation, given by treaty a distinct boundary which was never to be changed "while water runs and grass grows," he later found himself pushed beyond that boundary line, negotiated with again, and then set down upon a reservation, half captive, half protégé.

What could an Indian, simple thinking and direct of mind, make of all this? To us it might give rise to a deprecatory smile. To him it must have seemed the systematized malevolence of a cynical civilization. . . . Manifestly, the Indian has been confused in his thought because we have been confused in ours.

It is not surprising, therefore, that in the most searching survey of the American government ever made, James Bryce, the great English political writer who had three times visited America, undeceived by the declarations of the Secretary of State in 1913, had dismissed the Indian System with dignified forbearance. He wrote:

The Secretary of the Interior is chiefly occupied in the management of the public lands. . . . and with the conduct of Indian affairs, a troublesome and unsatisfactory department, which has always been a reproach to the United States, and will apparently continue so till the Indians themselves disappear or become civilized.

That the Indians were disappearing more rapidly than they were being civilized was manifest for at the end of a century and a half, of over five hundred thousand full blood Indians for whom the government at one time or another had assumed responsibility, less than one hundred and eighty thousand remained with a total Indian population including all degrees of Indian blood down to the sixty-fourth degree of but 265,683.*

While in theory the courts deal only with the law as it is written, after all those who frame their decisions are human beings whose words inevitably reflect both governmental policy and public sentiment. This has ever been particularly true in the case of the Indian decisions of the American judiciary. The Supreme Court had long since seen the havoc its decision of 1904 had created. Following Lane's bold semi-official condemnation of the government which, in a sense was an attack upon the judicial policy of the United States, in 1915 it did not hesitate at the instance of Charles Warren, Assistant Attorney General, to reverse itself again. It now held that a political guardianship over the tribal Indians and their property was not incompatible with national citizenship.†

In 1915 there occurred what might have given rise to another Indian war had it not been for the influence of General Scott. Trouble had occurred in Utah over the murder of a Mexican in Colorado with which a Paiute Indian was charged. The suspect had sought refuge among his tribe in Utah, which was related to the Utes of Colorado, and had intermarried with the Navajo tribe which boasted about 30,000 members. The Colorado authorities had made application for the extradition of the refugee but numerous posses had failed to capture him. Then the United States Marshal in Utah with a posse of about seventy-five gunmen had gone to Bluff and, being afraid to negotiate with the Paiutes, had surrounded a band of them. Believing that they were being threatened by hostile cowboys who were constantly invading their reservation, the Indians resisted arrest and the posse was defeated with bloodshed on both sides. Later an innocent Indian lad was arrested and murdered on the pretext that he was trying to escape. Early in February, 1916, a serious war with the Paiutes, Utes and Navajo was threatening when Secretaries Garrison and Lane

* Census 1920.
† *U. S. v. Nice* (1915), 241 U. S. 591.

called on General Scott to intervene. Finding that the Indians had been abused as usual and that a land-grabbing scheme was behind the whole thing, Scott had little trouble in quieting the Indians down. By giving the facts publicity he then compelled the authorities to drop the matter. He wrote:

Peace continued there seven years notwithstanding constant effort to get the Indians away from the land where they and their ancestors had been born. . . . The papers later recorded another difficulty with the Paiutes and various posses. I offered to go again and try to save the lives of those Indians but was told it was not an important matter, though I would be asked to go should it become important. It resulted, however, in the killing of Posey and some of his band. Later the Indian Department took the same measures for the prevention of such warfare that I had recommended seven years before, without result. I feel certain that if I had not gone in person with those Indians to Salt Lake City some of them would have been legally murdered by trial and conviction, with one side only represented. I am never afraid of what the Indian will do in such flare-ups—if I can reach him in time. I am always afraid of what the white men do—legally.

How sweet it is to look backward over the years and remember the gratitude of those simple, sincere and humble people, both red and white, for the salvation of the women and children of that country-side is worth far more than any military glory that might have been won in bringing about their subjugation and punishment by force of arms.

Although there is no doubt Secretary Lane was kindly disposed towards the Indians and really desired to do something substantial for their welfare, he was not without an eye to the Indian political possibilities. Statistics showed that in North and South Dakota, Wyoming, Montana, Oklahoma, Arizona and New Mexico there were enough Indians to decide the political complexion of those states if they were enfranchised and well organized. It might reasonably be expected that they would feel some obligation to support the Administration responsible for giving them the vote. But Lane was reckoning without his host. The West was not yet prepared to welcome the Indian vote. As the presidential campaign of 1916 drew near, however, and Lane undertook to organize the Department of the Interior with its army of federal employees in the West behind Wilson; he decided to encourage the Indians, who were already qualified for citizenship under the General Allotment Act, to exercise the right of franchise to which so far they had paid little attention. Thus, early in May, 1916, he proceeded to the Yankton Sioux Reservation in South Dakota where according to him he "admitted some one hundred and fifty competent Indians to full American citizenship."

In order to derive the fullest possible political advantage for the Administration the following patriotic ritual was devised by Lane which was well calculated to impress the Indians.

Indian Ritual Admission to Citizenship

The Secretary stands before one of the candidates and says:—

"Joseph T. Cook, what was your Indian name?"

"Tunkansapa," answers the Indian.

"Tunkansapa, I hand you a bow and arrow. Take this bow and shoot the arrow."

The Indian does so.

"Tunkansapa, you have shot your last arrow. That means you are no longer to live the life of an Indian. You are from this day forward to live the life of the white man. But you may keep that arrow. It will be to you a symbol of your noble race and of the pride you may feel that you come from the first of all Americans."

Addressing Tunkansapa by his white name.

"Joseph T. Cook, take in your hands this plough." Cook does so. "This act means that you have chosen to live the life of the white man. The white man lives by work. From the earth we must all get our living, and the earth will not yield unless man pours upon it the sweat of his brow.

"Joseph T. Cook, I give you a purse. It will always say to you that the money you gain must be wisely spent. The wise man saves his money, so that when the sun does not smile and the grass does not grow he will not starve."

The Secretary now takes up the American flag. He and the Indian hold it together.

"I give into your hands the flag of your country. This is the only flag you will ever have. It is the flag of free men, the flag of a hundred million free men and women, of whom you are now one. That flag has a request to make of you, Joseph T. Cook, that you repeat these words."

Cook then repeats the following after the Secretary:

"Forasmuch as the President has said that I am worthy to be a citizen of the United States, I now promise this flag that I will give my hands, my head, and my heart to the doing of all that will make me a true American citizen."

The Secretary then takes a badge upon which is the American eagle, with the national colors, and, pinning it upon the Indian's breast, speaks as follows:

"And now, beneath this flag, I place upon your breast the emblem of citizenship. Wear this badge always, and may the eagle that is on it never see you do aught of which the flag will not be proud."

These Indians, declared Lane, "never can become hyphenates." In letters written after the election Lane pointed out the decisive influence the Department of the Interior had been made by him to play in capturing the Western states. "Speaking of the election," he wrote the editor of the *New York World* on November 11, 1916, "I want you to bear distinctly in mind . . . that the states which the Interior Department deals with are the states which elected Mr. Wilson."

29

The Indians and the First World War

NOTWITHSTANDING its undoubted evils, modern war is an educative force by reason of the new contacts and experiences it affords the masses of the people whom it involves in its far-flung meshes. This was particularly true of World War I with respect to the American Indians.

It has been shown that the Indians fought in the Revolution and the War of 1812, not only as tribes, but as individuals in the armies of the United States and Great Britain. In the Civil War they also fought in both armies in special Indian military units and as individuals. It was in 1917, however, that the Indian, without regard to whether or not he had become enfranchised, was to take his place beside the white and black citizens of the United States and discover a world hitherto unknown to him while doing far more than his proportional part.

That the world at large was still thinking of the red race in terms of the past was illustrated upon America's entrance into the war by a great painting exhibited in Paris in which the artist depicted the many peoples now joined to overthrow the Central allies. The Indian in this impressive group was shown in full native war regalia when as a matter of fact he had donned the homely olive drab of the American army. The idealization of the artist was not without good effect, however, for Buffalo Bill and his Sioux warriors had not visited Germany in vain. There, too, the Indian who was known to be on his way to France was depicted in the popular mind bedight in war paint and feathers, while no other adversary aroused more fearsome interest. Over and over the Kaiser's veterans inquired with keen solicitation about this new enemy. Willing enough to match their unsurpassed skill and fortitude against the African and East Indian of whom they had had experience before 1914, thoughts of the blood-curdling war whoop and the scalping knife of the "barbarous savages of America" filled them with horror. More than one of them read the Declaration of Independence with added appreciation of the protest it contained against the employment in war of "savages."

To the everlasting credit of the Indians they were the most forward among the people of America in volunteering for military service. Even before the draft was put into effect, more than 2,000 had entered the American and Canadian armies. Upon the passage of the Selective Service Act it was arranged that the Commissioner of Indian Affairs should conduct the resignation of the tribal Indians in a manner similar to that pursued in the several states for the whites. A board was established on each reservation which consisted of the superintendent of the agency, the chief clerk, and the physician. The registration cards of citizen Indians were allocated to local boards having jurisdiction of the area in which the reservations were situated and the residents classified in the same manner as the citizens of the United States generally. Not being subject to military service, noncitizen Indians were exempt from the draft, but were allowed to volunteer.

In determining the citizenship of Indians, the rules laid down by the Bureau of Indian Affairs during the election of 1916 were followed. Generally speaking, an Indian born in the United States was deemed to be a citizen if he, or his father or mother, prior to his birth or before he attained the age of twenty-one, had been allotted land subsequent to May 8, 1906, and had received a patent in fee to his land; or if he were residing in the old Indian Territory on March 3, 1901; or had lived separate and apart from his tribe and had adopted the habits of civilized life. It would, of course, have been difficult to determine who was and who was not exempt from the draft had the Indians themselves not manifested the utmost good spirit in assisting. Furthermore, it is beyond doubt that many Indians voluntarily registered who were not bound to do so. The result was that a total of 17,313 Indians were registered. Of the 11,803 registered prior to September 11, 1918, 6,509, or 55.15 per cent, were inducted into the service. This is a percentage more than twice as high as the average for the 24,234,021 registrants, or all those enrolled under the law.

Despite the fact that hundreds of Indians had volunteered for service in the Canadian and United States armies prior to June 5, 1917, or the date of the first registration, but 228 of the 17,313 Indians registered claimed exemption—a negligible percentage. Most of them were forced to apply for exemption because of age, and some of them then volunteered. The Bureau of Indian Affairs declared that in fact more than three-fourths of those enrolled were actually volunteers. This is a record which should put to shame the thousands of persons who though citizens of the United States claimed exemption upon a multiplicity of grounds not recognized by the authorities.

While the limits of this work preclude more than the most casual reference to the racial contribution to the national effort, another fact must not be omitted. The Indians subscribed to 10,000 or more Red Cross memberships, and purchased a total of not less than $15,000,000 of Liberty bonds, making a per capita subscription to the national loan of approximately $50. The spirit animating them which made possible this astounding result is well illustrated by a single anecdote.

At a gathering of the Ute Indians, among the scores who spread all the fingers of one hand to indicate their subscription to the Red Cross, was an old woman of seventy-five years. Understanding that one finger meant $10.00, the presiding superintendent recorded her subscription for $50.00. A few days later when she limped to the agency to fill out her card she became indignant, explaining through an interpreter that she had given $500. "But," said the superintendent, "you have only $513 to your credit." Quickly came the answer, "$13 left? That's enough for me."

Knowing the admirable spirit with which the Indians assumed their share of the national burden, it remains to see what kind of soldiers they proved to be.

Innumerable anecdotes concerning them have appeared in the press and periodical literature, from which various conclusions have been drawn, some overhastily, many wholly without warrant in fact. It is human nature to base general conclusions upon special circumstances. Furthermore, the Indian being a romantic figure, his exploits naturally attracted a special attention which gave to him unusual notoriety. It is not, however, from the exaggerated reports of special cases of heroism, fortitude, dexterity, and peculiar individual ability, but from the carefully weighed testimony of reliable witnesses and investigators that a true estimate of his military character is to be formed. Always one must put to himself the question, is the testimony being presented evidence of the peculiar prowess of the Indian as a warrior, or merely of the prowess of an individual warrior. Thus tested, ninety-nine out of a hundred instances cited in good faith as evidence of peculiar Indian prowess will be disposed of, for invariably his reported exploits may be matched by similar ones on the part of the associates of the red man. Nor do the well-wishers of the Indians serve their interests by endowing them with virtues and qualifications which exist merely in the imagination of enthusiasts. The creation in the popular mind of an unreal Indian must in the end bring only bitter disappointment to all alike. It is much better for him that his white associates understand him as he is in reality, for thus only can they place a rational valuation upon him as a national asset, and frame their expectations accordingly.

Is the unparalleled enthusiasm with which they volunteered for military service to be construed to mean that still smouldering in the Indian character is the ember of savagery—an instinctive love of war?

His whole history shows that he has never loved war, but that when duty has called he has never sought to evade military service. Between the two there is a marked difference. It is very easy to confuse them. For instance, a number of years ago, in order to enliven the monotony of life on a certain reservation in connection with the annual agricultural fair that had been inaugurated to stimulate farming activity, there was arranged a mock foray, including a very realistic "old time" attack upon an emigrant train. To encourage the warrior braves the wagons were loaded to the "gunnels" with stores of good things, food, raiment and the like. Numerous pyres of inflammable material placed about the prearranged place of "massacre" were soaked with oil in order to illuminate the spectacle which was to be held late at night, while scores or more of local ranchmen willingly submitted to the unfamiliar role of being the victims. A huge gathering assembled to witness the realistic performance which the Indians conducted with superb *élan*, fortunately more hungry for the cherished prizes which the wagon train contained than for the blood of its defenders. So thrilled did the two hundred or more Indian women in the improvised grandstand become that when the "massacre" was complete, they surged out upon the lurid field chanting the weird victory choruses of their ancestors.

If this incident be considered without analyzing its true meaning, it is apt to be misleading as to Indian character, more often than not the case with unfamiliar customs. What did it really mean? Simply that even the women of the Indian race are possessed of a warlike disposition?

Not at all. Only that they honor the self-sacrificing spirit of the willing warrior; that manhood service is understood by them in its more noble aspects.

The surest way to test the meaning of the Indian victory choruses is to interpret the words. They set no store on fighting prowess or mere bravery as distinguished from the spirit of self-sacrificing courage. They do not exalt war or conquest. They praise those who fight in defense of home.

In the "strongheart fighting songs" of the Indians, which are usually sung by the women before and after battle, there is to be found much Indian psychology. During the war the various draft contingents were at times accompanied to the trains by their womenfolk who sent their warriors off to join the army with the stirring strains of their victory songs not only ringing in their ears but vibrating through their beings as well. The effect of this is not difficult to imagine.

There were others, too, who gave encouragement to the Indian conscripts. Among these should be mentioned the venerable John Grass, the last of Custer's scouts, who, with fervid eloquence, urged the men of his race to enlist.

The first soldier from South Dakota to receive a decoration in France was Chauncey Eagle-Horn, the son of a Sioux warrior who had fought against Custer. After he was killed in action, one of his associates wrote home: "I try to do everything they tell me, but some of it seems awfully blood-thirsty!" So, it is believed by those who know best, that it is a mistake to assume the Indians in 1917 were actuated principally by lust for adventure. A people do not contribute in the way they did out of their meager means to adventure alone.

When the Selective Service Act first cast its net over the Indians it was urged by Secretary Lane and various misguided enthusiasts that the Indians, like the whites and Negroes, be organized in distinct units.

The experiment of organizing Indian military units has not been untried. Indian frontier police had long been in use when in 1892 about 3,000 Indians were enrolled in the Regular Army in separate infantry and cavalry units, in addition to a small number of scouts. At the end of the first period of enlistment of several years the experiment was not repeated. This was due to the fact that the considerations which had dictated the policy were no longer of importance, and not because the Indian had proved unsatisfactory as a soldier. Enlistment as an antidote to open distaste for reservation confinement was no longer necessary since the last flicker of serious Indian independency had died down. Independent Indian enlistments, however, henceforth were accepted. The exploits of Arthur Bonnycastle, a sergeant in the 9th U. S. Infantry, in the Philippines and China will long be remembered in the Regular Army. In 1903 this heroic soldier returned to the Osage tribe, took the blanket, married the head chief's daughter and himself became head chief.*

The man who probably knew more about Indians in 1917 than any other was Major General Hugh L. Scott, the Chief of Staff. In his book, which is replete with valuable information about them, derived from a lifelong experience, he says:

Secretary Lane had a plan for raising a separate division composed of Indians and when my opinion was asked by the Secretary of War I had to say again, "It won't do." He asked, "What makes you talk that way? You have always been the friend of the Indian and once had an Indian troop of regular cavalry that did well. Why do you talk that way now?" I replied that I was just as good a friend of the Indian as Secretary Lane, or anybody else was, but that we must not build our army of special corps; that this time there should be no Polish, Armenian, or Russian regiments, as in the Civil War; no "fought mit Siegel," no "sons of Garibaldi"; nothing but homogeneous American troops. The separate Negro organizations we cannot avoid.

* He was persuaded by the author not to reenlist. He died about 1925.

"Suppose, Mr. Secretary," I said, "you had an Indian division decimated on the first line. You would have to pull that division back out of the way, and send back over here for new Indians to fill the ranks, and meanwhile your decimated division would be worse than useless. On the other hand, by my plan you could fill your divisions overnight. I know the Indian, Mr. Secretary. He can serve harmoniously as an individual, in any white organization, and will do himself credit and you, too. But please do not allow the Indian or anybody else . . . to be segregated in our army."

The position of General Scott was supported by the Commissioner of Indian Affairs so that the Indians were assigned to the divisions organized in their respective districts.

The decision of Secretary Baker in this matter

gave umbrage to Secretary Lane but the result turned out as predicted. The Indian was popular in his organization of white soldiers, wherever he served, no word of misprision ever reaching my ears. As a race he played a higher part in the war on the side of patriotism than the ordinary white man, notwithstanding the fact that it was but a short time since we were pointing guns at him. He put aside his long list of grievances against his white brother, without waiting to be drafted, and of about fifteen thousand men able to pass our military examinations, ten thousand served in the army and navy, a greater proportion than was furnished by the white man whose war it was primarily— this being one that the Indians did not make. Although an impoverished race, the Indians bought more than twenty-five million dollars' worth of Liberty bonds, subscribed liberally to the Red Cross and kindred societies, and would have done more had they known how to carry out our money-raising schemes. We may indeed all be proud of our Red Race and its record in the World War.

In his report the Commissioner of Indian Affairs, Cato Sells, said:

I think the best military status for the Indian is with the organization of white soldiers, where under the usual army discipline the benefits are measurably reciprocal, with a definite educational advantage to the Indians. The military segregation of the Indian is altogether objectionable. It does not afford the associational contact he needs and is unfavorable to his preparation for citizenship.

My personal observation when visiting cantonments and reports to me show that the Indians are making remarkably good soldiers, and I am gratified to learn that they were placed without regard to the fact that they are Indians. This mingling of the Indian with the white soldier ought to have, as I believe it will, large influence in moving him away from tribal relations and towards civilization.

From the standpoint here suggestively stated, to which other reasons might be added, I regard it as inadvisable to call a council for the purpose of arousing sentiment by agitational appeals to the Indians in the direction of separate military units, but that on all reservations and at Indian schools on and off reservations throughout the Service and among Indians everywhere, the spirit of patriotism and loyalty should be taught and emphasized and that all Indians acceptable under regulations should be encouraged to enlist in some organization of the regular establishment.

Military men were much surprised when the psychiatric test showed that in mental poise and the power to resist strain the Indian was superior to the white soldier.

After the actual fighting began it was not long before throughout the A. E. F. there were current many thrilling stories about Indian soldiers. In a November, 1918, issue of the *Stars and Stripes*, one finds evidence of the romantic interest that had early been imparted to the battlefield by the red man.

It was the Prussian Guard against the American Indian on the morning of October 8th, in the hills of Champagne. When it was all over, the Prussian Guards were farther on their way back toward the Aisne, and warriors of 13 Indian tribes looked down on the town of St. Etienne. The Indians—one company of them—were fighting with the Thirty-sixth Division, made up of Texas and Oklahoma rangers and oil men, for the most part. "The Millionaire Company" was the title that had followed the Indians from Camp Bowie, Wyoming, and there followed them also a legend of $1,000 checks cashed by the Indian buck privates—of privates who used to spend their pass in 12-cylinder motor cars—of a company football team that was full of Carlisle stars and had won a camp championship. Collectively they owned many square miles of the richest oil and mineral lands of Oklahoma, and back home there were thousands of dollars in royalties piling up every day for the buying of Liberty bonds. In the company were Creeks and Sioux, Seminoles, Apaches, Wyandottes, Choctaws, Iroquois, and Mohawks. It was a company with a roll of names that was the despair of the regimental paymaster, who never could keep track of Big Bear, Rainbow Blanket, Bacon Rind, Hohemanatube, and the 246 other original dialectic nomenclatures.

"Stuff" of this nature filled the press. It was popular if for no other reason than that it satisfied the national craving for a spark of romance in a war whose deep shadows seemed unrelieved by color and the higher lights. Imagine a company of Indians, even though composed of 250 Hiawathas, exerting an appreciable effect upon the sturdy and gallant Prussian guards!

In 1918, while in the midst of the great combat in the Argonne, Lieutenant J. R. Eddy, of Pennyslvania, formerly Superintendent of the Crow Reservation,* while serving with the 4th Division, recommended the organization and training of the Indians as rangers and their assignment in small groups throughout the Army for special service as scouts. This, he insisted, was the way to take advantage of their undoubted inherited char-

---

* Grinnel had dedicated one of his Indian books to Eddy.

acteristics. He was seconded by Lieutenant Red Cloud who, while attending a school at Langres, urged that they be formed into special Signal Corps units. The latter was killed, but after being incapacitated by gas, Eddy managed while being evacuated to America, to have himself assigned to General Headquarters in order that he might collect all the authoritative information possible to obtain concerning the Indian soldiers. When the Historical Section of the General Staff was formed, a questionnaire designed by him to elicit all the facts of military value was distributed to over fifteen hundred combat units by Brigadier-General Oliver L. Spaulding, the chief of the section.* A study of the evidence thus acquired compels one to regard with suspicion those who are willing to formulate definite conclusions from the necessarily limited and narrow basis of fact that is available to the ordinary person.

It was significant that though the Indian was subject to more than ordinarily careful observation by reason of his comparative rarity among the other troops, a fact which naturally caused his virtues in many cases to be over-generously extolled and his defects to be noted more readily, nowhere among the mass of early testimony was there a suggestion that he was inferior in moral stamina to the white associates with whom he was mingled From this it was fair to assume that under the conditions which obtained at least he measured up to the average white soldier.

How he would have comported himself if separately organized in large units, as were the East Indians of the British Army and the American Negroes, it is impossible to say with evidential precision. Would he have found within himself the moral resistance which modern European conditions of warfare demand? Who knows? Inured to extremes of climate he would probably have given evidence in the mass, as he did individually, of a resistance superior to that of the East Indian whose morale was adversely affected by climatic conditions. Some things, therefore, seem certain from the psychiatric test.

The military psychologist well knows that the physical bears an essential relation to the moral factors entering into the character of the soldier, nor fails to properly evaluate both in the problem of creating the fighting man. He cannot conclude from the collected evidence, however, that there is any peculiarity of Indian nature that might have reversed the history of the races had the Caucasian been opposed by the red man in Europe. History speaks loudly of the moral ascendancy of the Caucasian over all the others against whom throughout the travail of the recorded ages he has been pitted. There are innumerable authenticated instances in which others, when disciplined and led by the Caucasian, have measurably attained the latter's standards, none in which man for man with equal

* The author was a member of the section at the time and helped Lieutenant Eddy.

opportunities of training and armament, the Caucasian has manifested racial inferiority as a combatant. Nor have the intellectual endowments of the white man alone accounted for this. There have been times when the culture of the East has failed utterly under skilled leadership to vanquish those who were relative barbarians. The so-called Nordic must, therefore, still be taken as the foremost "killer" of the human race.

One of the first facts in connection with Indian soldiers to attract universal attention was their undoubted ability to communicate intelligibly with each other in a language that was wholly unknown even to the erudite "Professor Fritz." This is stated as a fact, and it may be safely accepted as such. Yet, it is not to be deemed peculiar in any way to Indian nature.

The Indians of the different plains tribes had developed a sign language of motions and symbols for the purpose of cooperation in the chase of the buffalo which the natives of the forest regions, limited in their range, did not find necessary. Although this sign language was of such narrow scope that at first it proved to be quite inadequate to the needs of modern military science, it was not long before the Indian soldiers extended the vocabulary. Thus, many lingual equivalents used by them were generally adopted for obvious reasons.

Decidedly the most interesting report upon this point obtained by Eddy was that of Colonel A. W. Bloor, commanding the 142nd Infantry, 36th Division, which, by reason of its special value, is given here in full.

In the first action of the 142nd Infantry at St. Etienne, it was recognized that of all the various methods of liaison the telephone presented the greatest possibilities. The field of rocket signals is restricted to a small number of agreed signals. The runner system is slow and hazardous. T. P. S. is always an uncertain quantity. It may work beautifully and again, it may be entirely worthless. The available means, therefore, for the rapid and full transmission of information are the radio, buzzer, and telephone, and of these the telephone was by far the superior—provided it could be used without let or hindrance, —provided straight to the point information could be given.

It was well understood that the German was a past master in the art of "listening in." Moreover, from St. Etienne to the Aisne we had travelled through a country netted with German wires and cables. We established P. C.'s in dugouts and houses, but recently occupied by him. There was every reason to believe every decipherable message or word going over our wires also went to the enemy. A rumor was out that our Division had given false coordinates of our supply dump, and that in thirty minutes the enemy shells were falling on that point. It was, therefore, necessary to code every message of importance and coding and decoding took valuable time.

While comparatively inactive at Baux-Champagne, it was remembered that the regiment possessed a company of Indians, who spoke twenty-six different languages or dialects, only four or five of which were ever written. It was hardly possible that Fritz would be able to translate these dialects, and the plan to have these Indians transmit telephone messages was adopted. The regiment was fortunate in having two Indian officers who spoke several of the dialects. Indians from the Choctaw tribes were chosen and one placed in each P. C.

The first use of the Indians was made in ordering a delicate withdrawal of two companies of the 2nd Battalion from Chumfilly to Chardeny on the night of October 26th. The movement was completed without mishap, although it left the Third Battalion greatly depleted in previous fighting, without support. The Indians were used repeatedly on the 27th in preparation for the assault on Forest Farm. The enemy's complete surprise is evidence that he could not decipher the messages.

After the withdrawal of the regiment to Louppy-le-Petit, a number of Indians were detailed for training in transmitting messages over the telephone. The instruction was carried on by the Liaison Officer, Lieutenant Black. It had been found that the Indians' vocabulary of military terms was insufficient. The Indian for "Big Gun" was used to indicate artillery. "Little gun shoot fast" was substituted for machine gun, and the battalions were indicated by one, two and three grains of corn. It was found that the Indian tongues do not permit verbatim translation, but at the end of the short training period at Louppy-le-Petit, the results were very gratifying, and it is believed, had the regiment gone back into the line, fine results would have been obtained. We were confident that the possibilities of the telephone had been obtained without its hazards.

A careful analysis of Colonel Bloor's report justified the conclusion that the special use made of Indians on the occasion described was possible more by reason of their unfamiliar tongue than of any peculiar military qualifications which they possessed. Would not some strange African tribesmen, speaking a language equally unfamiliar to "Professor Fritz," have been equally useful under the same circumstances?

There was one characteristic in particular, of the Indian, though more or less common to frontiersman of whatever breed, namely, that of stealth, which Eddy proposed to capitalize.

For uncounted generations Indians have been engaged in a conflict with the forces of nature, and from nature have wrested support for themselves. To find their way from place to place and to secure the wild animals which furnished them food and raiment, they have been obliged constantly to study nature, the habits of the wildlife upon which they depended, and the surroundings among which the animals lived. This has gone on until there has grown up among the Indians an inherited adaptability to deal with the conditions of nature and to draw conclusions as to how animals —including man—will act under certain circumstances. No doubt before the white man became civilized he possessed this same instinct. In the many generations that have elapsed since this natural mode of life was abandoned, this faculty, through disuse, has been lost by the latter. The Indian, however, who is only two or three generations distant from his ancient environment, still retains this inherited ability almost unimpaired.

In his case, as in that of other natives, an understanding of nature is still instinctive so that in many instances he will arrive at an accurate conclusion intuitively rather than by a conscious process of reasoning. Illustrative of this faculty and its common possession by natives is a story narrated by a prominent British officer who commanded a unit of Gurkhas in France. On one occasion he made a wager that half of a detachment of his men could crawl unobserved through a line of British sentinels. All twenty-five of the little Gurkhas entered the guarded sector to reapppear with astonishing rapidity in the rear of a line of patrols which may be assumed to have been more than ordinarily alert.

Major Glenn of the Intelligence Staff, British Army, an officer of wide experience with natives, after observing an extensive series of tests with American Indians, concluded they were far superior as military material to the ordinary southern natives by reason of their ability to adapt themselves to the rigors of northern climes. Relatively, however, the American Indian soldier was found lacking in physical stamina. This was only natural for while there are many Indians living today under equally as good conditions as those surrounding the whites, and but comparatively few in a primitive state, yet it cannot be denied that the race as a whole has suffered from the conditions under which it has been compelled to live so many years. In forced intimate contact from the first with the lowest elements of a frontier population with the blood of which they have been over-liberally infused, they have suffered intensely in many ways from the contaminating influence of civilization.

Illustrative of the relative lack of physical stamina of the Indian race, the late James Mooney cited to the author an interesting incident. In 1875 the United States government demanded of the Kiowas, Comanches, Kiowa-Apaches, Cheyennes and Arapahoes who had but recently gone on the warpath, sixty odd hostages who were transported to Florida and held there under restraint, under the custody of Lieutenant R. H. Pratt.* Among these were six Mexican captives whom the chieftains, with native cunning, had designated. Twenty years later three out of four of the Indians were dead, while all six Mexicans remained hale and hearty under the altered conditions of climate and living. There is no story that illustrates better than this one the susceptibility of Indian nature to the destructive defects of civilization.

It is frequently said that Indians, when subjected to an indoor routine of occupation, as a rule, soon develop a cough. If this be true, it might have been well had the government devoted itself in the past more to the medical than to the political treatment of the Indian, just as did the British medical relief societies to the natives in the East with almost instantaneous good results.

Frequently, the keen interest which Indians ordinarily display in the

* Carlisle School, as shown, was founded with some of these prisoners.

athletic exercises common among white men, is believed to be the equivalent of extraordinary physical aptitude for athletic games. Although we know that polo originated in Tibet, and that the ancient Indians played at a game suggestive of baseball, it is possible that their present interest in athletics is in large measure due to the novelty of the various forms of sport which are introduced to him by his more civilized associates. The Carlisle football team proved nothing to the contrary. It is not proved that Indians in their uncivilized state were peculiarly prone to engage in athletic exercises.

The evidence described having been obtained it was compiled with scientific accuracy by Eddy in the following digest of opinions, here printed for the first time:

Digest

1. Does the Indian stand the nervous strain of the soldier?
   He does.
2. Does he prove a natural leader in ranks?
   He does not.
3. Does he associate readily with white men?
   He does.
4. Is he regarded by the whites as an unusually "good" man?
   He is regarded as a very good soldier.
5. Has he demonstrated fitness for any special arm?
   Automatic Weapons.
6. What capacity has he shown under the following heads?
   a. Courage; endurance; good humor?
      Very good in all.
   b. Keenness of senses; dexterity?
      Good in both.
   c. Judgment and initiative?
      Fair in both.
   d. Ability to utilize mechanical methods, maps, buzzers, etc.?
      Poor in all.
   e. As night worker, runner, observer and verbal reporter.
      Good in all excepting the last: a poor verbal reporter.

When questioned as to their ability to maintain direction and to find their way about night or day in open country or in woods the almost invariable answer of many Indians interrogated was to the effect that they always felt quite sure of their direction.

The Indian ascribes his generally constant individual orientation to quick mental adjustment of terrain relationship resulting from habitual observation of wind, sun, moon, stars, landmarks, memory of country traversed and to knowledge of woodcraft.

Blindfolded and competitively in an intelligence section group under training in an open woods, Indians were the only scouts in the exercise able to reach previously indicated objectives 100 feet ahead. These were attained by careful crawling, feeling out the ground and growth along the course, and by memory picture of the intervening terrain. These tests were repeatedly made in intelligence training in the 36th Division.

Analysis of data covering reports on representatives of about forty tribes indicates that the Indian, apart from being a good soldier, possesses characteristics making him particularly valuable for scouting personnel.

He proves to be a good athlete, shows remarkable sense of direction, goes about his duties uncomplainingly, does not get lost, is a good runner, has unlimited patience and reserve, is a good shot, crawls habitually on night patrols, has non-light reflective countenance at night, is silent at work, stoical under fire, and grasps the significance and makes free use of signals.

In the battalion scout platoon the Indian soldier of average education may be trained to acquire facility in handling maps, mechanical methods, buzzers, etc. Here, too, he will receive the instruction given to all scouts, snipers and observers in the scout platoon.

The fact that the Indian has not shown facility with the above specified equipment of the battalion scout section is due largely to lack of sufficient training with these materials. The educated Indian soldier, given an opportunity to acquire knowledge of these devices, will prove able to handle them efficiently.

With respect to verbal message transmission, it may be said that in a Division where special exercises and training are given to develop ability to accurately transmit verbal messages, educated Indian soldiers showed up remarkably well in verbal message transmission tests.

If detailed to act as scouts and guides in problems involving attack, the example set by the Indians in making aggressive use of cover and low crouching advance will stimulate men of the ranks to assume similar methods of approach.

In war of movement Indians should be used as guides and scouts to orient the advance of platoons and platoon sections into attack. They should be detailed to this service from the battalion scout platoon.

The compass, upon which our officers now depend for direction in advance through woods and over open country necessitates frequent pauses for orientation, and during these pauses the officers are positive marks for enemy snipers. Indian scouts, moving from cover to cover guiding advancing groups and keeping direction without need of frequent reference to the compass will tend to hold direction for advancing units and avoid to a considerable extent the easy identification of officers and other advancing group leaders.

In war of position they should be used as guides for night patrols and as observers and snipers, and to guide troops into and out of trench positions at night.

Conclusion

The Indian is exceptionally qualified by natural characteristics and disposition as a scout for service in modern warfare.

Recommendations

That recognition of the scouting qualifications of the Indian be officially indicated with a view to having his services more generally made use of in the battalion scout platoons. Deserved recognition would stimulate the Indians now in the service and Indian cadets undergoing military training in the U. S. Government schools to show added interest in the Service and tend to provide excellent scouting personnel for the army.

Almost cryptic in its pointed brevity, this report embodied in a few sentences the pr celess results of years of sympathetic association with the red race, months of arduous labor on the part of Lieutenant Eddy. Although his invaluable services have been lost to the government both as a soldier and as an official in the Indian Service, his report is worthy of the most serious study and consideration. In it there is tersely epitomized the history of the suffering, the experience, and the splendid service of the American Indian in the ranks of democracy. May the nation profit by that record, of which the red man has every reason to be proud.* Who today shall say what was the color of the Unknown Soldier of America?

---

* Lieutenant Eddy died June 9, 1925. In 1921, the author, as a member of the Legislative Committee of the American Historical Society, delivered an address at the Carnegie Institute, Washington, on the Indian in World War I, embodying the substance of this chapter, which was also incorporated in a monograph furnished by him to the General Staff of the Army at the request of Major General Bryant H. Wells, Assistant Chief of Staff.

30

Indian Citizenship and Their Anomalous Politico-Legal Status

U<small>PON THE DISCHARGE</small> of the Indian servicemen from the reservations, many of them found no provision whatever for their return to civil life. In some cases the most outrageous injustices had been perpetrated in their absence. Like some other ex-servicemen they wondered if the patriotism which Lane had invoked with his ceremony just prior to the election of 1916, had not imposed a penalty.

By reason of the conspicuous part played by the Indians in World War I, however, it was inevitable that the question should arise, "Is the Indian veteran not as much entitled to citizenship as an Indian allottee?"

The political capital that was to be made out of enfranchising the Indian veterans had been made apparent by Lane. At any rate, ostensibly, at least, in recognition of the military service of the Indians, in 1919 Congress declared that every American Indian who had served in the military or naval establishments of the United States during the war against the Imperial German Government, and who had received or who should receive thereafter an honorable discharge, if not already a citizen, might if he desired, upon proof of his discharge, be granted by a court of competent jurisdiction, full citizenship without in any manner impairing or otherwise affecting his property rights, or those of his tribe. Neither party in Congress opposed, both supported the law. But it was decided not to reopen the Carlisle Indian School. Some of its graduates, unwilling to accept conditions on the reservations, had proved to be disturbing with their agitation of the tribes. A few, in fact, not unnaturally had been "bad actors."

It was now too that the Secretary of the Interior undertook in obedience to popular demands to expropriate as public land a tract owned by the Pueblo or the Navajo community of Santa Rosa in New Mexico. At once the Pueblo acting as a corporate unity sought to restrain him. The government argued that this could not be done by Indian wards. In this case which was carried to the Supreme Court, the Pueblo was sustained with a ruling that the remedy of injunction lay with it. "The existing wardship," declared Justice Van Devanter, "was no obstacle to the assertion by a tribal community of its constitutional legal remedies."*

Harding's Administration witnessed an increasing demand for oil; therefore, Albert B. Fall, the Secretary of the Interior, a rugged westerner who cared nothing for Indians, called on the Attorney General to pass on the legality of licenses to take oil from Indian reservations. Since the Indian right of occupancy had always carried with it the usufruct,† the Honorable Harlan F. Stone, now Justice of the Supreme Court, promptly rendered an able opinion against the licenses contemplated by Fall.

An increasing number of philanthropic institutions were now seeking to aid the Indians. Being indisposed to rest content with the mere enfranchising of the ex-servicemen, they too, like the Mohonk Conferences, were pressing for radical reforms. Lane's bold confession of the national guilt was recalled. The fact was cited that notwithstanding the annual boast of the government as to the condition of the Indians, over 100,000 according to its own figures, had disappeared since 1910, that on the reservations there were still Indians who spoke only their native language, no more fitted than were their aboriginal ancestors to enter the white man's society. It was also contended by some that the reservations in effect had become burial grounds for the red race. Harsh as such criticism may have been, it was manifest that if the sunlight of civilization were to be excluded from the reservations, what had been designed to be havens for the education and transformation of the race, could in fact be no more than dungeons for those whom the Congressional neglect was imprisoning.

Inasmuch as the Indians and their friends because of Lane's enterprise were disposed to credit a Democratic Administration with the Indian Veteran Franchise Act of 1919, it was seen by those in power that something must be done to meet the demands of the government's critics. Accordingly, Dr. Hubert Work, immediately upon succeeding Fall as Secretary of the Interior, appointed a committee composed of one hundred publicists, or persons of eminence supposed to possess more than ordinary interest in and knowledge of the Indians, to investigate the conditions existing among them, and to recommend to the government such reforms as were found to be desirable. At the same time Congress passed a Juris-

---

* Lane v. Santa Rosa Pueblo (1919), 249 U. S. 113.
† United States v. Cook, 19 Wallace 541.

dictional Act under which the Sioux might litigate their monumental claims. In 1923 the so-called Committee of One Hundred, rendered a report in which, after a full review of the Indian policy of the United States, no doubt was left as to the plight of the race. Part of it is here quoted.

The Continuing Problem

We shall not attempt to appraise the achievements for the Indian in the past nor to measure the mistakes or inabilities of the present. We find ourselves beset by many of the same problems which have faced the Government for nearly 50 years. Regardless of progress actually made, the great objectives of our benevolent desires have not been attained. This situation and this history show the extravagance of all efforts which are not directed by the best ability, supported by adequate funds, or maintained by sufficient consistency.

Early Ending of Government Activities

We believe that temporary increases in financial operating through the ablest men obtainable under just compensation would speed up the work of the Indian Bureau and result in large ultimate savings to the Government through an earlier ending of many if not all of the Government activities for Indian welfare.

Education

To this great end of achievement of good for the Indian and of ultimate cessation of the Government's paternal activities for him, we call attention to the paramount importance of education among all the functions of aid now properly committed to the care of the Government. We recognize that the quality and value of education is largely determined by the quality and character of the teachers. We urge with all earnestness upon the Indian administration, and particularly upon Congress, the granting of appropriations of education sufficient to secure, through largely increased salaries, teachers of highest ability and training competent to achieve the mighty task committed to their care.

A School for Every Child

The Government should assure educational facilities for each and every child among the wards of the Nation. Among these tribes suffering from insufficient schools the Navajo is particularly needy.

Admission to Public Schools

The public-school system of our country should be fully open to the Indian as an effective means of preparing him for good citizenship.

Help for the Ambitious

Because of the hindrances affecting Indian youth in their elementary education, they should not be debarred from Government aid when they become of legal age, but if possessed of desire and ability they should be allowed to pursue and complete the courses prescribed in the Government schools. Furthermore the Government should, where necessary, provide scholarships for able students who desire further education in high schools and colleges with a view of fitting them for positions of native leadership.

The Fullness of Life

The committee wishes to place on record its sympathetic approval of every effort to bring the wards of the Nation, whether that be by the incorporation of the genius of the Indian in music, literature, and the decorative arts, through instruction by the most competent in these lines in the several tribes, or by bringing the youth into the familiar phases of Caucasian civilization through a discriminating use of the outing system, or through some form of community service at the various schools which brings adult and youth into a sympathetic understanding of what life and living mean. In this connection this committee heartily commends the attitude of the Government, the Secretary of the Interior, and the Indian Commissioner in the encouragement that they give to the efforts of all religious denominations and sects to bring religion into the thought and life of the Indians.

Health and Sanitation

We recognize the national responsibility to combat such evils as tuberculosis, pyorrhea, and trachoma, and we favor the use of whatever means will quickly and effectively meet the situation represented by them. We urge earnestly upon Congress the appropriation of the sum of $100,000 asked by the Bureau of Indian Affairs for this purpose, and join with the Secretary of the Interior in his request that the National Health Council proceed immediately with its projected survey of Indian health. We further urge that every possible aid of State boards of health be enlisted in cooperation with the National Government in this health campaign.

Peyote

We urge that the National Research Council be requested to undertake an immediate and definite study of the effects of peyote and if this investigation shall show that this drug is fundamentally detrimental to the health and morals of its users, then Congress shall be asked to pass appropriate legislation to prohibit its use, sale, and possession.

Trained Doctors and Nurses

As a major part of the program for health and social welfare we urge upon Congress adequate appropriations for securing a sufficient number of trained physicians and nurses on the several reservations. We heartily commend the experiment of the Indian Bureau in securing trained workers in field matron positions and urge here also more adequate appropriations for this work so essential in improving health conditions in the individual homes.

Indian Dances and Ceremonies

The Committee of One Hundred cordially commends the substance and the spirit of the letter of February 24, 1923, and of Circular No. 1665, of the Commissioner of Indian Affairs relating to certain Indian dances and customs, and desires to make clearer the following positions with which it believes the Indian Office and the majority of the friends of the Indians are in accord.

First. That cultivation of all lawful ancient ceremonies, rites, and customs of the Indian race and the various tribal and community divisions of the race, is the privilege and liberty of the Indians. This is not to be curtailed or infringed.

Second. When any of these contravene the laws of the land, or the interests of morality, they are manifestly, by concurrence of Indians and whites, to be discontinued and discouraged.

Third. The encouraging of the characteristic native arts and crafts, and the development to the highest artistic and commercial values of Indian basketry, pottery, and blanket weaving, many of which are ceremonial and religious, and are to be thoroughly commended.

Fourth. The economic and social aspects of the question as referred to in the commissioner's letter and message should receive due consideration. And when elaborate ceremonies, prolonged dances, "give away" customs, and commercialized Indian shows and fairs interfere seriously with Indian welfare and interests, they should be discouraged,

Admission to the Court of Claims

This committee finds that the Indians of the country, unlike any other citizen or class of citizens of the Nation, are denied access to the Court of Claims of the United States, except in special instances under special enactments, to have their claims against the United States arising out of violations by administrative officers of the United States of their treaties or agreements reduced to final judgment, and this committee recommends the enactment to suitable legislation that will enable every group or tribe of Indians to obtain through the Court of Claims a judicial and final accounting with their guardian, the United States.

Executive Order Reservation

We recommend that the Secretary of the Interior suspend all departmental proceedings touching the sale or lease of oil, gas, or minerals on or from Executive Order Indian Reservations pending action by the Congress to vest the title of said reservations in the Indians occupying them.*

This was a terrible indictment of the Indian system by a large group which included some of the highest intellects of America. Such a group was, of course, not prejudiced, not apt to exaggerate. Moreover, it was only natural that the "Russian bureaucracy" whose early abolition was now recommended, should have continued to devote its energy to opposing the suggested reforms. All bureaucracies are the same in some respects. Losing sight of the purpose of their existence, they are not prone to self-elimination. In its view the Committee were but tyros. This itself is sufficient to condemn the Bureau. Meantime a policy of governmental economy had been instituted which forbade an endorsement of the Committee's recommendations by the Secretary of the Interior who instead actually recommended the reduction of appropriations for Indian welfare work. The net result of this last effort at reform was, therefore, the transmission to Congress of the Committee's resolution with comments by the Secretary of the Interior designed to offset the criticisms it contained.†

In August, 1923, Calvin Coolidge, of Vermont, the direct descendant of Chief Crawford, a Connecticut Indian, succeeded to the presidency upon the death of Warren G. Harding. Leases to Indian oil lands ended therewith. Though still Congress was not disposed to deal with the Indian

* "The Indian Problem: Resolution of the Committee of One Hundred . . ." presented by Mr. Snyder January 7, 1924. Government Printing Office Dec. No. 149, 68th Cong.

† Comments on the Recommendations of the Advisory Committee on Indian Affairs, Hubert Work, Secretary of the Interior, Pamphlet, Government Printing Office, 1924.

problem in a serious way, the Commissioner of Indian Affairs, the Honorable Chas. H. Burke, like Lane had been, was aware that there were sufficient Indians in Oklahoma, South Dakota, Wyoming, Montana, Arizona and New Mexico to determine the political complexion of those states in the next election. At any rate it was deemed wise to extend the franchise to all the Indians. Accordingly, on June 2, 1924, largely at the instance of the Commissioner, Congress passed the following law:

Be it enacted by the Senate and House of Representatives of the United States of America in Congress assembled: That all non-citizen Indians born within the territorial limits of the United States be, and they are hereby, declared to be citizens of the United States; Provided, That the granting of such citizenship shall not in any manner impair or otherwise affect the right of any Indian to tribal or other property.

Thus, 135 years after national citizenship was created, and fifty-seven years after it was extended to the Negroes, it was conferred upon the Indians as a race.

The mere enfranchising of the Indians, being no more than a political gesture, by itself could not, of course, better their lot in any material way. Moreover, it is doubtful if in the entire range of the Indian policy of the United States there is to be found a more ill-considered measure. Not only did it confer national citizenship upon all Indians without regard to individual capacity to exercise the rights incident thereto but, failing to relieve the competent Indians from the burdens of the old federal guardianship, it perpetuated an ill-defined politico-legal status for which in the law of human relations there is no precedent. Moreover, since tribal Indians residing on federal reservations were still not deemed citizens of any state, it added to the complexity of their status. Now was extended to them the peculiar form of citizenship which the allotment acts had first made possible. The old regulatory statutes inconsistent with citizenship such as the limitations on their right to sue upon the contractual obligations of a treaty, to recover interest as an essential element of just compensation, to employ counsel free of the dictation of the government were not abolished. Thus, these new citizens were left without the remedies available even to visiting aliens and the possibility of securing the willing aid of eminent and able counsel in contests with the government which at best are unequal. Was there any wonder that after the first furor of excitement over their enfranchisement, the Indians found that citizenship meant little to them but a name?

At the same time, largely at the instance of Commissioner Burke, Congress at last empowered the Court of Claims to hear the claim of the Yankton Sioux arising out of the misappropriation of the famous Pipe Stone Quarry.

31

Reaction and Continued Neglect

W HATEVER THE MOTIVES of Congress may have been, however sincere it was in its purpose to do the Indians honor, hardly had they been enfranchised when, obedient to the rigid executive policy of economy that had been inaugurated, an executive rule largely nullified their constitutional rights as citizens. Measures for their relief, including legislative acts necessary to confer jurisdiction upon the Court of Claims to pass upon their treaty rights, were now required to be approved by the Bureau of the Budget before they received the approval of the President! In other words, even the legal remedies which it was in the power of Congress to provide were made dependent upon an executive fiscal policy which clearly violated the prerogative of Congress. But most extraordinary of all, in 1924 while the Solicitor General of the United States was arguing before the Supreme Court that the Tejon tribe of California was possessed of a right of occupancy of which it could not be deprived even by Congress without its consent, an assistant to the Attorney General was arguing coincidently in another case in the Court of Appeals of the District of Columbia—the second highest court of the land—that where Congress saw fit in the exercise of its plenary power to take the property of the Karok tribe of California without compensation, no legal remedy was available to the Indians!

The explanation of such inconsistency is, of course, that in the one case where third parties were trespassing, it was to the advantage of the government to sustain Indian rights; in the other, where the government itself was the transgressor, to override them.

The truth is, the policy of the Department of Justice is in no sense that of a judge advocate between the tribes and the government. For instance, in 1925 the Six Nations caused to be instituted a test case designed to recover possession of a part of the St. Regis Indian Reservation which the state of New York had undertaken to acquire by treaty contrary to the law, and granted to private persons. Inasmuch as the fee in the property was in the United States, the Indians called in the Bureau of Indian Affairs for assistance. The matter was referred to the Secretary of the Interior who in turn referred it to the Attorney General for appropriate action. One statute provided that reservation Indians should not be ousted, another that they should be represented by the U. S. District Attorney in all cases affecting their reservations. After instructions were prepared for intervention, a representative of the Attorney General of New York appeared at the Department of Justice following which intervention was refused. Thus, the Indians were compelled to prosecute a case unaided by the government to recover from the present corporate holders land and water rights *prima facie* the property of the United States. Despite the statutes mentioned, and the fact the St. Regis Indians derived their right of occupancy from the treaty of 1796 with the Seven Nations of Canada, the lower court on a technical plea to its jurisdiction held that no federal question was involved. This it could not have held had the United States joined with the plaintiff and insured a hearing on the merits, or sought to recover its own property.*

The results of the reactionary policy described, are indicated by the following letter:

A committee meeting was called at Greenwood last Saturday, the 22nd, for the purpose of laying before you the awful conditions that some of the Indians are to face this coming winter.

As you will remember, we had a short crop last year, and many suffered from lack of food and fuel. In fact, all the willows that grew along the river and on the little islands are gone, and soft coal is selling for $14.00 a ton. The Reservation is going through one of the worst droughts since '94. Up until the 1st of July everything looked promising but we haven't had a decent rain since.

The Indians will have to do without their dried sweet corn as we haven't got a roasting ear.

I understand the Superintendent has written the Indian Office in the matter. I know he will do all he can.

Antelope and Standing Bull were strong in their speeches, Saturday, that now is the time that help is needed. Antelope thought it might be possible that two or three months of rations could be gotten.

I understand that credit has been shut off at the Agency stores, and as many

* *James Deere v. St. Lawrence River Power Co., et al. and the State of New York*, 22 F. (2d), 851; 32 F. (2d) 550.

342 THE RED MAN IN THE NEW WORLD DRAMA

Indians were living on the strength of share rents, you can imagine conditions with such a season.

Lawrence just returned from Cheyenne River and says the country was blessed with lots of rains, but the Indians were living on horse flesh.

A complaint is made that a resident doctor had been promised after July 1st and instead we have a contract doctor who is paid $1200 per year and figures he can't serve the whole tribe for that amount, and says he will serve just those who have no money. He'll find out that there are a whole lot without money. These are some of the things that I was asked to lay before you.*

(Always friendly, this tribe had received and entertained Lewis and Clark most hospitably during the winter of 1804. Upon the birth of one who was to become their head chief he was wrapped by Lewis in the American flag. The Yanktons came to honor and love that flag, and gave timely warning of the Minnesota Massacre of 1862. They took no part in the Sioux uprisings of 1876 and 1891 and have never raised a hand against the government. In 1917 they sent forth their young men in a body to fight side by side with their white brothers in France. Among them there was not a single case in which exemption from military service was sought. In 1925 they were peacefully litigating their rights in the Supreme Court against the United States looking to the recovery of the famous Pipe Stone Quarry. They were not complaining. They were not asking for charity. They were only asking for aid in a cruel struggle with the elements.

An appeal on their behalf to the Bureau of Indian Affairs was made in vain. No funds were available for their relief. It was intimated that if they had not been improvident in one way or another, particularly in the leasing of their lands to white farmers incapable of cultivating them properly, they would not be in their present plight. The aid of the Red Cross was then sought. The Director, the Honorable John Barton Payne, formerly Secretary of the Interior, tactfully arranged to have an investigation made by his agents into the condition of the Yanktons. This disclosed the fact that only about ten per cent of the tribe were destitute and in need of charitable aid.)

But what of a guardian who permits one in ten of his wards to perish whether through positive neglect of them on his part or their own improvidence?

If such conditions existed among the relatively advanced Yankton Sioux who for a generation had been tilling the soil, it was only reasonable to assume that they also existed among the less enlightened Indians.

Almost a century and a quarter after Jefferson penned the Declaration

---

* Letter dated August 26, 1925, from business committee of the Yankton Sioux to J. C. Wise, who in 1924 succeeded the Honorable Charles Evans Hughes as counsel for the tribe, and in 1925 was made an honorary chieftan (Wamdi Kasapa).

of Independence, there died in Weimar a mad Prussian. The philosophical child of Darwin and the political brother of Bismarck, in his relatively lucid moments he wrote much that mankind may well ponder. "The process of evolution involves the utilization of the inferior species, race, class, or individual by the superior; all life is exploitation, and subsists ultimately on other life; big fishes catch little fishes and eat them, and that is the whole story of life."*

The idea is not an uncommon one. Even an American statesman within the lifetime of those living declared with that "disgraceful tact which his people called 'honesty,' that the world is ruled by might and not right."†

"But," says Van Loon with terrible irony, "we go to our little white schoolhouses and our pretty white churches in Idaho and Wyoming and Montana and Nebraska, and we sing our hymns and thank a merciful Heaven that we are not like those tribal foreigners who publicly avow that the strong will inherit the earth and the fullness thereof, and that the weak will be deprived even of the little that is their own. We detest such an idea. Impatiently we shout 'No'! It may be true of others, but our own people, we feel sure, will never be guilty of such a thing."

Yet, here was that very thing going on in the good year 1925 just as surely as it had in New England in the days of John Winthrop, though now it was better concealed.

At this juncture resort was had to a scheme to arouse public sympathy for the Indians who were everywhere being represented by agents of the government as unreasonable. In March, 1925, a group of six Yankton Sioux chiefs, including Standing Bull, Hollow Horn and Antelope were brought to Washington and presented to several members of the Cabinet and the diplomatic representatives of eighteen countries at a private luncheon. The British and Spanish Ambassadors, and six Ministers were present besides General Miles, General Parker, General Bliss, the Adjutant-General and other senior Army and Navy officials, Senator Du Pont, John Hays Hammond—eighty persons of eminence in all. The Honorable James M. Beck welcomed the Indian guests whereupon Standing Bull and his compatriots responded with an oratory that amazed their hearers. As pre-arranged by the writer, or the host, they now told the true story of Custer's death in a way that explained vividly the Indian point of view. It was a dramatic incident—this appeal of the venerable Sioux chieftains for under-standing of their race. It was quickly followed up.

Now and then in the martyrdom of man an epical plea is made, so convincing of the woes of a people, of some great moral wrong being done them, that even the most callous will pause to harken to the cry that rises

---

* *Thus Spake Zarathustra*, Nietzsche, E. P. Dutton, Inc.
† *Story of America*, Van Loon, Hart Publishing Co., Inc.

to the high heavens. So it was in the case of Euripides' imperishable tragedy—*The Trojan Women*—which shook to its foundations the Greek civilization, introducing into the moral philosophy of his time a new humanism. Longfellow but reiterated in *Evangeline* the protest of Euripides.

Sometimes the wrong is so indisputable, the right so overwhelmingly on the side of the reformer, that even the most craven of men will not take issue. Then, the potential energy of the moral conviction which has been stored in the conscience of men translates itself into the moving energy of some great reformative act. Thus it was when Granville Sharp climbed over the side of an English slaver in the port of Greenwich with Lord Mansfield's writ of habeas corpus in his hand, leading directly to the early abolition of slavery in the British Empire. So it was later with Harriet Beecher Stowe's *Uncle Tom's Cabin*. A picturesque romance it was, and one well calculated to inflame the passions of men, yet it was compounded of pathetic though isolated truths, each of which was recognized by some man here or there as that which could not be denied before God. The travail of the Negro was a present fact. It was beyond the power of the most facile imagination to relegate his wrongs to the past. It was, therefore, but a question of time before the potential energy of the moral conviction which Wilberforce, Sharp and Clarkson had generated was to be translated by Garrison and Stowe, however violently, into the moving energy of abolition.

But in the case of the American Indians it has been different. In their case the reformer has been unsupported by any motive of political or economic gain. There being no one with whom to wage a conflict over the Indians, no factional passions were to be aroused by periodic revelations more dreadful than any made by the abolitionists. Therefore, in vain, Washington, Franklin, Monroe, Harrison, Houston, Grant, Schurz, Arthur and Cleveland had appealed to the conscience of the nation on behalf of a dependent people for whom the young American republic had assumed before God and the nations full responsibility, and to whom it owed far more than it did to the other race which it held in its bondage. Yet, no man stood forth to deny the appalling indictment brought by Helen Hunt Jackson in 1881 against the American people. Her *Ramona*, like her *Century of Dishonor*, was not to prove the *Uncle Tom's Cabin* of the red man. The oft-published speech of Logan, the old Cayuga chieftain, the heroic stories of Lamotachee, the Creek, and Osceola, the dauntless Seminole martyr, may have touched the conscience of the nation, but they had not moved it. No more had Zane Grey's *Vanishing American* which brought the dreary sordid story of wrongs to the Indian down to date. Still the smug complacence of the American people remained unshaken.

But why did the big fish continue with appetite unappeased? Surely not because the red man was still deemed the "spawn of Hell," nor because of any ill will for him, but simply because, looking at the Indian problem in a superficial way, the friends of the Indians were content to translate their good will for the race into indictments of the governmental agency charged with administrative responsibility for them. Upon a moment's reflection, however, it should have been obvious that railing at the Bureau of Indian Affairs, like recurring threats on the floor of Congress to investigate this or that commissioner, was utterly useless since these agencies, however bad, were but the expression of the system for which the American people themselves were responsible. Moreover, the many excellent philanthropies which had been organized to alleviate the Indians' lot were not striking at the root of the evil but contenting themselves with applying the balm of their charity to the surface sores of the race. Each had a hobby, its own scheme. Attempts to coordinate their efforts were futile.

But were the Indians to be left forever in the cycle of Congressional indifference?

Was the nation to continue to pour its wealth and aid into Armenia, Russia, Turkey, China and Japan, leaving its own Indian citizens to starve upon inhospitable and blighted reservations, still ignorant even of the language of their guardian government and those with whom they were compelled to contend?

Were foreign policies, battleships, submarines, airplanes, road-building, harbors, drainage, irrigation schemes, canals, post offices, and the further economic developments designed to benefit the white man to take precedence over the duty of the nation to these helpless people?

Such were the questions which arose in the minds of those who had not despaired of reforming the system responsible for the continuing ills of the Indians, and who saw that something must be done and done at once if the remnant of the red race was to be saved. Accordingly, appeals were now made by the writer to most of the great private philanthropies—Carnegie, Rockefeller, Scheppe, Wanamaker—to whom the opportunity to create a great foundation for the betterment of the Indians was pointed out. What a wonderful thing it would be to bring about those reforms in the law and in the system of governance to which an entire race—300,000 helpless people—were subject, this not as a matter of charity for them but for the mutual benefit of both the red and white races; an institution that would advocate with compelling influence the amendment of obsolete and the enactment of new laws, the coordination of the several departments of the government dealing with the Indians, including the Bureau of Indian Affairs, the Land Office, the Department of Justice, and the Smithsonian Institution; that would discourage unnecessary litigation, insure competent counsel when needed, exercise a general surveillance of charities, encourage the more worthy aspirations and traditions of the race.

But the Indian problem was everywhere deemed essentially political, and even the founder of the Rodman Wanamaker Historical Expeditions feared to become involved in it. "Go to the Bureau of Government Research," was the general reply. This was done but it had many tasks ahead of it. Such was the situation when in 1925, as the basis of a better organization of public opinion, a plea for the Indian citizens of the United States was presented at the opening session of Congress. After the anomalous situation of the Indians had been fully set forth, the appointment was urged of a commission, whose membership should include a minority of Indians, to investigate and report upon the facts with recommendations to Congress.*

"A copy of this document should be in every household in America," said the honorable Senator who presented the plea,† while ten thousand copies were printed at the expense of another Senator, and circulated under his frank.‡

Everywhere the proposal was endorsed. College presidents, statesmen, publicists, scholars, editors, and all the leading Indians gave it their unqualified approval. Not since the time of Wendell Phillips, it was declared, had such a plea been made for a dependent race.

"It offers," said Philip Alexander Bruce, the historian, "one of the most striking examples known to me of that burning feeling of indignation which Swift summed up in the famous words—'Saeva Indignatio.' It is a remarkable document. . . ."

"This plea," said Senator Bruce, "is borne out, on the whole, by every thoroughly dispassionate observer who has ever had a word to say about the relations of the whites in the United States to the Indian. No less a person than Benjamin Franklin . . . expressed the opinion that Indian outrages were but the natural sequels of the wrongs done the Indian by our race. . . ."

"I am in thorough sympathy," wrote John Grier Hibben, President of Princeton Univerity, "with the project to petition Congress to create a Commission to study the Indian problems in a scientific manner, and suggest to Congress the laws necessary to relieve the present chaotic condition of the Indian situation."

"However strongly I speak," wrote Ellen Glasgow, "I cannot speak strongly enough on this subject. It is not only pitiable, it is deplorable that this impoverished people should be forced to spend its substance in contending for its Constitutional rights."

"Without exception," declared Joseph W. Latimer, the noted Indian

* "Plea for the Indian Citizens of The United States," by Jennings C. Wise, spread on the Congressonal Record, Dec. 15, 1925, at the instance of Senator William Cabell Bruce, of Maryland.
† The Honorable William Cabell Bruce, of Maryland.
‡ The Honorable Thomas F. Bayard, of Delaware.

philanthropist, "I consider this the most effective argument for the free-dom of the American Indian ever written," while Zitkala-Sa, the noted Indian authoress and welfare worker, said: "The proposition of a Com-mission composed of the ablest obtainable men, including Indians, to make a survey and offer their recommendations, has a strong appeal to me. May I add that these men be chosen, strictly eliminating politicians, white and Indian. I have no special favor to ask for the Indian people, only that intelligence rather than political expediency be used in Indian affairs."*

Finally, public opinion was well summarized by the *Scientific American* in the following words:

Plainly the old political system must be revised. How should this be done? Not by the passage of new and disconnected statutes. The problem must be solved in an intelligent and scientific way. First, the needs of the Indians, which vary greatly with localities, must be studied. This can only be done, as pointed out by President Cleveland, by a body of able, unprejudiced men, similar to the Dawes Commission of 1893. On the Commission should be placed only men of vision and ability—trained sociologists, economists, and political scientists. It should include several Indians in order to afford to its white members a lens of understanding, through which the true nature and aspirations of the Indian may be viewed in a sympathetic way. Upon the findings and recommendations of such a jury alone can Congress devise the necessary laws to put into effect a policy designed to solve the Indian problem. Until the information that such a commission can acquire is forthcoming, let the well-wishers of the Indians direct their efforts solely to the support of its work and cease to urge upon Congress special legislation that alone cannot accomplish the emancipation of the race. For that an entire new and well coordinated political system is necessary.†

From all this it was manifest that the heart of the country was in the Indian cause so that appeals were now made to the more important uni-versities to include a study of Indian Affairs in their courses of political science and sociology.‡

Voltaire declared that nothing enfranchises a people like education; that when once a people begin to think it is impossible to enslave them. Perhaps it was for these reasons that the Bureau of Indian Affairs did not undertake to extend its educational work among the Indians. At any rate, the plan of disciplinary control which it now undertook to substitute for the education which the Committee of One Hundred had recommended was, to say the least, unusual. Most certainly it could but retard the eman-cipation from its control of a people already enfranchised.

* Robert Lansing, President Farrand of Cornell, John Hayes Hammond, Fairfax Harrison, Charles Warren, also were among those who endorsed the plan.

† *Scientific American*, January, 1926.

‡ Harvard, Yale, Princeton, Cornell, University of California, Stanford University, and others.

For many years there had existed what are known as reservation courts of Indian offenses—petty disciplinary tribunals administered by lay appointees of the Bureau. In the law there was no sanction whatever for these disciplinary agencies except where the Indians had voluntarily submitted to them in treaties or executive agreements providing that they might exercise through such means a control over the conduct of their own people. The so-called judges were ordinary Indians whose stipend was $10 a month, which alone indicates the capacity of those who presided over the courts. These courts were, of course, no part of the judiciary system of the United States or of the states, nor could Congress itself, much less the Bureau, confer upon them powers in excess of the Constitutional power of the federal judiciary since judicial power cannot be conferred upon an administrative agency. Nevertheless, it appeared that the Bureau of Indian Affairs had taken advantage of the situation to erect upon the foundation of the tribal courts what was in effect a judicial department of its own!

An Indian citizen who had been imprisoned under the sentence of one of these courts now had the temerity to sue on a writ of habeas corpus, and established in the proceeding the lack of power on the part of a reservation court to commit him to penal servitude. To this interference with its system the Bureau had no idea of submitting. Consequently, in 1926, it brought forth a measure which, with the unqualified endorsement of the Secretary of the Interior, was introduced in the form of a bill by the Honorable John W. Harreld in the Senate, and the Honorable Scott Leavitt in the House, Chairmen of the respective Committees on Indian Affairs.* To the everlasting shame of its proponents Section 2 of this iniquitous bill read as follows:

The reservation courts of Indian offenses shall have jurisdiction, under the rules and regulations presented by the Secretary of the Interior, over offenses committed by Indians on Indian reservations, for which no punishment is provided by Federal law. Provided, that any one sentence of said courts shall not exceed six months' imprisonment or labor, or a fine of $100, or both.

Thus, it was proposed in the good year 1926 to sentence citizens of the United States to penal servitude, contrary to the Constitution, without presentment or indictment before a grand jury, trial by jury, the sworn testimony of confronting witnesses, or due process of law in any other respect, solely at the will of a layman whose knowledge of and capacity to administer justice is indicated by his stipend. And this for offenses expressly declared not to be punishable offenses under the law of the United States!

It was, of course, beyond the power even of Congress to endow such an

* H. R. 7826, 69th Cong., 1st Sess.

agency with power to deprive a human being of liberty since the judicial power of the United States is limited by the Constitution to the punishment of offenses against the United States and cannot be enlarged by Congress. Nevertheless, the measure was urgently supported by the Bureau of Indian Affairs. Unbelievable as it may seem, in the extended hearings before the Committees of Congress, it was developed that the reservation courts of Indian offenses had long been accustomed to impose penal sentences and on occasions had placed culprits in chains.* A search of the laws of other civilized countries will disclose but one similar tribunal in existence, and that in French African Congo among the bush Negroes. It is doubtful, however, if even there such a tribunal exerts a civilizing influence so that the Bureau's bill was finally rejected. Meantime, however, the great Sioux claims filed under the Act of 1920 had been sidetracked, more or less indefinitely, while the Court of Claims again overruled the Yankton claim to the racial shrine, or the Red Pipe Stone Quarry. On the other hand, during the 69th Congress a bill was also introduced designed to give the oil interests licensed access to Indian reservations despite the opinion of Attorney General Stone in 1924, during Secretary Fall's incumbency as Secretary of the Interior, holding such licenses unconstitutional.†

Manifestly, even if the Attorney General were wrong, and the oil within their reservations were not subject to the usufruct of the Indians, agriculturists and drovers could not wrest from their holdings a livelihood if the surface of their farms and ranches were turned over to oil operators.

For years it had been customary to charge the trust funds of Indian tribes with the cost of governmental inspections of their property. But it also developed before the 69th Congress that large amounts had been habitually charged by the Bureau of Indian Affairs against Indian tribal funds without the knowledge or consent of the tribes, as arbitrary contributions to the construction of bridges, highways and other public works shown to have been of no practical benefit to the Indians, including the famous Navajo bridge which the Governor of Arizona testified not a score of Indians would cross in a year.‡

Such things as these were possible only because the Indians were disintegrated as a race, untutored, poor, and patient. They had learned by a

---

* Hearings, H. R. 7826, 69th Cong., 1st Sess., Congressional Record, March 4, 1926; March 23, April 23, 1926; Speeches of Honorable James A. Frear, in opposition. For his uncompromising stand against this vicious measure efforts were made to remove Mr. Frear from the Indian Committee. Not being in tune with the system he was branded as a radical. When the nature of the bill was exposed Senator Harreld was quick to disclaim responsibility for it.

† H. R. 9133, Congressional Record, March 4, 1926; speech of the Honorable James A. Frear, in opposition.

‡ Ibid.

century and a half of sorrowful experience that their rights were deemed
by the government to be more or less on a parity with those of the
buffalo—to be ignored when they stood in the way of the white man's progress. "It is useless," said a Senator, "for an individual or any small group in
Congress to fight the Indian System. It is not a party matter. It is one thing
in which an overwhelming Congressional sentiment is united. Indian matters are looked upon as exclusively within the non-partisan disposition of
Congress. Only when the Indian wields his vote effectively will the system
be reformed."

Here, indeed was the key to the solution. It was only too true. The
Indian citizens must make their political influence felt by Congress. Accordingly, it was now decided to furnish them with a leadership which
might fan the last ember of hope among them. Followed the organization
in 1926 in Washington of the National Council of American Indians,
representing many tribes. The Indian authoress and welfare worker, Red
Bird, or Zitkala-Sa, a Yankton Sioux and granddaughter of Sitting Bull,
was made its president.

At once resort was had to the old processes to discredit the Council.
Nevertheless, the declaration of the United States to Great Britain in the
Cayuga Case was to prove a boomerang. As a last resort, under the
provisions of the First Amendment of the Constitution, guaranteeing the
right to petition for redress of grievances, early in 1926 the Council
presented to Congress a petition in which, after referring to the statements
of the Secretary of State, it said:

Are these solemn international declarations to be given the weight of truth?

Or will the Government go on talking two ways, and tell our people when
they call for the fulfillment of its pledges that it meant one thing to the Indians
and another to the King of Great Britain and the world at large?

Our people speak but one language. They may be ignorant of all the ways
and cultures of mankind other than their own, but the language they speak to
those with whom they have smoked the peace pipe has but one possible
meaning, and in the simple candor of their nature they have never sought to
give it another. Untutored in the reservations of governments which speak two
languages, they cannot love or respect a government that speaks more than one
to them. Yet, although they ask for nothing that the Government of the United
States has not in its might voluntarily declared belongs to them, when they
appeal to the Government it tells them Congress would never pass a bill giving
them that much, and it is useless to ask Congress for their own property.

If that much, along with the protection of its laws and the human liberties
guaranteed all men by its great writing, the United States will not yield to
them, they can only regard its declarations and the decisions of its courts as
meaningless, and the citizenship that has been conferred upon them as a sham
to increase their liabilities without in fact according them the rights of human
beings, much less those of citizens.

In presenting this petition to the Senate of the United States on behalf of the Indian citizens of the United States the National Council of American Indians assumes a responsibility for which it must answer both to the Senate and to the Indian citizens. The justice of its demands is its answer.

The council has but one purpose, the organization of a constructive effort to better the Red Race and make its members better citizens of the United States. These objects it cannot attain unless the Indians are accorded the rights essential to racial self respect and a spirit of loyalty to the United States. It is for that reason alone that it presents their grievances.

The council is well aware that in the laudable effort which it proposes to make it will not have the encouragement of certain agencies of the Government. Fearing the power that comes of union, even now agents of the Government advise the tribes not to join it. The blind support of things as they are is made the one test of loyalty to the United States. Those who do not indorse this idea or that or seek to prove wherein one of them may be wrong, even though the program for the Indians includes their imprisonment and the administration of their estates without due process of law, are branded as malcontents. Such is ever the case in a struggle between progress and the forces of bureaucratic action. We know that it is only to be expected that those forces will persist in that diplomacy which proved so effective in breaking up the Indian National Confederacy and which has ever since kept the Indians disrupted. But though there may be malcontents among the Indians the National Council of American Indians will not lend itself to an attack upon anyone or any agency of Government that does not stand in the path of progress. Those who seek to resist legislation designed to benefit the Indians will themselves create the opposition.

Until now our people have been interpreted by an agency or Government that by its own program has shown that it is entirely out of sympathy with them. Against that agency we make but one charge, and that is that it is no different from any other bureaucracy. It was inevitable that it should become inflexible; that in its effort to sustain itself it should have largely forgotten its true purpose—to emancipate the Indians from the guardianship committed to it and thereby render its own function unnecessary.

There are fundamental characteristics of human nature which may be denied but which cannot be destroyed. In seeking to overcome the inertia of the present Government we do not attack individuals but a system which we believe must be reformed, and no fear of the temporary loss of favor will deter the council from voicing the legitimate aims and aspirations of our race.

We do not pretend to say how all the existing evils with respect to the Indians and Indian affairs are to be corrected, and the Indian problem eventually solved. We do know, however, that our race is entitled to the redress of its grievances and relief from its present intolerable situation; that it is not charity that it requires, nor the overhasty distribution of its estate, but adequate education, practical guidance in the utilization and enjoyment of its property, personal liberty commensurate with the dignity of a free people, and the fair and efficient administration of their estate by the guardian-trustee thereof, and a clarification of the multiplicity of laws dealing with them and their property.

This, then, is our program and one in support of which reason must unite all Indian citizens and their well-wishers.

With these conditions of life assured to them, the Indians who remain and their posterity will take their place in the social and economic life of the Nation just as our young warriors took their place in the embattled ranks of 1917.

We ask the Senate to decide in all fairness if there is anything in such demands that should arouse opposition to our aims; that can justify for us the brand of malcontents which reactionary governmental agents would place upon us.

A time there was when the protest of our race against injustice was voiced in the war cries that rose from the primeval forest. No less audibly shall this protest resound through the hills and vales of our Fatherland, echoing the far-carrying appeals of justice and reason, never to be silenced until the pledge of the Nation, made to us by the Great Grandfather, and sealed by our blood on the fields of France, is redeemed.*

One of the principal difficulties to be encountered in any effort to secure cooperation among the tribes is they have been preyed upon and deceived so long they have little faith in anyone. Therefore, it is always easy for hostile interests to set one working against another. Accordingly Congress simply ignored the threat contained in this petition. Therefore, with the aid of funds contributed by several friends—namely, the late Senator Henry A. Du Pont, John Hayes Hammond, the Honorable Perry Belmont, Mrs. Anne Archbold, Henry W. Anderson, of Virginia—at the writer's instance the Council dispatched representatives to the West during the summer of 1926 to organize the Indian voters. The result was, in the November elections Senator Harreld, long Chairman of the Senate Committee of Indian Affairs, was defeated in Oklahoma, while the Congressional ticket supported by the Indians in South Dakota was elected. Thus did the Indians, forced into politics at last, manifest the power with which the politicians unwittingly had but recently endowed them. The cry which now arose from the defeated candidates was long and loud!

Immediately after the Congressional elections of 1926 the Institute of Government Research, a nongovernmental agency, was requested by the Secretary of the Interior to prosecute an independent survey of the economic and social conditions of the Indians.

During the summer of 1927 President Coolidge established the "Summer White House" in the Black Hills among the Sioux where all the honors which it was possible for the Indians of the West to confer were

* Petition of the Nation Council of American Indians to the Senate, printed in Congressional Record, April 24, 1926, at the request of the Honorable Thomas F. Bayard.

bestowed upon him. There he saw their true condition. It was during the autumn of 1927 that in a noble opinion Justice Sutherland handed down the unanimous decision of the Supreme Court reversing the Court of Claims in the case of *Yankton Sioux v. The United States* and holding that the tribe was entitled to just compensation for the appropriation of the Pipe Stone Quarry. Moreover, it was also held that "just compensation" included interest on the value of Indian property from the time of its taking. Thus, at the end of a struggle which had continued seventy-six years and litigation which had lasted forty-three years, the original contention of the Sioux was upheld though meantime they had been irrevocably divested of their shrine.*

In his annual message the following December, although the President made reference to Indian needs, he recommended a continuation of the policy already adopted. So that the National Council of American Indians with reason could not conclude that much was to be expected of the Administration. Therefore, it appealed again to the Indian Committees who were now alive to the importance of the Indian vote.

Straightway, there was introduced by Senator King, of Utah, Democrat, who had refused to present to Congress the petition of the Council, a resolution calling for a full investigation of Indian Affairs. In the preamble the plight of the Indians was even more fully described than by the Committee of One Hundred, or in the petition of the Council.

Whereas there are two hundred and twenty-five thousand Indians presently under the control of the Bureau of Indian Affairs, who are, in contemplation of law, citizens of the United States but who are in fact treated as wards of the Government and are prevented from the enjoyment of the free and independent use of property and of liberty of contract with respect thereto; and

Whereas the Bureau of Indian Affairs handles leases, and sells Indian property of great value, and disposes of funds which amount to many millions of dollars annually without responsibility to civil courts and without effective responsibility to Congress; and

Whereas it is claimed that the control by the Bureau of Indian Affairs of the persons and property of Indians is preventing them from accommodating themselves to the conditions and requirements of modern life and from exercising that liberty with respect to their own affairs without which they cannot develop into self-reliant, free, and independent citizens and have the rights which belong generally to citizens of the United States; and

Whereas numerous complaints have been made by responsible persons and organizations charging improper and improvident administration of Indian property by the Bureau of Indian Affairs; and

Whereas it is claimed that preventable diseases are widespread among the

---

* For facts of this celebrated cause see *U. S. v. Carpenter*, III U. S. 350; *Yankton Sioux v. U. S.*, Case No. 31253 Ct. Claims; and same title, 270 U. S. 637; 272 U. S. 351.

Indian population, that the death rate among them is not only unreasonably high but is increasing, and that the Indians in many localities are becoming pauperized; and

Whereas the Acts of Congress passed in the last hundred years having as their objective the civilization of the Indian tribes seem to have failed to accomplish the results anticipated; and

Whereas it is expedient that said Acts of Congress and the Indian policy incorporated in said Acts be examined and the administration and operation of the same as affecting the condition of the Indian population be surveyed and appraised: Now, therefore be it

Resolved, That the Committee on Indian Affairs of the Senate is authorized and directed to make a general survey of the condition of the Indians and of the operation and effect of the laws which Congress has passed for the civilization and protection of the Indian tribes; to investigate the relation of the Bureau of Indian Affairs to the persons and property of Indians and the effect of the acts, regulations, and administration of said bureau upon the health, improvement, and welfare of the Indians; and to report its findings in the premises, together with recommendations for the correction of abuses that may be found to exist, and for such changes in the law as will promote the security, economic competence, and progress of the Indians.

Said committee is authorized to send for persons and papers, to administer oaths, to employ such clerical assistance as is necessary, to sit during any recess of the Senate, and at such places as it may deem advisable. Any subcommittee, duly authorized thereto, shall have the powers conferred upon the committee by this resolution.*

Almost coincidentally with the passage of this resolution, the Bureau of Government Research, having completed an investigation extending over two years, rendered an impartial report which took the matter of government neglect out of the field of speculation. No longer could the real issue be submerged in feuds between the government and Congressional investigators, or whitewashed by both parties.

The Secretary of the Interior said in his annual report for 1928:

This was the most thorough and comprehensive survey of Indian affairs ever undertaken. The report, which is entitled "The Problem of Indian Administration," comprises 872 pages. It contains many constructive suggestions and recommendations for the betterment of all branches of the service. The officers of the department have been studying it intensively, and some of the recommendations have already been incorporated in the Indian program.

As the inadequacy of the educational system for the Indians was one of the reasons for the department's request for the survey and report, the following summary of the findings of the investigators on this subject is of especial interest:

The survey staff finds itself obliged to say frankly and unequivocally that

* Senate Resolution 79, 70th Congress, 1st Sess., referred to Committee of Indian Affairs December 17, 1927.

the provisions for the care of the Indian children in boarding schools are grossly inadequate.

The diet is deficient in quality, quantity, and variety.

The great protective foods are milk and fruit and vegetables, particularly fresh green vegetables.

The diet of the Indian children in boarding schools is generally notably lacking in these protective foods.

The boarding schools are overcrowded materially beyond their capacities.

The medical service rendered the boarding-school children is not up to a reasonable standard.

The medical attention given children in day schools maintained by the Government is also below a reasonable standard.

The boarding schools are supported in part by the labor of students.

The service is notably weak in personnel trained and experienced in educational work with families and communities.

With this open confession of conditions in hand, the Council next appealed indirectly to Herbert Hoover, Secretary of Commerce, and prospective nominee for the presidency on the Republican ticket, who was much impressed by the need of effective reforms. With his approval there was incorporated in the platform upon which he and Charles R. Curtis, the grandson of a Caw Indian, were to stand for election as president and vice-president, respectively, the following unequivocal pledge to the 300,000 Indian citizens of the United States:

Our Indian Citizens

National citizenship was conferred upon all native-born Indians in the United States by the general Indian enfranchisement act of 1924. We favor the creation of a commission to be appointed by the President, including one or more Indian citizens, to investigate and report to Congress upon the existing system of the administration of Indian affairs and to report any inconsistencies that may be found to exist between that system and the rights of the Indian citizens of the United States. We also favor the repeal of any law and the termination of any administrative practice which may be inconsistent with Indian citizenship, to the end that the Federal guardianship existing over the persons and properties of Indian tribal communities may not work a prejudice to the personal and property rights of Indian citizens of the United States. The treaty and property rights of the Indians of the United States must be guaranteed to them.*

* This plank was drafted by the author and approved by the National Council of American Indians, whereupon it was offered by Henry W. Anderson, member of the Sub-committee of the Committee of Resolutions, of which Senator Smoot was chairman, and unanimously adopted with no material change.

Immediately after President Hoover's election the Senate Committee on Indian Affairs instituted the investigation for which provision had been made in the King Resolution the preceding December, the first hearing being held November 12, 1928. The facts which at once began to come to light were appalling.*

With the retirement of Hubert Work as Secretary of the Interior and Charles H. Burke as Commissioner of Indian Affairs, Ray Lyman Wilbur and Charles J. Rhoads were appointed by President Hoover to succeed them. Both were peculiarly fitted for their great tasks. Their appointment was hailed as an omen of promise to the Indians. At once they began reforms in the Bureau of Indian Affairs, especially with respect to the old bureaucratic ideals prevailing there, which quickly gave it an entirely different tone. As declared by Secretary Wilbur in a series of articles explanatory of President Hoover's enlightened policy, it was his desire that the Bureau of Indian Affairs should "work itself out of a job."

But the Administration soon found itself in no position to fulfill its promises. The Great American Depression doomed the new program. There were no funds available for innovations and the Meriam Report languished for want of ability of the Bureau of Indian Affairs to institute reforms it had suggested. Nevertheless some of its recommendations were effected. Indian education which had received severe criticism was revitalized by the creation of day schools on reservations where they were feasible thus eliminating the traditional boarding school as the only institution for Indian primary and secondary education. It remained for the new Administration of Franklin D. Roosevelt to develop a new Indian policy.

---

* "Survey of Conditions of the Indians in the United States," 70th Cong., 2nd Sess., S.R. 79, printed hearings.

32

The New Deal and the Indian Reforms

THE ADMINISTRATION of Franklin Roosevelt began with ominous portents since his campaign for President had stressed fiscal responsibility and differed little from previous promises to Indians made by other candidates in the years since Indians had received the voting franchise. Roosevelt was sincerely interested in assisting the Indians, however, and his years as President were probably the best years in American history for Indians. Reform that had appeared promising in the 1920s suddenly bloomed forth when Roosevelt appointed John Collier as Commissioner of Indian Affairs. Collier had been a well-known anthropologist specializing in Indians and led in the struggle over the Pueblo lands by opposing the Bursum bill which sought to deprive the New Mexico Indians of their lands. Collier was directed to create a new Indian policy and to develop the necessary programs to put it into effect. He did his job with devotion and spectacular results.

Realizing that the assimilationist viewpoint had created numerous failures whenever it had been written into programs for Indians, Collier developed the concept of cultural survival as a means of bringing the Indian tribes into the modern era. In 1934 he introduced the idea of reservation self-government via the Wheeler-Howard Act which was officially known as the Indian Reorganization Act. The I.R.A., as it became popularly known, recognized Indian institutions as having a valid place in American life. It provided for self-government by allowing the reservation people to organize federal corporations, adopt constitutions and by-laws to govern themselves, and to embark on economic development projects to produce tribal income to operate these corporations. Originally

Collier advocated the idea that all allotments could be traded in for shares in a tribal development corporation with each person receiving shares in the corporation comparable to his land holdings on the reservation. This device was an effort to solve the complicated land problems that had resulted from allotment of the reservations some half-century before.

In 1887 and succeeding years the vast tribal estates had been divided into small farming plots and the remainder had been declared surplus, sold to the government, and opened to white settlement. As each of the original allottees had died, their lands were thrown into a complicated heirship status since few of them left wills and the interests in the estate required that at least symbolically the lands that they had owned should be divided according to certain formulas. Over the years with additional deaths, births and marriages the allotments came to be owned by an increasing number of Indians. Some tracts were owned by as many as a hundred Indians, none of them owning more than a symbolic fraction of the land or the income derived from it.

The plan was rejected during hearings on the bill and when the Indian Reorganization Act was passed into law no provision was made to solve this problem. Instead a revolving loan fund was established from which tribal governments could borrow to repurchase allotted lands to rebuild their reservation land base. This program did not prove satisfactory, however, since Congress never did appropriate sufficient funds to make program more than a token gesture at land reform.

For those tribes who accepted the provisions of the Act and organized tribal governments under it, further allotment of Indian lands was prohibited. This section saved the lands of some tribes that had not yet been allotted but caused other tribes to reject the law so that they could continue their allotment programs. Not only was allotment stopped but the constant necessity of petitioning the Secretary of the Interior to extend the period of trust was remedied by making all lands of I.R.A. tribes, as those who accepted the Act came to be called, indefinite trust lands until changed by Congressional directive. Prior to the Indian Reorganization Act, as the original twenty-five-year trust periods came to an end, Indians had to petition the Secretary of the Interior to continue the trust status of their lands. This preserved the tax exemption on them and gave the Indians some protection against white encroachment and unwise land sales to speculators.

More important, perhaps, for the long-term effect on Indians was the provision legalizing Indian religions on the reservations. Shortly after the tribes were placed on reservations the native religions were banned. Holy shrines, as we have seen in the case of the Yankton Sioux, were destroyed or confiscated. All practice of traditional social customs was forbidden. In destroying the religion the government felt that it would be easier to solve the "Indian problem."

But the native religions did not die out. The rites and ceremonies were passed down by word of mouth, practiced secretly in the remote regions of the reservations, and wherever possible the sacred medicine bundles, pipes and other religious objects were kept hidden. Collier recognized that religious sensitivity in communities bound them together more firmly than any other factor. He had fought side by side with the Pueblos, the most traditional of Indians. He knew well the religious orientation of the Taos Pueblo toward its sacred Blue Lake area and the part that land played in Indian beliefs.

Incorporated in the Indian Reorganization Act was a provision allowing the practice of native religions on an equal basis with the Christian religions that had been superimposed on the different tribes half a century before when the respective reservations were allotted to the various missionary societies. This provision caused great concern among the Christian churches who saw it as a backward step in the great task of civilizing the Indians whom they now considered their special religious wards. Much opposition to the new program was stirred up on the reservations by missionaries who viewed the move toward self-government and religious freedom as direct threats to their own programs of changing the Indians to meet their image.

A special preference in hiring Indians for positions within the Bureau of Indian Affairs was also made part of the legislation. The Indian Bureau had been the exclusive domain of non-Indians for nearly a century. With the exception of a handful of well-educated Indians such as Carlos Montezuma, an Apache, and Charles Eastman, a Sioux, both doctors, few Indians had been employed in government service. If the Bureau of Indian Affairs was to "work itself out of a job" as both the Hoover and Roosevelt administrations had advocated, then it was apparent that the Indians would have to be prepared to provide whatever services their own communities desired. The Bureau of Indian Affairs was visualized as the training ground of future tribal and governmental leaders and administrators.

The New Deal also had the insight and wisdom to include Indians in its regular programs of economic recovery that had been created for its white citizens. On many reservations Civilian Conservation Corps camps were established and Indians worked alongside non-Indians in the various projects under this program. From this beginning a number of tribes soon established buffalo and antelope herds and restocked their tribal lands with game for hunting.

Part of the New Deal program was the construction of public buildings as a means of providing employment. Most of the Indian reservations received new schools, hospitals, roads and administrative buildings. For many tribes it was the first time that any significant capital investment in community facilities had been made. Specific funds were set aside for home improvement plans and a goodly number of family homes were built for people who had previously lived in tents and leanto shacks. The tragedy was that these houses, many of them established as temporary shelters, proved to be permanent homes when the programs were cut back during World War II.

In order to encourage economic development of individual Indians the government set up large tribal herds from which individuals could borrow cattle to begin their own herds. When an Indian's herd had increased to the point that he could return the cattle borrowed he was then free to pursue ranching operations on his own. A number of Indian people on the larger northern plains reservations in Montana and the Dakotas achieved some measure of financial independence under this program. But more tragic consequences also attended it. Whenever some Indian killed a head of borrowed cattle to eat (and many Indians were starving), then it became an act of destruction of government property and the Indian was liable for a criminal charge.

Nearly two thirds of the Indians on federal reservations chose to organize under this legislation. The new governments were not, in most cases, comparable to the traditional governments which the tribes had used. Of the tribes rejecting the Indian Reorganization Act, most did so because the provisions appeared to forbid traditional chiefs and headmen. While some were able to work their chiefs into the new I.R.A. structure, the majority of tribes wishing to keep their traditional leaders drew up constitutions and by-laws independent of I.R.A. and had these approved by the Secretary of the Interior. The I.R.A. had proven one thing and that was that Indian tribes had to formally organize themselves in new corporate forms in order to deal with modern society. It was not possible to return to the old days.

Within a few years the Indians had shown the I.R.A. to be a very successful way of handling their problems. Large numbers of young people attended college or began work in the Bureau of Indian Affairs under the new Indian preference hiring system. Tribal governments began substantial programs in land consolidation and economic development that were, for the times, advanced and sophisticated. The traditional religions, now equally protected in their practice with the Christian denominations, flourished. The reservation became less a prisoner-of-war camp and more a home. The greatest days of Indian life in the twentieth century were, strangely enough, in the midst of the greatest depression this nation had ever experienced.

But the horizon darkened quite rapidly. Reactionary Republican senators who viewed the innovations of the New Deal with suspicion were not long in finding a communistic attitude lurking within the new tribalism. Both the Senate and House exercised a continuous overseeing function on Indian Affairs, waiting for Roosevelt's popularity to wane so that they could repeal the programs his administration had begun. In 1944 a House investigating committee made strange noises about the subversive activities in Indian country and when the Republicans began their post-war comeback, the Bureau of Indian Affairs ranked high in their sights.

In 1948 the Hoover Commission, which was investigating the feasibility of restructuring the federal government, gave the I.R.A. concept of tribalism a strong boost but nevertheless recommended more involvement by state governments in Indian Affairs and a gradual phasing out of federal responsibilities.

Perhaps the last reform made by the New Deal philosophy was the Indian Claims Commission established in 1946. Since the pre-Civil War era Indian tribes had not been given standing to sue the federal government for violation of treaties and agreements. If a tribe desired to go to court, usually the Court of Claims, it had to seek Congressional authorization in the form of special legislation. This legislation spelled out precisely what procedures could be used, what topics could be covered, and what appeals the tribe might have once a decision was rendered. This tedious procedure, while it protected the United States against its Indians, who were regarded as mere wards of the state, meant that each session of Congress had to consider numerous Indian bills giving the tribes access to the Court of Claims.

Part of the hearings in the late 1920s and early 1930s had recommended that all Indian claims be settled before any further adjustment was made in the tribal-federal relationship. Thus reform in this area of Indian Affairs was not a new development, it was something that had been growing for decades. By the middle 1940s the stage was set for dramatic action.

In 1946, partially growing out of a desire of Congress to "solve the Indian problem," the Indian Claims Commission Act was passed. This Act allowed all tribes to file their claims against the United States in a special court known as the Indian Claims Commission. The Act contemplated a mere five years to settle the outstanding complaints that the tribes might have had. At the end of that time any further causes of action which might arise against the government had to be taken to the Court of Claims in the same manner as non-Indians could take their disputes to that court.

Some 800 claims were filled within the five-year period. These covered a great many subjects and were in many instances not simple land cession revisions as everyone had been led to anticipate. Thus the Indian Claims Commission began to sift through the wrongs of a century and a half and it clearly became impossible to settle very many of the claims within the original time limit. Some tribes sued for misuse of their tribal funds, others for land cessions, some for loss of hunting and fishing rights, others for minerals and rights to natural resources. The Claims Commission has had to be extended a number of times with little end in sight. Like other government agencies it has been able to perpetuate itself beyond any conceivable term envisioned by its creators.

One reason that the claims have dragged on has been the attitude of the Justice Department in its handling of the defense of the actions of the United States. It was originally thought that the cases could be handled easily, with each side presenting a few facts, a quick decision made based on those facts, and the award, if any, would be given to the tribe concerned. At that point the United States would be free to sever its relationship with the tribe having corrected any wrongs that it had historically committed against the Indians.

But Justice saw its job as militant defender of United States interests. It was rarely prepared to go to court against the tribes. Accountants searched the century-old records for any payments in goods or services that the United States had made to the tribe which could be set off against the tribal claim. The anticipated claims cases lapsed into struggles between sets of accountants preparing massive memorandums listing every blanket, kettle and plow that was ever exchanged between the Indians and the government.

In the nearly twenty-five years of litigation since the creation of the Indian Claims Commission less than half of the claims have been settled. The tribes have been denied interest on their claims even in those cases where funds instead of lands have been taken. Some claims have been settled by payments that would be unfair even at the prices of yesteryear. The most ludicrous example is the claim of the California Indians in which the price awarded them for the taking of California is a mere 47 cents an acre, certainly less than they could have received had they been allowed to sell it a century ago.

The practice grew up that only money could be offered for the lands taken. Thus, with the exception of Taos Pueblo, no Indian tribe has been given any of their land back, whether the lands in question were being used by the government or not. To a people desperately in need of an adequate land base for an expanding population, the cash settlements have been a poor redress for centuries of exploitation.

In 1971 the Indian Claims Commission is due to expire. Congress has previously allowed only five-year extensions and the last extension for a five-year period was in 1966. At that time there was some sentiment among Indians and Congressmen alike that the Commission should be allowed to expire and all pending cases transferred to the Court of Claims. The feeling was that the Commissioners were not making sufficient progress in settling the disputes to justify keeping the Commission alive. In spite of criticism, the Indian Claims Commission appears to be similar to other government agencies and continues to perpetuate itself while apparently fulfilling its appointed task.

[V. D., JR.]

33

The Great Indian War of the Twentieth Century

THE INDIAN REFORMS of the New Deal had barely gotten started before suspicions arose that John Collier was increasing federal responsibility instead of lessening it. Congress had quite often been informed that the Indian problem was just on the verge of disappearing when some report shocked it into action. But the underlying feeling that somehow things should one day come to a conclusion was always present in the minds of Senators and Congressmen who worked on Indian legislation.

Although the new reservation governments appeared to be making substantial progress in developing the human and economic resources of the Indian tribes, people in Congress felt they had been led to expect even more spectacular results. Therefore there was continual surveillance by the various committees concerned with Indian problems all through the Collier years.

In 1947 the Senate Civil Service Committee held hearings on ways to cut government expenditures. Since the Bureau of Indian Affairs had always been accused of lavish expenditures and abject failure to perform its appointed tasks it naturally came under the inquiring eye of this committee. Then Acting Commissioner William Zimmerman was asked to testify before the committee on ways that appropriations could be reduced by ending federal services and supervision over Indian tribes.

Zimmerman was reluctant to testify since he realized that whatever he said could be used in a number of ways and might prove detrimental to Indians. Nevertheless he gave a carefully drawn presentation of how he foresaw ending federal supervision in the field of Indian Affairs. Dividing the existing federal tribes into three classes, Zimmerman thought that the first group could end federal supervision almost immediately providing certain safeguards were written into the legislation protecting their treaty rights and land resources. This class was composed of a few tribes having rich natural resources and a stable reservation population.

The second group contained the majority of tribes served by the Bureau of Indian Affairs. Zimmerman felt that these tribes, with proper training and development, could sever their federal ties within a definite period of time, providing that they continued to make the progress they had been making during the previous decade.

The last class contained those tribes that suffered abject poverty and had not as yet developed any significant tribal programs. These tribes were generally isolated from the rest of American society and had not as yet compromised their traditional ways of life with the world outside the reservation. Prominent in this list were the Hopi and Navajo, Arizona tribes who at that time were almost totally isolated from the rest of Indian Affairs and lived pretty much as their ancestors had lived, by herding sheep in the deserts of northern Arizona.

The tribes that Zimmerman had picked for the first category were for the most part providing their own social and governmental services from the income accruing to their tribal enterprises. Therefore even if federal services were denied these tribes, if the basic economic base which they had achieved was not disturbed, they could maintain themselves as self-sufficient communities. When the Senate Civil Service Committee discovered how little would be saved from severing the federal relationship with these tribes it abandoned the project and turned elsewhere in search of unnecessary expenditures.

The following year the Hoover Commission Report on the reorganization of the federal government was issued. It contained a good evaluation of the impact of the Indian Reorganization Act on the development of Indian resources but still advocated that federal services to Indians be turned over to the respective states and that the federal government close out its responsibilities to Indian people. The impact of the report was to convince a number of men in Congress that the federal government could terminate its ancient responsibilities to Indian tribes without severely changing the status of Indian treaty rights.

Coinciding with these developments in government was the desire of some of the church groups, who had little knowledge of the nature of the Indian relationship with the federal government, to "make the Indians first class citizens." Thus little pressures began to develop from outside the field of Indian Affairs to "do something." This vague feeling was buttressed by developments in the field of civil rights as more and more people became concerned about the problem of the American Negro community. People began to reason that while the black problem could not be solved in the foreseeable future the Indian problem might be speedily resolved to their satisfaction.

By 1952 both political parties pledged to free the Indian from his shackles and make him a first class citizen equal to any other citizen of the nation. These were fine words but they had no basis in historical reality. In treaty after treaty the chiefs and headmen of the different tribes had refused to touch the pen until they were assured that the tribe would be given the special protection of the federal government as long as the grass should grow, the sun should shine, and the rivers ran down to the sea. Indians were fully aware of Andrew Jackson's cynical pronouncement that the Supreme Court could enforce its own decrees. They meant to have the special assurance of the Great White Father if they were to cede their lands peacefully.

"Freeing" the Indians, in the campaign slogans of the early 1950s simply meant that these longstanding protections would be denied to the tribes in violation of the United States' pledged honor that it would never leave them at the mercy of private citizens and state courts. In 1924, when Congress was considering the Indian Citizenship Act, there was much discussion about whether making Indians citizens would not abrogate longstanding treaty agreements. A clause was added to that legislation specifically disclaiming any change in tribal rights as a result of granting the Indians citizenship.

In August of 1953 a House resolution was passed stating that it was the intent of Congress to terminate all federal responsibilities for Indians at the earliest practicable time. The resolution passed both Houses of Congress without a dissenting vote as obscure resolutions are wont to do. The following year Indians learned to their sorrow exactly what the resolution meant.

Early in 1954, Senator Arthur Watkins of Utah, chairman of the Senate Interior Subcommittee on Indian Affairs, and E. Y. Berry, Representative from South Dakota and chairman of the counterpart committee in the House of Representatives formed a special joint subcommittee on Indian Affairs and began a survey of federal Indian tribes that might be ready for termination. Might is hardly the word to describe the criteria that Watkins and Berry used. Digging up Zimmerman's old testimony as proof that termination had been discussed by Congress for years and that therefore everyone, Indian and fellow committee members alike, understood what it meant, they began to write legislation unilaterally terminating the treaty rights of various tribes.

What infuriated Watkins almost immediately was the fact that the Menominees had recently won a case against the United States and were due to receive a substantial sum from the federal government. The idea that alleged wards of the government could sue their trustee for what had been blatant violation of its role as trustee gave Watkins nightmares. He felt that it was his duty to sever all Indian relationships as quickly as possible before the United States incurred more liability. He was thus totally without mercy or understanding when it came to the solution of social problems that had plagued the respective tribes for decades. His only solution was one of informing the tribes that they would do better if completely denied any assistance from the United States government.

The conditions which Zimmerman had carefully described as prerequisites to any alteration of the federal relationship disappeared in smoke. The sole criterion became whether or not the joint subcommittee could threaten harass or dupe the unsuspecting Indians into agreeing to give up their treaty committments and let the United States out of its century-old contractual agreement. Thus Watkins told the small bands of Paiutes in southern Utah that the legislation was only to validate their tribal marriages as legally recognizable in Utah courts. When the small groups agreed to this interpretation of the legislation he then wrote a bill terminating their federal trusteeship and placing them under the guardianship of a bank in his home state.

The Klamaths, Menominees, Catawbas, Alabama-Coushattas, Grande Ronde, Siletz, mixed-blood Utes, and California tribes fell under the onslaught. When tribes refused to agree to the legislation Watkins told them that he would pass it anyway and in the process deny them their funds in the federal treasury. By agreeing to the legislation the tribes understood they could at least get their own money which was being held in the federal treasury for them. It was apparently the case of the lesser of two monstrous evils.

The genius of the joint subcommittee then became apparent. Traditionally all legislation went from one House to the other and received consideration by both of the committees. Thus, when one house passed legislation that did not have the same phrasing as that being considered by its counterpart, a conference would have to be called to iron out the differences in the two versions. Indians therefore were able to present their views to members of both Houses and if they could convince either House to support their desires they could have the legislation tied up until a sounder solution was found to the problem.

With the joint subcommittee there was no need to have two pieces of legislation. Watkins and Berry simply wrote one version which was submitted to both houses in identical form with pious pronouncements of its careful consideration and no one dared to question the viability of the proposal. Bills simply flew through the two Houses of Congress and were ready for the President to sign before the tribes knew what was happening. It was almost as if Senator Watkins had been given a blank check by other members of Congress to do whatever he wanted to the Indians.

The destructive impact of the legislation was not immediately evident because some of the bills provided for a certain period of time before they were finally effective. Thus a bill could be passed into law and no one would discover that it was poorly written or violated constitutional safeguards until years later when problems began to arise.

One of the major reasons given for passing the legislation was that Congress felt that Indian people should have full control of their own property. Lack of this right of absolute control, many Congressmen thought, had hampered the economic and industrial development of Indian reservations. But the terminal legislation did not give the Indians control over their natural resources.) The mixed-blood Utes of Utah were organized as a corporation and the assets of this corporation were placed in trust with a Salt Lake City bank. Instead of having the right to determine what would become of their lands, the Utes were more often told what the bank intended to do with them. So swiftly and carelessly was the Ute termination bill driven through Congress that there was little thought given to proper allocation of the reservation resources between the withdrawing members and those full bloods who were allowed to keep their federal rights. By 1970 a number of cases attempting to straighten out this legislation were already in court, one case alone was for $141,000,000 for deprivation of mineral rights.

The Menominee legislation was also rushed through Congress and it likewise lacked sufficient clarity and definition. The Menominees had been promised that their hunting and fishing rights would receive special attention when the legislation was proposed but no mention of them ever was made. Thus in the years after termination continual disputes arose between Menominees hunting on the former reservation lands, now incorporated as Wisconsin's seventy-second county, and state fish and game officials. After years of confusion on this subject the tribe finally arrived at the Supreme Court where, in the early summer of 1968, the court decided that the tribe had not even been terminated, it had merely been handed over to the jurisdiction of Wisconsin. With this decision questions immediately arose concerning the taxability of tribal income, the aboriginal rights to fish and game regardless of state laws, and the deprivation of federal services during the period between the passage of the legislation and the decision of the Supreme Court.

In southern Oregon the terminated Klamath tribe also found itself in a state of confusion. The terminal legislation provided for the continuance of the tribal entity but made no provision for amendment of the tribal constitution. Close to 78 per cent of the tribal members chose to withdraw from the tribe. The remainder were placed under the trusteeship of a bank in Portland and constituted the tribe after termination. Since the tribal constitution survived the terminal legislation intact and it called for a certain number of Klamaths to be present at any valid meeting of the tribe this meant that practically the whole group of remaining members had to be present at any meeting to do business. No clear solution has ever been reached on what now constitutes a legal meeting of the tribe which enables it to deal with its trustee. There is a real question, therefore, whether anything that has happened to the remaining Klamaths since termination has been legally binding on the trustee bank, the government or the tribe.

A few tribes were marked for extinction but managed to survive application of the policy for the strangest reasons. The Flatheads of Montana, listed originally by William Zimmerman, were saved through the personal intervention of Mike Mansfield, then a newly elected Senator from Montana. Mansfield shamed the members of the subcommittee by reminding them that the United States had considered the Flathead tribe an equal at the time of the treaty-signing and that it was a breach of good faith for the United States now to renege on the agreement.

The Seminoles of Florida were recommended for termination even though only one fifth of the tribe spoke English and most of them were so poor that they didn't own a pair of shoes. Somehow the subcommittee felt that if they simply tried a little harder to get along they would come through the experience unscathed. The Daughters of the American Revolution took up the cause of the Seminoles and were able to stop the legislation before it could be passed. The patent absurdity of the proposition that a group of uneducated non-English-speaking Indians who lived primarily in the swamps could succeed in the contemporary, highly competitive American economic struggle should have been self-evident.

After the first rush of legislation, the tribes, led by the National Congress of American Indians, fought back against further implementation of the policy. They were able to stop the introduction of further terminal legislation as the expression of a policy to which Congress and the executive were firmly committed. Then things took on a new light even more frightening to the Indian community. The threat of termination had proved a useful weapon against any tribe that asserted its right to govern itself. The fact that Indians were cowed and fearful of losing their treaty rights was not lost on anti-Indian members of Congress and bureaucrats jealous of Indian self-government of the reservations which had traditionally been regarded as belonging exclusively to the Bureau of Indian Affairs.

Instead of arbitrarily choosing tribes for termination, therefore, the Senate Interior Committee used the threat of termination to force compromises by tribes who wanted other legislation passed. If a tribe wanted to program its funds received in a judgment given to it by the Indian Claims Commission, termination was threatened so that the tribe would agree to a per capita distribution of the award moneys—thus precluding development of Indian industries on that particular reservation. Or if a tribe were being difficult on some minor point of a land restoration bill, the suggestion of termination might cause the tribe to take the Senate's version of the legislation instead of pressing its claim for a fair law.

In 1958 the Colville tribe of eastern Washington asked for the return of lands which had been taken away from it at the end of the last century for homestead settlement (which had never been settled). Under the Indian Reorganization Act the Secretary of the Interior had the power to return lands in this category to the tribe by administrative action. But Colville was not an I.R.A. tribe and therefore the tribal council had to seek enabling legislation to get title to the lands. The price of getting their own lands back was agreement to develop a plan for termination of the reservation within five years.

The tribe agreed and developed a plan comparable to the original plan suggested by William Zimmerman in 1947. It called for the gradual phasing out of federal services as the tribe became more experienced in the management of its own affairs. The Colville plan was sensible and provided safeguards so that property rights would not be lost nor would the social and cultural life of the community be harmed. When the plan was presented to Congress, the Colvilles were shocked to learn that the agreement to present a *plan* for termination was being interpreted by Henry Jackson, chairman of the Senate Interior Committee and Senator of their home state, as agreeing to terminate within five years.

Many of the people gave up and refused to consider any alternatives feeling that the government was going to do what it wanted anyway. They refused to vote in the tribal elections, thus enabling the pro-termination forces to capture the tribal government. The new council promptly passed a resolution stating that it was the desire of the majority of the people to terminate federal services as quickly as possible. Many of the new council looked only at the possible distribution of the assets of the tribe and not at what the results of sealing their federal rights would be. But the resolution was accepted by the Senators as indicative of the real desires of the Colvilles.

The Colville tribe faces continual pressures from the Senate Interior Indian subcommittee to accept the termination bill during every Congress. The bill is poorly written with many trick sections that, if detrimentally administered, would effectively deprive the Indians of any fair price for their lands. Most of the Senators on the committee have little knowledge of how the legislation originally developed. It has been proposed within the committee for so long that everyone assumes that somewhere in the distant past the decision to terminate the tribe was made irrevocable. Only James Haley and Wayne Aspinall, influential members of the House Interior Committee which handled Indian legislation in the House of

Representatives, have prevented the Colvilles from being terminated. They have refused to pass unfair legislation.

A similar situation exists with the Senecas. In 1962 President John Kennedy allowed the Kinzua Dam to be built on the Seneca reservation in New York State; thus breaking the oldest extant treaty in American history, the Pickering Treaty of 1794, which, as we have seen, was personally guaranteed by George Washington. When the Senecas went to the Senate to get compensation for the destruction of their reservation by flooding they were told that they would have to accept termination in order to be paid for the taking of their lands! With winter coming on and the majority of the tribe driven from their homes by the Corps of Engineers the tribe had little recourse but to agree to the terms laid down by the Senate Interior Committee. It was that or endure the harsh New York winter without homes. Thus the allies of the United States at the time of Little Turtle's drive to exclude whites from the Ohio valley were shabbily deprived of their very homes.

By the middle 1960s it was easy to see that the Buffalo Party had not given in, it had merely changed its tactics from outright theft of vast areas of untamed wilderness to executive sessions of the Senate Interior Committee where smaller pieces of land were taken with ease, cloaked only in the presumption that Congress would always act in good faith towards Indians as the United States had originally declared in the Ordinance of 1787. There was no basic change in attitude from Andrew Jackson to Henry Jackson—both felt that Indians have no more rights than the buffalo.

[V. D., JR.]

34

The Rise of Indian Organizations

I N HIS CLOSING YEARS as Commissioner of Indian Affairs John Collier could see the storm clouds gathering on the horizon. He realized that the pendulum of policy was swinging against the Indians and that another struggle over the Indian lands was imminent. Collier also understood that no longer could white friends exert pressures on the government of sufficient magnitude to prevent disaster. For nearly eight years sympathetic whites had organized pressure groups and influenced Indian policy, sometimes for the Indians' good, sometimes to fulfill their own desires. But in many ways they had worn out their welcome with Congressional committees who had responded to numerous schemes put forth to solve the Indian problem. After decades of following the advice of sympathizers, Congress was in no mood to heed the sentimentalists, as they came to be called. It was time for Indians to speak for themselves.

Collier was indeed correct. The field of Indian Affairs had shrunk over the previous decades. Where Indian treaties had once occupied the attention of the whole Senate, now Indian legislation was given to an obscure Congressional subcommittee in the Interior committees of the two houses. Indians had become an exotic and remote commodity fit only for perusal by tourists and those few people in Congress who had some local concern with them. World problems had now come to dominate the thinking of the American public. Indians, constituting an increasingly smaller percentage of the population, simply could not bring their problems to the attention of a world filled with wars and crises that affected the whole planet. If Indians were to survive they must find within themselves the strength, unity and stamina to do so. They could no longer rely upon popular sympathy to plead their case.

Thus in the early 1940s Collier urged the younger Indians within the Bureau of Indian Affairs to create a national organization which could represent the interests of the scattered Indian tribes of the country before Congress and the administrative agencies of the government. For a number of years he carried this message to his Indian employees so that by 1944 there was a considerable concern among the educated Indian people across the country to establish a national organization. In November of that year over a hundred Indians from different parts of the nation met in Denver, Colorado to explore the possibility of having their own organization. Among them were the outstanding leaders from the tribes, religious bodies, and the Bureau of Indian Affairs, probably the most representative group of Indian leaders of that generation.

Meeting for two days they recalled the numerous attempts that had been made over the years to create a national organization to speak for the tribes. Pontiac had attempted such an alliance, as had Joseph Brant and Tecumseh. Perhaps the closest that any group had come was the old intertribal alliance formed in Oklahoma to fight the allotment bill in the closing decades of the last century. Sporadic attempts had been made earlier in the twentieth century with the creation of the National Council of American Indians in the late twenties and the alliance of Charles Eastman, Carlos Montezuma and others during the first World War.

When it came time to decide what to call the new organization, the committee nearly revived the old council started by Archie Phinney and others two decades earlier but they finally chose to start a new group formally entitled the National Congress of American Indians because they felt that the Indian people should also assemble in Congress as the whites did. Accepting this name they then proceeded to draw up a constitution and by-laws, elected officers, and planned to carry on two functions— work in legislation affecting the various Indian tribes and a legal rights program to handle litigation on Indian matters such as voting rights, welfare and civil rights problems. Widespread discrimination, particularly in the field of voting rights, had been reported by delegates to the founding convention.

Operating with a volunteer lobbyist, Mrs. Ruth Muskrat Bronson, a Cherokee from Oklahoma, as their first director, the new organization was able to accomplish a number of things in its early years. In the years immediately after its founding, the National Congress of American Indians was able to exert significant pressure to get the Indian Claims Commission established and proposed special legislation to handle the problems of the Navajo and Hopi tribes of Arizona. Indeed the struggle of the N.C.A.I. to get the Navajo-Hopi Rehabilitation Act passed was indicative of its devotion to the Indian cause, since the Navajo tribe, largest and now richest Indian tribe in the nation, did not once acknowledge the existence of the organization and did little on its own behalf to get the legislation enacted. Much of the work on the legislation was done by volunteers of the N.C.A.I.

It has always been the regret of the other tribes that they devoted so much time to assisting the Navajos and neglected the establishment of their legal services program. When the Navajo finally received income from their oil and gas and had achieved some measure of financial independence, they constantly derided the N.C.A.I. and refused to assist the other smaller tribes that had once assisted them. Without the early work of the N.C.A.I. there would probably be no Navajo tribe today. But memories are as short among Indians as they are among any other group.

The early years of the N.C.A.I. also resulted in the emergence of a number of Indian leaders who would be very important in later years. Two future Commissioners of Indian Affairs served as program director of the N.C.A.I. in the post-war years, Louis Rooks Bruce and Robert LaFollette Bennett. Judge Napoleon B. Johnson, later a Chief Justice of the Oklahoma Supreme Court served as president the first eight years to be followed by W. W. Short, also an Oklahoma Indian.

By 1954 the National Congress of American Indians had achieved national recognition as the voice of the Indian people, although competitive groups of non-Indians organized in the East years earlier continued to represent themselves as spokesmen for the Indian cause also. It was in 1954, however, that the crisis envisioned by John Collier occurred, and the N.C.A.I., which he had strongly supported, was on hand to meet it. In that year Congress began its policy of terminating the federal relationship with Indian tribes. In unilateral actions a number of treaties were abrogated and the tribes were denied their federal rights. Before long it was apparent that the services and trusteeship gained by land cessions of the last century might all be destroyed and the Indian tribes with them.

Quickly responding to the threat, the National Congress of American Indians, then under the able leadership of Mrs. Helen Peterson, Oglala Sioux, who was the Executive Director, and Joseph Garry, a Coeur d'Alene from Idaho and president of the organization, called an emergency conference of the tribes in Washington, D.C. Indians from all over the continent swarmed into Washington, D.C. where they spent nearly eight weeks protesting the unwarranted attack on Indian rights and attempted to get the hated "termination" policy stopped. As Senators and Congressmen were cornered by angry and frightened Indians from their home states it became apparent that the policy was misdirected and even more important it became apparent that Indians were alerted to their rights and willing to fight for them.

The conference was organized without funds, primarily by word of mouth communications, and under the threats of certain Congressmen who saw "subversive" elements present in the Indians' effort to protect themselves. So effective was the N.C.A.I. defense that the two Interior Committees carried out their threat to "investigate" Indian conditions with the hopes that they could find something to smear the Indian organization with —thus destroying its effectiveness. The result was a biased report, House Report #2680, which ripped the lid off Indian Affairs and attempted to overturn such longstanding programs as the Indian Reorganization Act.

Congressional committees threatened to take away the self-governing powers of the Indian tribes who fought back against the termination policy. While this threat frightened the weaker tribes, it had no effect on the N.C.A.I. At its annual convention in Omaha, Nebraska in the fall of 1954 the organization issued a stinging rebuke to the Congress of the United States by passing a resolution that demonstrated the inaccuracies and biases in the House Report. By unmasking Congressional rhetoric and showing the injustice in the position taken by the House Interior Committee, the N.C.A.I. made Senators and Congressmen take a stand. Some did not want to be associated with the report since it clearly demonstrated that the Congressional policy had been unjust. The united Congressional front melted as various Senators and Congressmen were quick to disassociate themselves from it. Only those tribes that had been tricked or browbeaten into accepting termination were hurt by the policy.

The decade-long struggle against the arbitrary policy served to build up the National Congress of American Indians and tribal membership climbed steadily during that period. But with increasing membership came unanticipated problems within the Indian community. The group split into two political factions which spent almost as much time fighting each other and they did fighting the government policy. The annual election of 1959 held portents of rough seas ahead. The president of the organization won his office on the basis of disputed votes by the slenderest of margins.

In 1961, the next time the office of president was vacant, the scene was dramatically set for a palace revolution. In June of that year a group of anthropologists, anxious to present the new Kennedy administration with a united Indian voice, held a conference of Indians in Chicago. The incumbent N.C.A.I. group worked hard planning the program yet suffered severe criticism by their 1959 opponents as having "rigged" the conference. No one had suspected that the conference provided an extra political forum in which the competing groups might spar for the showdown later that fall.

Chicago proved to be the undoing of the N.C.A.I. administration that had valiantly led the fight against termination during the 1950s. In the N.C.A.I. elections that fall they were soundly defeated by a coalition composed mainly of tribes from South Dakota and Montana. The fight was so bitter in 1961 that it set off an intertribal war that lasted the decade and thus Indians were effectively split and fighting among themselves during the years of greatest social movement in America.

But the Chicago conference did have another effect. For some years prior to Chicago, the Indian youth had been searching for a vehicle by which they could enter the national Indian arena. The National Congress of American Indians had student memberships and a fairly respectable number of Indian youth held memberships in it. But the voting process within the organization was largely controlled by older Indians representing tribal governing bodies. Thus participation by Indian young people in the electoral processes of the organization was slight.

These young people met at Chicago and made plans to initiate their own organization, drawing mostly from the southwest area. Later that year they had a preliminary meeting in Gallup and founded the National Indian Youth Council. By the following year, the N.I.Y.C., as it came to be called, was well on its way to becoming the second organization founded and conducted by Indian people and having a national program. By late 1964 it was conducting activist programs in the northwest in defense of Indian treaty fishing rights and had become an influence in national Indian affairs politically.

The N.I.Y.C. worked behind the scenes at the N.C.A.I. conventions and put two of its members into the position of Executive Director of the N.C.A.I. during the 1960s. More important, it provided for the radical position in Indian Affairs by always seeking a more liberal and activist position with respect to issues than did the older N.C.A.I. The young Indians thus opened up Indian Affairs as it had never been opened, dividing the traditional government by consensus into a progressive movement with a majority and dissenting philosophy always opposing one another.

In 1966 the popular Philleo Nash, then Commissioner of Indian Affairs, was fired because of his persistent opposition to the termination of the Colville tribe of eastern Washington state. The Colvilles had been one of the last tribes affected by the termination era. In the late 1950s they had sought the return of lands taken from them during the allotment era. Since they were not organized under the Indian Reorganization Act, the Secretary of the Interior could not return these lands to them by administrative action. They therefore had to seek Congressional action to get their lands back. The price of getting their lands was a termination rider in the legislation requiring them to submit a plan for termination within five years. Nash sided with those Colvilles who refused to terminate and for this he lost his job.

After Nash's departure. Stewart Udall, then Secretary of the Interior, planned to take a personal interest in the administration of the Bureau of Indian Affairs and called a conference at Santa Fe, New Mexico to discuss his plans with the chief administrative people in the bureau. The tribes learned of Udall's plan and came to Santa Fe to demand a voice in the decisions that would be made about the future programs of Interior that affected the reservations. From this meeting the tribes were promised major legislative action by the Johnson administration in the form of an omnibus bill incorporating all problems of Indian administration into one gigantic legislative package. And the Indians were promised a major voice in developing the project.

Since there had never been any efforts at consulting with Indians or their tribal governments when legislation concerning them was proposed, the Interior Department proceeded to draw up legislation to introduce to Congress. But the Indians would not accept benign neglect any longer. In what must be regarded as a major coup of Indian history, the N.C.A.I. smuggled copies of the proposed legislation out of the Interior Department and informed the tribes of Udall's proposed legislation.

That fall Robert Bennett, the new Indian Commissioner, was sent out on a hopeless task of pretending to consult with tribes in order to draw up the legislation as Udall had promised at Santa Fe while the tribes already had copies of the legislation which Interior officials declared was nonexistent! Even before the round of consultation was finished the legislation was introduced, thus belying the official administration line that it did not exist. In the winter of 1967 and later that spring, Interior again tried to fool the tribes into believing that they would have some voice in the proposal to Congress. Once again the N.C.A.I. got copies of the legislation and sent them out to the tribes.

When the bills were finally introduced formally in Congress in May of 1967, not one single Senator or Congressman would sign his name to them. They had to be introduced "by request" of the administration. For the first time in Indian history a group of tribes had beat the United States at its own game. The storms that John Collier had foreseen in the early 1940s had been weathered and the modern Indian movement had been created.

Even more important was the development of Indian thinking as a result of the Santa Fe meeting and the struggle over the Omnibus Bill legislation. For almost the whole twentieth century Indians had been leaving the reservations and going into the large cities to work. During the Eisenhower administration the official policy was to move the Indians off the reservations as quickly as possible. For those tribes who refused to terminate their federal relationships a special program was planned called relocation.

Relocation meant moving individual Indian families into the urban centers on the West Coast and in the Midwest where they were to be employed with the hopes that they would permanently make their homes there. By the late 1960s nearly a majority of the Indians in the nation had moved to the cities and for this population there were no federal programs. With the national Indian sentiment moving quickly toward more meaningful involvement of local people, it was not long before each city with a major concentration of Indians had its own urban organization. These began to coalesce by 1966 into regional and national groupings.

An abortive attempt was made in 1968 to organize the urban Indian centers into a national organization at a meeting in Seattle, Washington. But the effort was ill-defined and lacked sufficient leadership to grow into an effective force. No one as yet had the key to organizing urban Indians. Where the National Congress of American Indians could concentrate on defense of Indian lands and the National Indian Youth Council could appeal to young people, American Indians United, as the urban group called themselves, could find no common ground among the urban Indians other than their disenfranchisement from programs enjoyed by reservation Indians.

While American Indians United failed to formally organize existing urban Indian clubs and centers into an organization, a group of Indian activists succeeded in finding the answer to the problem. Beginning in Minneapolis in the late fall of 1966 the American Indian Movement— A.I.M.—was born. This group closely followed activist techniques of other minority groups but concentrated on problems suffered by the relocated Indians of the cities.

By late 1968 they had organized an Indian Patrol in Minneapolis that conducted a surveillance of police activities showing that there was definitely discrimination in the policing of predominantly Indian neighborhoods of that city. With this success under their belts the activists of A.I.M. quickly expanded into other cities in the Midwest, notably Cleveland, and became the leading Indian activist organization in the nation.

A similar movement began in late 1969 on the west coast. In November of 1969 the San Francisco Indian Center, which served the Indians of the Bay area, burned down. It had provided a home away from home for thousands of Indians coming to the San Francisco area for a long time and people were desperate to find another place where they could carry on their activities. And the island of Alcatraz looked abandoned, forlorn and inviting.

On November 19, 1969 a group of students from nearby colleges swam ashore and occupied the island calling themselves the Indians of All Tribes. Throughout 1970 other groups in the West Coast area, also calling themselves Indians of All Tribes, took over abandoned army posts and isolated islands as the Indian movement turned from mere defensive postures of the past to an aggressive feeling shared by all Indians that the struggle of the future should be over the return of lands illegally taken from their fathers and grandfathers a century ago.

The development of Indian organization was not wholly activist oriented. In 1964 a small group of Indians began the American Indian Historical Society in San Francisco as a means of improving the Indian image in the textbooks and getting a fair hearing for Indian people in the academic community. Beginning with a small mimeographed paper the group produced extremely fine materials challenging the Indian community and the white scholars to examine the whole content of Indian history.

By 1970 the American Indian Historical Society had founded an Indian press, developed their newsletter from a mimeograph into a fine quarterly magazine with a subscription list of thousands, and had sponsored the first meeting of Indian scholars. In March of 1970 at Princeton, New Jersey over a hundred Indians gathered to discuss what could be done by Indian scholars in the field of scholarship. It was the first time that so many Indians had gathered to consider the intellectual and theoretical implications of having a national Indian community. The conference served notice to Indian and non-Indian alike that the obscurity in which Indian people had spent the twentieth century had passed and a new day of intellectual respectability had dawned.

In the years since 1944 a great many things had happened. While the National Congress of American Indians was the oldest all-Indian organization, it was certainly not the only one. Indians had proved, perhaps better than they wished, the variety of thinking and concerns that flowed through the various tribal groups. Although few of the organizations had achieved financial stability or the overwhelming loyalty of Indian people, taken together the organizations proved one lesson. Indian people were learning that old tribal enmities must be overcome, and that unification and organization were the most important weapons in their arsenal if they meant to survive.

[V. D., Jr.]

35

Social Programs of the 60s

A S THE RED MAN entered the decade of the 60s the future did not look bright. The effects of the termination policy of the previous administration still lingered. The Klamath termination became final in 1959 and the Menominees of Wisconsin, even with their extension of time, were finally terminated from federal services in 1961. The Tuscaroras had just lost a major case involving their lands and the tiny New York reservation was being flooded by a holding dam to provide more power for New York State.

Further south along the Pennsylvania–New York border the Senecas were in the final process of losing the fight to prevent the construction of Kinzua Dam. In 1794 the United States had signed the Pickering Treaty with the Iroquois as a means of pacifying the powerful eastern League prior to the war with Little Turtle and the Indian nations of the Ohio Valley. Part of the Iroquois Treaty provided that no lands would ever be taken from them by duress and that they should always have their reserved lands to live on.

But industrialists in Pittsburgh downstream from the Senecas needed Kinzua built in order to reduce the costs they would otherwise have had to pay for polluting the river. So the Army Corps of Engineers proceeded to push the legislative authority to build Kinzua through the Congressional committees and the tearful Senecas could only remind the nation of George Washington's pledge that they would forever have their rights respected. But there was little outcry by the public which was convinced that no price was too high for progress even though the nation had given its word that Indians would never have to pay that price.

There were other threats to continued tribal existence. In 1954 the black community had won the crucial case of *Brown v. Board of Education of Topeka, Kansas* and the following year demonstrations began as the disenfranchised Afro-Americans attempted to make the nation live up to its constitutional promises. The white community was increasingly accepting integration of minority groups into American society as the only viable and just alternative to continued local oppression of racial minorities.

Integration, as you will recall, was the justification used to create the hated allotment process that had resulted in the loss of most of the Indian lands in the short period between 1887 and 1934. And a variant ideology of equality before the law, thoroughly integrationist although not practically so, had been at the root of the termination policy of the late 40s and 50s. As far as social ideologies were concerned, the beginning of the decade of the sixties was the darkest hour of the century for the original Americans. They were definitely out of phase with every important idea accepted by contemporary society as a fundamental premise of social reality.

Yet even in this time there were inherent in the emerging ideas of the New Frontier several concepts that were to be of utmost importance for Indians. During the primary campaign in West Virginia John Kennedy had been shocked at the poverty and deprivation of the people of Appalachia. Promising that he would do something special for the depressed mining areas of the east, Kennedy rode to victory in the 1960 Presidential campaign and prepared the way for the new concept of regional development commissions.

During his time in office John Kennedy was able to assist the Appalachians by creating a special multi-state commission to work on the problems of that region. Pushing the concept farther the Area Redevelopment Administration was created that gave special grants to depressed areas. The grants covered the development of industries and building basic municipal service buildings in places that had never had the capital to furnish themselves with community facilities. The pattern of special help for special problems, set early in the days of the New Frontier, became the pattern for the decade.

Every argument that could be mustered in support of special programs for depressed areas such as Appalachia could be eloquently illustrated on the respective Indian reservations. If West Virginia had bad roads, the Navajo Reservation had practically no roads and yet was larger in size and distance between settlements. If the people of the West needed county courthouses, the Oglala Sioux needed every kind of community building. If special schools were needed to solve the dropout problem of the urban areas, what happened to the Indian reservations where the schooling was averaging half of the national average?

As the various pieces of legislation were lobbied through a reluctant Congress, Indians and their few friends in Congress managed to write in Indian tribes as eligible sponsoring agencies. The tenacious fight for tribal sovereignty that the red man had waged during the last century unexpectedly paid off. As Indian tribes were recognized as local governments capable of receiving program funds under the new laws, the old sovereignty of domestic nations changed and tribes became the vanguard and perhaps archetypes of the later community action agencies authorized during Lyndon Johnson's administration under the War against Poverty.

Most tribes in the nation took advantage of the Area Redevelopment Administration and applied for grants to build community halls, tribal headquarters, and service buildings to attract industry to their lands. The only capital community facilities on most reservations were brick buildings built during the depression by the make-work projects of the New Deal. Many had barely been adequate at that time. By the early sixties they were completely outmoded. The smaller reservations still had the old wooden buildings first constructed by the government when the reservation was established. Most agencies had long since been closed for the smaller tribes and the old buildings had been abandoned or had fallen into such a state of disrepair that they were virtually worthless.

Soon new tribal offices began taking shape. The provisions of the A.R.A. allowed the tribes to contribute land and services in place of matching funds. Thus it was possible for an Indian tribe to receive a substantial grant of funds for a new building and the only requirement was that it contribute the land to put the building on and promise to provide upkeep of the new facility. Some tribes ambitiously built large buildings and rented space to the Bureau of Indian Affairs, thereby creating a rental income for themselves. On the whole the Area Redevelopment Administration was the most progressive program ever made available to the Indian tribes.

Following John Kennedy's death the social legislation of a generation began to move through Congress. Lyndon Johnson's landslide victory in the election of 1964 gave him such a majority in Congress that he was able to get a substantial amount of new legislation passed in 1965 and 1966. Even while he was campaigning during 1964, Johnson was moving new social programs into American life. The Economic Opportunity Act of 1964, the famous War on Poverty, was an election-year program although it marked his first departure in the domestic realm from the Kennedy ideology of the New Frontier.

In May of 1964, realizing that the Poverty bill might soon become law, the Council on Indian Affairs sponsored a major conference on Indian poverty in Washington, D.C. The Council on Indian Affairs was a traditional coalition of the old groups that had always been influential in

Indian Affairs, the major missionary church bodies and the private Indian-interest groups. Led by the able Reverend Clifford Samuelson of the Episcopal Church, who raised substantial sums to provide travel for the tribal delegates, the Council held a three-day meeting at which the major topic areas of Indian poverty were discussed. Tribal delegations from all over the nation attended and willingly discussed their problems with anyone who would listen, and during the election year a great many Senators and Congressmen, aware of the combined voting power of the Indians and their friends, did listen.

When the Poverty package passed Congress during the summer of 1964, Indians were written into the law in a special program that enabled them to take their grant applications directly to a special desk in Washington, D.C. and avoid the tedious and exhausting route of regional offices to which other groups had been consigned. By eliminating the administrative levels between the reservations and the funding sources in the nation's capital, the law enabled the tribes to embark immediately on reservation-wide programs of their own choice.

The significance of the Poverty program for Indian communities can scarcely be underestimated. Ever since the tribes had been driven to the reservations, their social and political life had been rigidly controlled by the federal government. Where there were strong and able chiefs the Interior Department officials deliberately undercut their dignity and influence so that the tribes would be more pliable toward the reshaping of Indian lives in the manner in which federal policy dictated. Religious ceremonies had been banned and the medicine men of the tribes had been ridiculed and taunted for their beliefs.

The Indian Reorganization Act was a major step toward reestablishing a sense of community on the reservations. But by and large it proved useful only for those tribes with income enough to support their own programs. Where a tribe had won a large settlement from the government and had used the money wisely for social and educational programs tribal self-government proved effective. But the number of tribes that had been able to financially support their own programs was minimal. Perhaps the only tribes ever to do so on a significant scale were the Osage of Oklahoma, who had a substantial annual income from oil royalties, and the Uintah and Ouray tribes of Utah who won a $31,000,000 suit against the government. Outside of those two tribes reservation programs had been sporadic and restricted to small land consolidation programs.

Suddenly the world was turned upside down. Where the government had blocked community self-government it now provided millions of dollars to encourage it. Where Indian children had been kidnapped from their homes and taken thousands of miles to eastern boarding schools such as Carlisle, now tribal governments were given large grants to develop Head-start and Neighborhood Youth Corps and were encouraged to take over the education of their children on the reservation.

Tribes that had previously strained to make a $25,000 budget cover all facets of reservation life were quickly forced to program and spend a million dollars or more on sophisticated developmental programs. On many reservations the highest salary for a nongovernmental job prior to the Poverty program had been that of tribal chairman, paying anywhere from $5,000 to $9,000 a year. These jobs had been eagerly sought as a means of working in the Indian community. This scale of salaries was soon swept away in the rush to develop programs.

Early in the development of the Poverty program the decision was made to fund Indian community action programs at the same salary scales and program rates as those programs being developed in the urban areas. Thus salaries quickly shot out of sight on the reservations. The pay scale rapidly went to $12,000 and then to $14,000 and above. Even jobs far down the administrative chain of command were still higher than the tribal chairman's salary which was dependent upon tribal income.

The result of this rapid escalation of job opportunities was twofold. First, it encouraged people who had not previously considered working in the Indian community to do so. For nearly a decade Indians had been trooping off to college but few returned to the reservation communities. There were simply no jobs for college educated Indians other than teaching jobs in government day schools. Thus young Indians with new ideas and a knowledge of the world outside the reservation often drifted into the cities and took jobs outside the Indian community, having grown accustomed to the conveniences of modern urban life and being skilled enough to command decent salaries.

With the development of community programs on the reservation a number of these educated young Indians returned and entered tribal life as administrators of programs. Indian reservations were able to tap an immediate source of strength that they had not previously had access to. A strong sense of tribal identity began to build as people who had long since forsaken the tribe returned to claim membership and regain their place in the reservation communities. The trend of gradual dissipation of young people which had plagued the tribal groups for over a century was reversed. In this aspect the Poverty program was critical in establishing a new sense of identity among Indian people.

But there was a bad aspect to the program also. Very quickly the competent leadership was taken up and still high-paying jobs were being created. Programs were designed to place a maximum involvement in local people but the planners did not reckon with the time when all of the capable people would be hired. As programs continued to expand, people with little to recommend them but their Indian blood were hired. An

Indian student graduating from college could be assured of a job paying $15,000 a year within two years of graduation as the continual expansion of programs created a vacuum which lifted people up the administrative and salary pyramid in spite of themselves.

The effect of this expansion was that programs quickly became filled with people who had little knowledge or experience of Indian communities. Young people were torn out of the student context and became national leaders overnight. The medium of job hunting quickly became the message of Indian life and the entire concept of community development quickly went by the wayside. Instead a tragic pyramid-climbing process of hunting ever better and higher-paying jobs was introduced into national Indian life. People would barely be settled in a job when they would be offered another and better-paying job two states away. Off they would go in search of additional power and prestige, leaving the program to drift into limbo as other temporary people sifted through on their way to new jobs.

As the community development programs expanded and state-wide organizations were developed, Indians thought less and less about the crucial issues that affected them. Instead all eyes were drawn toward achieving a position of total power on a national basis. Thus, with the rise of programs intended to solve basic reservation problems, less and less attention was paid to the actual conditions of the people. The early programs had emphasized problems that everyone understood. Indian people had been wanting to develop these particular programs for years but had been handicapped for lack of funds. Once funded, the early programs were understood better and were more relevant to reservation needs.

But the later programs were developed because funds existed in certain government agencies and because these funds could be made available to Indian groups that knew how to write grant applications. Thus the later programs were created for their own sake and not to serve reservation needs. Like Mount Everest, people went after the money because it was there. By the end of the decade millions of dollars were being poured into Indian organizations without any visible results. It was apparent that something was happening in Indian country but nobody knew exactly what.

In an effort to correct this situation most of the government agencies created advisory groups or task forces to help them direct grant money to the most useful expenditure possible. But this maneuver in turn created additional turmoil in the Indian community. People working their way up the job pyramid had to also compete for positions on the respective committees advising how grants were to be given and thus an additional competitive factor was introduced into what had been a process of community development. Almost every program designed to assist people to meet their immediate local needs was politicized nationally among competing groups of Indians. The availability of programs contributed proportionately to their failure.

The situation was considerably more complex however. It was not simply a case of having a plentitude of government programs from which to choose. Churches and foundations also made funds available to local groups. Funds from these sources were considerably smaller and more concentrated in certain areas. But after 1967, when the larger churches made significant commitments to raise millions of dollars to assist minority groups, funding from these sources increased in importance.

Churches and foundations created confusion among Indian people completely out of proportion to their actual funding ability for a number of reasons. First, both groups looked at Indian Affairs as a series of unfortunate bumblings by people of years past. They were convinced that there was one key problem which, if solved by a specific grant, would miraculously resolve the Indian problem into a comprehensible model of the average non-Indian community. Thus many grants were given to groups who had played the grantsmanship game and had convinced church and foundation people that they had found the answer to Indian problems. When the program was funded and did not work, the churches and foundations would continue their search for the correct solution.

Then there was the problem of ideological considerations by churches and foundations. In order to justify granting sums to minority groups they had to adhere to the contemporary rhetorical slogans of the social climate of the country. Thus funds were given to support integrationist programs in the early 60s and to support the power movements of the later 60s. There was little apparent consideration of the actual effect of programs on the reservations by people approving the funding of programs. Rather primary emphasis was on the code words used to justify the project under consideration.

When the government agencies began the task force and advisory process, the churches rapidly followed their lead, so that in addition to the government committees, each church had its own committee to give it advice on expenditure of funds. Often the task forces gave conflicting advice depending upon the agency requesting advice and the availability of funds. Indian Affairs became almost totally existential and basic philosophies and serious program planning had to be discarded to meet the crushing demand for advice and information.

There were two basic and fundamental flaws in the community development schemes of the 60s. First, there was little analysis by either public or private funding sources of the actual situation of reservation people. Few people realized that the most profound inadequacy of Indian people from the establishment of the reservations had been their inability to translate a hunting or farming economy to an industrial economy or to a type of community existence where the tribe as a whole could compete with other groups in a modern economic context.

During the previous century the majority of the western tribes had their basic economic capital in the wild game of their homelands. The buffalo, antelope, salmon and deer provided a floor economically so that tribes could maintain themselves indefinitely at a subsistence level on the game available to them. With the settlement of the West and the destruction of the game, a new economic floor had to be built under the reservation communities to enable them to survive.

Many of the tribes adapted to reservation life by becoming ranchers or by holding their land communally as farmers with the amount of land determined by a family's ability to use it beneficially as a tribal member. The allotment process effectively divided the reservations into units of land so small as to preclude efficient farming or ranching. After allotment, the best an Indian family could do was to lease its lands to a white farmer or rancher with sufficient capital to operate on a larger scale and to seek relief from the government.

From the time of allotment until the early 60s, Indian economic self-sufficiency had gone down hill simply because the cost of living had risen enough to absorb the little income from the leasing of allotments. Where $150 a year could support a small family that had a garden, some horses and cattle, and lived in an isolated area of the West in 1900, by the late 1940s it was impossible for a family to live in the old ways. The cost of basic commodities such as sugar, bacon, flour, coffee and beans totaled many times the annual lease income and the family was thrown into poverty.

The Poverty programs of the 60s overlooked the fact that no replacement of the economic substructure was being built that would give to the reservation communities a stability from which a formal community could later be developed. Instead, almost every program except the capital development grants for community buildings given by the A.R.A. and its successor, the Economic Development Administration, was built upon the premise that Indians had to be motivated to do better than they had been doing.

With motivation as a built-in factor in the grants, it was painfully apparent to the skillful observer why the programs were having no visible effect on the economic status of the reservations. People could be motivated forever, but if they had no funds to invest, then there was really no way that they could ever improve their financial status and escape the effects of grinding proverty. It was curious that with all the funds made available to Indian communities and the massive funds being spent on research into the causes of Indian poverty that no one took careful note of what the tribal leaders were actually saying during those years. Whenever a reservation leader was asked to describe the progress being made on his reservation, he would invariably start with a description of the houses being built, the new community building, and the potential factories and businesses that his tribe was planning. Little was said about motivation.

The other fatal flaw of the community development process was the inability of administrators to conceive of things in the world of experience. Grants were written on a yearly basis. Administrative positions were mapped out according to functions. For example, a project might have an executive director, a director of research, a director of training, a director of public relations, and a project planner. The entire grant was funded with the assumption that all components would be immediately staffed by experienced people who could immediately take up their tasks with skill and knowledge as soon as the funds were in the bank. There was no acknowledgment that most organizations are the product of a maturing process or that natural growth creates certain functions in one organization and not in others.

Instead of a project growing naturally and being able, through experience, to handle increasingly complex problems, funds were simply given and the local community was cast adrift to staff and operate the program as best it could. This process created the scramble for jobs described above. No Indian was allowed to grow with his job and learn from developing a program how communities actually develop. Rather, staff positions were filled politically and the people hired were expected to be immediately expert in the position for which they had been hired. This was the philosophy on which the grant was developed and everyone was expected to follow it to its logical conclusion.

The result of this hidden philosophy or analysis of the workings of organizations was absurd. Leaders were forced to give inadequate or flip answers to problems that could not conceivably be resolved except by the most intimate knowledge of Indian communities and the most sophisticated education in corporate and organizational structure. Anyone considering the manner in which the various components of a program would actually fit together would have been horrified at the philosophy inherent in the development of grants. But no one took the time to consider this aspect of program management.

On the whole, the 60s were a good decade for Indian people. After nearly a century of oppression and efforts to destroy the institutions of the red man the government finally gave overwhelming support to local efforts to develop relevant institutions. There can be no doubt that with a plentitude of opportunity available for the first time to the different tribes, a mighty effort had been made to redress the wrongs that previous generations had endured.

The tragedy of the decade, however, was the failure to provide any lasting economic institutions on the reservations. There were scarcely any more jobs in the reservation villages by 1970 than there had been in 1960, if one eliminated the administrative jobs created by the government programs. In the meantime the cost of living had skyrocketed on the reservations because the many Poverty jobs had created a demand for additional consumer goods and the large salaries had provided cash with which the goods could be purchased. It was yet undetermined in 1970 whether the Poverty program had not become a permanent part of reservation life. Withdrawal of the programs would have meant economic catastrophe of the most severe kind, since in the decade communities had become almost wholly dependent upon the Poverty programs for their economic existence.

[V. D., JR.]

36

The Rise of Indian Activism

As the Civil Rights movement turned from integration to the development of Black Power in 1966, a corresponding shift in emphasis was made in the Indian community. Formerly people had depended upon the tribal councils to initiate movements and generate issues. The tradition had been set during the previous decade when tribal councils led the fight against the termination policy. And in many instances city-dwelling Indians and young people waited for the tribes to define vital issues before they became active in pushing for reform.

As early as 1963, however, the climate began to change. In that year, the National Indian Youth Council, a group of younger college-trained Indians who had organized following the Chicago conference of 1961, led a demonstration in the northwest in an effort to highlight the problem of Indian treaty fishing rights. Marlon Brando offered his support to the movement and the Indians went out on the Nisqually and Puyallup rivers south of Seattle, Washington to protest the treatment of Indian fishermen by the departments of Fish and Game of the state of Washington.

Fishing was a way of life for the Indians of the Puget Sound. They had formerly lived along the numerous rivers and streams emptying into the sound. In 1854 and 1855 when the United States was desirous of obtaining a peaceful settlement of the region, Isaac Stevens was sent to the coast to negotiate treaties for land cession. Stevens received title to the entire northern coastal area in a series of six treaties. The small tribes agreed to go to selected reservation areas but they rigorously maintained that they should have the right to hunt and fish in all of their accustomed places since their entire economic structure was built upon the salmon runs of the rivers of the territory.

At first the Indians were left in relative peace and they provided the new population with almost all of its seafood. In fact at one time there was considerable concern that the Indians would not fish and that the territory would be left without adequate food supplies. But as the territory of Washington became a state and more people poured into the land, tensions began to mount. The discovery of efficient methods of canning meant the development of a commercially profitable fishing industry and soon Indians were being pushed aside from their traditional fishing sites by whites.

During the 1920s, the development of sports fishing and the tremendous state income deriving from the sale of fishing licenses meant virtually the end of Indian fishing. Quickly the state began moving in against the Indians, confining them within the reservation boundaries and often challenging their right to fish, even on the reservations.

Nineteen sixty-four began the modern Indian war in the northwest over Indian fishing rights. The state of Washington, in violation of the edicts of the Supreme Court of the United States and a Congressional statute of 1954 specifically exempting Indian fishing rights from the jurisdiction of the states, began a relentless harassment of the Indian fishermen—particularly on the Nisqually, Green and Puyallup rivers. The state Game Department was particularly oppressive since it represented the sportsmen (being almost entirely financially dependent on the revenue of fishing licenses for its income).

The Fish and Game officers would swoop down on an Indian settlement, break up the Indian boats, cut or confiscate their nets, and after numerous acts of brutality, arrest the Indians for "disturbing the peace," "unlawful assembly," or "inciting to riot," thus effectively preventing the helpless red men from getting the issue of fishing rights into the federal courts where they could receive protection.

By the middle of 1970 the Indians had begun to get sufficient publicity on the struggle so that the public at large was becoming aware of the unequal situation. So the Game Department officers lessened their harassment and gangs of vigilantes, consisting sometimes of their sons and relatives, took to cowardly tactics of ambush and night raids to destroy the Indian fishing gear and to frighten off the Indian fishermen.

Finally the inevitable happened. On January 19, 1971 in the early morning mists, two white vigilantes caught Hank Adams, leader of the Indian fishermen, alone in his car sleeping by the riverbank. They threw open the car door and shot Adams in the stomach. It was a near miss. Fortunately Adams survived the assassination attempt, since the sportsmen were not the best of shots. But the crisis had undoubtedly reached its climax. Later that day in the hospital, police questioned Adams and implied that he had shot himself in an effort to gain publicity for his cause. Neither the police nor the F.B.I., reluctantly called in by the hesitant Nixon administration, which had been elected on a platform of law and order, followed up the assassination attempt very rigorously.

Perhaps the worst part of the fishing rights struggle was the almost complete news blackout in the state of Washington itself. The newspapers were almost unanimously against the Indians, and what stories were published reflected the sportsmen's side of the controversy. In Tacoma, the nearest large city to the Indian settlements, the television stations were controlled by the newspapers—thus effectively blocking out any news whatsoever on the fishing troubles. Numerous Indian leaders across the nation tried to bring the issue to the American public via national television programs, but these programs were never shown in the state of Washington. When it came time to rerun them the stations of the state would substitute other programs. Thus the white population of the state that might have been sympathetic toward the Indians was never allowed to hear the Indian side of the story.

If the fishing rights struggle showed anything, it was that the Indian wars had not ended and that in all probability they would never end so long as there were any Indians on the continent. The spectacle of 200 Indian fishermen constantly harassed by two state departments representing some 3,000,000 whites and aided behind the scenes by federal officials who were sworn to protect the Indians demonstrated the basic inequality of the American way of life more eloquently than had Wounded Knee or Sand Creek. America had not come far in the intervening century.

Tribal councils were rather reluctant to engage in the struggle to protect treaty rights. The National Congress of American Indians came to be dominated by a few larger tribes that lived rather comfortably while their brethren from the smaller tribes were pushed around. But for one period in 1966-67 even these lethargic political leaders saw the danger that was confronting them.

By early 1966 the Poverty program was spending substantial sums in Community Action Programs on the respective reservations. The basic theory of the Office of Economic Opportunity was to involve the poor in all decisions affecting their lives. Thus tribal councils had early become sponsoring agencies for the various Poverty programs operating on the reservations. This development meant that suddenly millions of dollars were being subcontracted to the tribal governments for operation. With funds of their own to operate programs, many tribes began to show remarkable independence from the Bureau of Indian Affairs. And hence the trouble began.

After nearly a century of complete domination of tribal life the career bureaucrats of the Interior Department found themselves in a drastic position. In previous years they had only to submit their budgets, pacify the reservation leaders, listen to the continual complaints of a rural people who had no knowledge of the workings of the government, and await their retirement date. By 1966 these civil servants were extremely nervous and irritable. Congressmen and Senators were asking pointed questions about the inability of the Bureau of Indian Affairs to solve basic problems and the efficiency of the Bureau was being unfavorably compared with the apparent success of the Poverty program.

Hence, in April of 1966 there was an effort by the officials of the Interior Department to convince Congressmen on the appropriations committee to change the operation of the Poverty program and to give the funds now being diverted to tribal governments directly to the Interior Department for programming. Contending that the Interior Department had extensive experience with Indians and that the reservation people could not operate a large program successfully over a longer period of time, they asked that all funds used in the Poverty program for Indians simply be included in Interior's annual budget.

The impact of this plan would have been to deprive the tribal councils of some $40,000,000 in community action funds which they were then contracting from the Office of Economic Opportunity. The various action projects on reservations were already achieving more true community participation with that small amount than Interior officials had allowed even with their massive budget totaling close to $320,000,000 a year. It was a real crisis for the tribal governing bodies who had tasted their first real freedom since the Great Depression.

In April of 1966 a meeting of the tribes was called to coincide with a meeting of the Bureau of Indian Affairs in Santa Fe, New Mexico. The Bureau meeting was advertised as a means of letting the high Bureau officials gather with Stewart Udall, then Secretary of the Interior and directly responsible for Indian programs, to decide what was best for reservation development. For three days, two meetings were held simultaneously. The Bureau of Indian Affairs met at a little green Episcopal church in town while the assembled tribes met at the federal building three blocks away.

Sixty-two tribes answered the call for meeting. With their already considerable experience operating the community action programs and sensing a heady success when the bureaucratic front began to melt, the tribal leaders planned out a series of policy resolutions that would come to fruition in later years. The foremost of these was subcontracting of agency functions to the tribal governments along the same guidelines as the Poverty program. An old law, the "Buy Indian Act," allowed the Bureau of Indian Affairs to subcontract services and purchase of materials to any

Indian or Indian tribe. Yet this law was rarely used. It was the consensus of the tribes at Santa Fe that the tribes gradually take over the operation of the Bureau of Indian Affairs by subcontracting as many of its functions as possible.

At first the Bureau officials would not allow Indian delegates into their conference. This was particularly galling to the Indian meeting, since representatives of the national churches and a representative of one of the old-line Indian interest groups were attending the meeting. Breaking through this ancient alliance of white friends and career civil servants was not originally a goal of the tribal meeting. But it became apparent to many delegates that they might again be squeezed out of all considerations and the policies of the old alliance would once again be put into effect. It was this coalition that had made the allotment process so destructive and had resulted in the loss of the Indian land base.

The result of the Interior meeting was the announcement that the Secretary of the Interior would consult with the various tribal leaders and, on the basis of this consultation, present to Congress the following year, a massive piece of legislation tentatively entitled the "Omnibus Bill," which would update and solve all the tribal problems. But when the meeting of tribal leaders asked for an advisory committee to the Secretary of the Interior to assist in drawing up the legislation, they were turned down, thus indicating that Interior officials had their own ideas of what was meant by consultation.

By the summer of 1966 it was apparent that there would be no effort to consult with the tribal representatives. Rumors began to float through the various government agencies that legislation was being prepared in spite of the promise by Udall that Indians would share in the creation of the Omnibus Bill. The National Congress of American Indians managed to secure a copy of the proposed legislation and in September of that year passed out copies to every tribe in the land. From that point on, it was downhill for the new program. In a series of meetings covering the major areas of Indian population, the tribes stood firm for the principle of consultation and the legislation, when it was finally introduced, received only perfunctory consideration by Congressional committees and then died.

The first meeting of the consultation tour taken by then Commissioner Robert Bennett was indicative of the rising expectations of the Indian people. In Minneapolis when the tribal leaders were called together to discuss the new legislation, a group of Indians living in the twin cities demonstrated against the policies of the Bureau in that urban area. Some minor changes in procedure were effected but the success of the demonstration foretold the manner in which techniques were changing. Following the Minneapolis demonstration, the Indians of that city organized the American Indian Movement, an activist group that was prepared to challenge the arbitrary manner in which the Bureau of Indian Affairs administered its program for off-reservation Indian people.

In the years following Minneapolis, the American Indian Movement, nicknamed "A.I.M." expanded its operations over most of the Midwest. It became the most sophisticated Indian activist group, with branches in Milwaukee, Chicago, Denver, Rapid City, South Dakota, Cleveland, and Spokane, Washington. In its home city of Minneapolis, A.I.M. concentrated on eliminating the discriminatory practices of the local city government. The chief practice they opposed was enforcement of city ordinances against the resident Indian population.

For nearly a century the Chippewa living in the northern part of the state had come to the Twin Cities to seek employment. Gradually little Indian neighborhoods formed, usually around certain bars where the men would go at night after work. During this time the practice had grown up of arresting the Indians for drunkenness and disturbing the peace whether they were doing so or not. Technically it could be called "preventive detention." In practical terms it meant that the police could pad their records by inflating them with statistics representing Indians unfairly arrested.

A.I.M. organized an Indian Patrol to conduct surveillance of police activities in areas predominantly Indian. Choosing areas which showed the highest arrest records, A.I.M. relentlessly followed the police and checked every arrest they attempted to make. After nearly a year of surveillance there were no arrests of Indians made. A.I.M. proved beyond a doubt that the police practices had been discriminatory and that once proper procedures were followed in determining arrests that Indians were no more lawbreakers than any other group.

While the American Indian Movement was expanding its influence throughout the Midwest, another movement was beginning on the West Coast. Centered in the San Francisco Bay area were the universities of California and Stanford with their thousands of activist-minded students. The Bay area also contained a substantial number of Indians who had moved to the coast in the last twenty-five years. When the ideologies of the student began to influence the resident Indian population, Indian activism took on a new twist.

Sometimes identifying with the Third World Movement, sometimes as an offshoot of the activism and ideology of Berkeley, small groups of Indian students began to make themselves felt. They began to develop a militancy and a willingness to go farther than any other group of Indians had ever gone to make their point. The problem they faced was that few of them had ever lived on reservations and there was an inability of some to identify with the philosophies of the tribal leaders who controlled the reservations. With this split, based primarily on residency and experience, the urban Indians began to appear as a threat to reservation existence.

In November, 1969 a crisis developed in the Bay area that radically changed the nature of the national Indian community. The San Francisco Indian Center which had served the Indians of the Bay area for a number of years burned down following a meeting of Indian centers. With no center to which they could relate, the Indians of the Bay area were cast adrift in confusion. Appeals to the Bureau of Indian Affairs were useless since the Bureau had already determined that urban Indians would not receive services from its regular budget.

In casting about for a new location for an Indian center, eyes began to turn toward a small island in San Francisco Bay—Alcatraz, the abandoned federal prison. The rush was on.

In two different invasions the island was secured and the young Indian activists achieved world-wide attention overnight. Alcatraz was the master stroke of Indian activism. It was a symbol with which anyone in the world could identify. A stark forbidding island topped with a crumbling penitentiary, barren and useless. It was, the invading activists declared, exactly like an Indian reservation. And Americans quickly got the point.

Alcatraz inspired young Indians everywhere and its organization, Indians of All Tribes, was quickly copied all over the country. Almost instantaneously the nation was blanketed by groups calling themselves Indians of All Tribes and they meant business. An activist roll call comparable to the Civil Rights roll call of Selma, Birmingham, and Memphis was quickly made up. "Indianness" was judged on whether or not one was present at Alcatraz, Fort Lawton, Mount Rushmore, Detroit, Sheep Mountain, Plymouth Rock, or Pitt River. The activists took over and controlled the language, the issues, and the attention that other Indians had worked patiently and quietly to build.

By late 1970 the tribal leaders were cringing in fear that the activists would totally control Indian Affairs. Old issues of taxation, treaty rights, tribal sovereignty, land consolidation, and economic development were being completely neglected as the federal government and the American public were responding to the demands of the Indian activists. But it was the tribal leaders' own fault that the activists had parlayed Alcatraz into a national phenomenon. For decades the N.C.A.I. and other national groups had cast aside young people and urban Indians in favor of local reservation politicians. They had no reason to expect that these people, cast aside and unwanted, would suddenly rally to their aid.

The situation was deeper and more serious than that however. Almost every Indian active on the national scene had missed the significance of Alcatraz. They all looked at Alcatraz in terms of land restoration, and when balancing seventeen acres of rock in San Francisco Bay against the possible loss of the Colville Reservation in Washington state which was in danger of being terminated, or against the return of Blue Lake to Taos

Pueblo, 44,000 acres of sacred land, no one could see why Alcatraz should have any importance.

If the tribal leaders had rallied to the cause of Alcatraz immediately and used it correctly as a symbol of land restoration legislation that was badly needed by the tribes they could have redoubled the effectiveness of the activism while using the movement to project real needs of reservations into the area of policy consideration. But the tribal leaders' first inclination was to avoid the issue and withhold support and this doomed them thereafter from raising their issues in a context in which they could have gained public support.

Nor did the young activists take the time to examine the issues that affected the reservations. They were centered primarily in the colleges and young working classes. Heirship land problems, taxation of Indian allotments, and economic development had little meaning for them. There was little available literature to inform them had they been interested. And tribal chairmen did nothing to inform or cooperate with the activists to bring them into Indian Affairs as an action arm of national Indian politics.

Thus this unfortunate combination of circumstances resulted in activism for activism's sake and irrational fears on the part of reservation political leaders that their influence was being undermined. Indian Affairs disintegrated into a confused state and no one could conceive of any way to bring the diverse groups into one unified front for any kind of action or program. Conferences split between activists and tribal chairmen, each suspicious of the other and neither willing to grant the validity of the other person's ideas of programs.

The situation was further marred by the appearance of pseudo-organizations of Indians. Charismatic Indians who had a spectacular style of speaking began to make their appearance. By simply publishing a newspaper, claiming an unusually large membership, and appearing at every Indian conference, a number of individuals were able to influence events far beyond their actual individual following, if indeed they had any following. After all, who would check a nonexistent membership list in the middle of a conference to determine if the speaker represented any Indians at all? Once started, the contemporary Indian movement expanded beyond any boundaries ever conceived at the start of the decade by the youth in Chicago in 1961. There were no ground rules and no channels by which ground rules could be constructed.

One thing that the Indian activist movement did prove was that competent leadership existed off the reservations as well as on the tribal councils. People such as Stella Leach of Alcatraz, Bernie White Bear of Fort Lawton, Fred Lane of Seattle, Dennis Banks, Clyde Bellcourt and Russell

Means of A.I.M., and other activists could easily have chaired the larger tribes with outstanding results. Leadership was not confined to reservation councils. A great deal of it existed in the activist groups.

And conversely the movement showed that some respected tribal leaders could only function in an atmosphere in which there was no competition. If confined only to the internal politics of the National Congress of American Indians or regional groups such as the Northwest Affiliated Tribes these people were adequate. But once faced with a new situation and intelligent young rivals, long-standing Indian leaders were unable to function and resorted to tactics of fear in an effort to buttress their reputations.

When the activist years are considered in a historial perspective the situation has little novelty. The confusion of the present was mirrored in the efforts of Little Turtle, Pontiac, and Tecumseh to create a coalition of tribes and their subsequent failure to maintain a united front in the face of constant opposition by the invading white men. Within the currents of historical movement it has been exceedingly difficult for anyone to properly assess his actual situation in time to respond effectively. Only in those rare moments when religious fervor enables a people to coalesce for action have any significant changes been made.

Insofar as the activist movement had triggered off a return to ancient tribal customs, then, it could be counted as an important development in the red man's historical drama upon this continent. For by returning to Indian religions, by adopting the traditional customs by which tribal members related one to another, by forming useful and efficient alliances with forces in contemporary society, by these means alone could the red men ensure their survival.

The beginnings of such a movement were clearly evident in a meeting of the traditional medicine men of the tribes, which was held on the Crow Reservation in Montana in the summer of 1970. Originally organized by a group of anthropologists from Monteith College in Detroit against the advice of supposedly knowledgeable Indian political leaders, the conference was an unqualified success. After a week's meeting in which ideas and evaluations of the contemporary scene were made, the rapid movement of Indian emotions toward the traditional values was evident to everyone. No one had dreamed that the offshoot of activism had been to revive the inherent strengths of the basic tribal beliefs to the point that they were beginning to dominate every decision made by Indians.

A long time ago Chief Seattle had warned that the white man would never be alone in his civilization and that the Indian dead were not really dead but still roamed and walked the land. Even Wovoka had foreseen the day when the irresistible movement of Indians would reclaim the continent.

The meeting at Crow had made clear that this time was approaching. It remained only for the religious leader to arise and integrate all of the diverse strands of political, economic, religious and social movements into one strong contemporary Indian structure. For the reᴅ man, at least, the drama was taking on a special significance.

[V. D., Jʀ.]

37

Conclusion

THE DRAMA OF THE RED MAN began a long time ago. After nearly four centuries of conflict between the original inhabitants of the continent and the invading European settlers, the drama is far from ended. Indeed, we may be entering the most fascinating era of North American history. With the sudden advancements of technology during the last half century, the various segments of American society have achieved a parity of communication. It is no longer possible for any group to remain forever outside the experiences of all other groups or of society as a whole. We are thus all linked inevitably with one another for better or worse.

The important aspect of the story of the red man is his stubborn refusal to give up his tribal identity and become simply another American citizen. While the years have shown a partial assimilation of other groups, only the red man has stood firm, resisting all efforts to merge him with the groups that surround him. Whether the battle is on the banks of the rivers of Washington state, in remote reservation mission stations, on the streets of the large cities, or the lonely island of Alcatraz, there is something uniquely different about Indian people that sets them apart from other Americans.

Indian organizations defy the ordinary laws of organizational life. Where other groups would merge, go bankrupt, or transform themselves, Indian groups remain peculiarly impervious to change. Regardless of their fortunes they manage to land right side up. In fact, the entire conception of organization is changed in an Indian context, belying the traditional

American contention that efficiency and relevancy are key components of organizational life.

The vision of today is colored quite radically by the decade behind us. Thus, to many people the situation is not radically different from the beginning of any previous decade of Indian life. Many influential Senators and Congressmen still believe that they can remake red men into white men by adjusting the laws that protect Indian property, although they would not attempt such a program with any other comparable group of citizens.

Many Indians believe that for the first time the federal government is listening to the voices of Indians. This is not true either. The government always listened to Indians but it rarely did what they wanted it to do. This principle will probably hold true for the future. The government will continue to do what it thinks is best for Indians and somehow the Indians will absorb whatever program is devised for them and remain in much the same state as they formerly were.

The mood of the nation is now determined by the manner in which the communications industry presents the respective problem areas of national life. In this sense Indians will probably continue to suffer reverses. With the vast majority of Indians living in the remote western states and the television and publishing industries entrenched on the two coasts, Indians remain the one group with the potential for becoming an exotic commodity projected by the communications media. As such Indians will probably remain a people of fantasy and never achieve the credibility or reality that other minority groups have achieved.

At present there has been a reactionary movement to place decision-making and consultation strictly within the group of elected tribal leaders and to exclude all Indians of whatever persuasion who are not elected tribal officials. Little do the proponent. of this idea understand of the nature of elective processes, for they assume that they will always occupy the highest offices of each tribe. Thus we cannot expect the reactionary forces to capture the spotlight of attention for very long.

By the same token the younger and more liberal activists feel that they should determine what paths the national Indian community should follow. The conception of contemporary Indian Affairs is thus blown far beyond the original ground rules set down during the days of George Washington. The United States has responsibilities to distinct tribal groups with which it has signed treaties and for whom it has accepted certain responsibilities. Individual Indians and nontreaty groups have no more claim on the United States than does any randomly chosen group of people without historical relationships with the federal government.

Between these two extremes the drama of the red man must be worked out in the years to come. Whether or not anyone likes it, the future, as the past, revolves around the continued sovereignty of Indian tribes. Thus the principle for which the red men fought over the centuries has continued to be the important building block in any solution of the conflict between red and white. The coalitions of the past were mere rehearsals of what must come in the future—a new type of activity in the political arena—a new definition of citizenship.

What is certain in one generation is rigorously questioned in the next and the reverse holds also. The Buffalo Party has continued to declare the Vanishing American as finally vanished while always fighting the latest manifestation that the red man has not quite vanished. In every generation there will arise a Brant, a Pontiac, a Tecumseh, a Chief Joseph, a Joseph Garry, to carry the people yet one more decade forward.

At present the national Indian scene is conspicuous by the number of capable organizations that characterize it. Which group will ultimately prove to have built its future on important issues and which will have fallen by the wayside is yet to be determined. The exciting part of being alive in these days is to watch how the various pieces of the puzzle come together and what pattern they will eventually make.

If you should desire to join in the enactment of the drama of the red man in the new world you have a variety of groups to which you can give your allegiance. It is in the interaction of these groups with the forces outside the Indian community that the final scene of the red man will be written. In the appendix that follows, a list is provided which contains the major Indian groups at work today. They are writing the script for tomorrow.

[V. D., Jr.]

Appendix A

AMERICAN INDIAN MOVEMENT
Minneapolis, Minnesota

AMERICAN INDIAN HISTORICAL SOCIETY
1451 Masonic Avenue
San Francisco, California

AMERICAN INDIAN LEADERSHIP CONFERENCE
Manderson, South Dakota

AMERICAN INDIAN DEVELOPMENT, Inc.
Zook Building
Denver, Colorado

AMERICANS FOR INDIAN OPPORTUNITY
1825 Jefferson Place, N.W.
Washington, D.C.   20036

ARROW, Inc.
1346 Connecticut Avenue, N.W.
Washington, D.C.   20036

AMERICAN INDIAN PRESS ASSOCIATION
1346 Connecticut Avenue, N.W.
Washington, D.C.   20036

ALASKA FEDERATION OF NATIVES
Anchorage, Alaska

CALIFORNIA INTERTRIBAL COUNCIL
Sacramento, California

OKLAHOMANS FOR INDIAN OPPORTUNITY
University of Oklahoma
Norman, Oklahoma

NATIONAL CONGRESS OF AMERICAN INDIANS
1346 Connecticut Avenue, N.W.
Washington, D.C.   20036

NATIONAL INDIAN YOUTH COUNCIL
Albuquerque, New Mexico

SURVIVAL OF AMERICAN INDIANS
Tacoma, Washington

SOUTHWESTERN INDIAN DEVELOPMENT, Inc.
Gallup, New Mexico

SMALL TRIBES OF WESTERN WASHINGTON
Puyallup, Washington

UNITED SCHOLARSHIP SERVICE
Denver, Colorado

Appendix B

Official Indian Wars
(War Department Compilation)

1790-1795   War with the Northwest Indians: Mingoe, Miami, Wyandot, Delawares, Potawatomi, Shawnee, Chippewa, and Ottawa—September 19, 1790 to August 3, 1795. Included are Harmar's and St. Clair's bloody defeats and Wayne's victory at Fallen Timbers, which compelled peace.

1811-1813   War with the Northwest Indians, September 11, 1811 to October, 1813. General Harrison defeated the Confederate tribes at Tippecanoe, Indiana. Tecumseh was killed at the battle of the Thames, in Canada, October 5, 1813.

1812   Florida or Seminole war, August 15 to October, 1812. Spanish Florida invaded by Georgia Militia under General Newman, and the Seminole under King Payne defeated. These disturbances never ceased until Florida was ceded by Spain to the United States. In fact, one band of Seminole were never conquered and reside in Florida to this day.

1813   Peoria Indian war in Illinois, September 19 to October 21, 1813.

1813-1814   Creek Indian war in Alabama, July 27, 1813 to August 19, 1814. It was in this war that General Andrew Jackson first attracted attention as a commander. He defeated the Creeks in a bloody engagement at Talladega, November 9, 1813, at Emuckfau January 22, 1814, at Enotochopco January 24, and finally at the Horseshoe Bend of the Tallapoosa River March 27, 1814, which humbled the Creek pride completely. At this battle 750 Creeks were killed or drowned, and 201 whites were killed or wounded. In this war the brave Creeks lost 2,000 warriors. Ten years afterward the tribe still numbered 22,000.

1817-1818   Seminole war in Georgia and Florida, November 20, 1817 to October 31, 1818. It was during this war that Jackson took possession of the Spanish territory. He seized St. Mark's and Pensacola, Florida, hanged two Englishmen, Arbuthnot and Ambrister, for inciting the Indians to hostilities, and brought the Indians to terms.

1823   Campaign against Blackfeet and Arickaree Indians, upper Missouri River.

1827   Winnebago expedition, Wisconsin, June 28 to September 27, 1827; also called La Fevre Indian war.

1831   Sac and Fox Indian war in Illinois, June and July, 1831.

1832   Black Hawk war, April 26 to September 21, 1832, in Illinois and Wisconsin. Black Hawk escaped from General Atkinson, but surrendered at Prairie du Chien August 27, 1832. He was taken to Washington to see the "Great Father," and ever afterward lived at peace with the whites. He was but a chief of a secondary band. He settled upon the Des Moines River, in Iowa, where he died October 3, 1838.

1834   Pawnee expedition, June to September, 1834, in the Indian Territory.

1835-1836   The Toledo war, or Ohio and Michigan boundary dispute.

1835-1842   Seminole war in Florida, December 23, 1835 to August 14, 1842.

1835-1837   Creek disturbances in Alabama, May 5, 1836 to September 30, 1837.

1836-1838   Cherokee disturbances and removal to the Indian Territory.

1836   Heatherly Indian troubles on Missouri and Iowa line, July to November.

1837   Osage Indian troubles in Missouri.

1849-1861   Navajo troubles in New Mexico.

1849-1861   Continuous disturbances with Commanche, Cheyenne, Lipan, and Kickapoo Indians in Texas.

1850   Pitt River expedition, California, April 28 to September 13, 1850.

1851-1853   Utah Indian disturbances.

1855   Winna's expedition against Snake Indians, Oregon, May 24 to September 8, 1855.

1855-1856   Sioux expedition, Nebraska Territory, April 3, 1855 to July 27, 1856.

1855   Yakima expedition, Washington Territory, October 11 to November 24, 1855. Commanded by Maj. Gabriel J. Rains, afterward a Confederate general. Composed of a small body of regulars and a regiment of mounted Oregon troops. The expedition was a failure. The following year, under command of Col. Geo. Wright, U. S. A., better success was had against the Indian allies, and peace subsequently compelled. Lieutenant Sheridan, afterward lieutenant-general, greatly distinguished himself at the Cascades.

1855-1856   Cheyenne and Arapaho troubles.

1855-1858   Seminole war in Florida, December 20, 1855 to May 8, 1858.

1857   Sioux Indian troubles in Minnesota and Iowa, March and April, 1857.

1858   Expedition against northern Indians, Washington Territory, July 17 to October 17, 1858.

1858   Spokane, Coeur d'Alene, and Paloos Indian troubles.

1858   Navajo expedition, New Mexico, September 9 to December 25, 1858.

1858-1859 Wichita expedition, Indian Territory, September 11, 1858 to December, 1859.

1859 Colorado River expedition, California, February 11 to April 28, 1859.

1859 Pecos expedition, Texas, April 16 to August 17, 1859.

1860 Kiowa and Commanche expedition, Indian Territory, May 8 to October 11, 1860.

1860-1861 Navajo expedition, New Mexico, September 12, 1860 to February 24, 1861.

1862-1867 Sioux Indian war in Minnesota and Dakota. The Sioux killed upwards of 1,000 settlers in Minnesota. They were pursued by General Sibley and General Sully, with about 5,000 men, scattering in Dakota. The operations against them were successful. Over 1,000 Indians were made prisoners, and 39 of the murderers were hanged after a fair trial. In 1863 the Minnesota Sioux were removed to Dakota.

1863-1869 War against the Cheyenne, Arapaho, Kiowa, and Commanche Indians in Kansas, Nebraska, Colorado, and Indian Territory.

1865-1868 Indian war in southern Oregon and Idaho and northern California and Nevada.

1867-1881 Campaign against Lipan, Kiowa, Kickapoo, and Commanche Indians, and Mexican border disturbances.

1874-1875 Campaign against Kiowa, Cheyenne, and Commanche Indians in Indian Territory, August 1, 1874 to February 16, 1875.

1874 Sioux expedition, Wyoming and Nebraska, February 13 to August 19, 1874.

1875 Expedition against Indians in eastern Nevada, September 7 to 27, 1875.

1876-1877 Big Horn and Yellowstone expeditions, Wyoming and Montana, February 17, 1876 to June 13, 1877. Three converging expeditions under Generals Gibbon, Custer, and Terry were sent against the hostile Sioux who had previously repulsed General Crook in the Little Big Horn country. Custer divided his command when in the vicinity of the Indians, and he with 250 of his men were surrounded and killed to a man by at least some 3,000 Sioux warriors. The bands of Sitting Bull, Crazy Horse, and other hostiles afterward fled into Canada, from whence they did not return for some years. Eventually all came into the agencies.

1876-1879 War with Northern Cheyenne and Sioux Indians in Indian Territory, Kansas, Wyoming, Dakota, Nebraska, and Montana.

1877 Nez Perces campaign and Chief Joseph, in Idaho.

1878 Bannock and Paiute campaign, May 30 to September 4, 1878.

1878 Ute expedition, Colorado, April 3 to September 9, 1878.

1879 Ute campaign.

1879 Snake or Sheepeater Indian troubles, Idaho, August to October, 1879.

1878-1887 Apache Wars, Geronimo.

1891-1892 Sioux Ghost Dance disorders and the death of Sitting Bull.

1913 The Navajo war in New Mexico.

1915 The Paiute war in Colorado.

Index